Sustainable Development in Environment

Sustainable Development in Environment

Edited by **Kane Harlow**

Callisto Reference

New York

Published by Callisto Reference,
106 Park Avenue, Suite 200,
New York, NY 10016, USA
www.callistoreference.com

Sustainable Development in Environment
Edited by Kane Harlow

International Standard Book Number: 978-1-63239-578-8 (Hardback)

Printed in the United States of America.

Contents

Preface

The world is advancing at a fast pace like never before. Therefore, the need is to keep up with the latest developments. This book was an idea that came to fruition when the specialists in the area realized the need to coordinate together and document essential themes in the subject. That's when I was requested to be the editor. Editing this book has been an honour as it brings together diverse authors researching on different streams of the field. The book collates essential materials contributed by veterans in the area which can be utilized by students and researchers alike.

The issue of environmental management has gained frequency in recent years. In sustainable development circumstances, central and local governments more often observe the necessity of acting in ways that decreases negative impact on environment. Environmental management occurs at different levels- commencing from global level; for instance, climatic changes, by way of national and regional level (environmental policy) and concluding on micro level. This book displays many examples of environmental management. The multiplicity of presented aspects within environmental management and approaching the subject from the prospects of various countries contributes considerably to the development of research fields of environmental management.

Each chapter is a sole-standing publication that reflects each author´s interpretation. Thus, the book displays a multi-facetted picture of our current understanding of application, resources and aspects of the field. I would like to thank the contributors of this book and my family for their endless support.

Editor

Chapter 1

Overview of Past and Ongoing Experiences Dealing with the Environmental Management at Cluster Level

Francesco Testa, Tiberio Daddi, Fabio Iraldo and Marco Frey

Additional information is available at the end of the chapter

1. Introduction

Small and medium-sized enterprises make up a large part of Europe's economy, representing some 99% of all enterprises and 57% of the economy's added value (COM (2005) 551 fin). Therefore they also play a primary role in shifting the European economy to more sustainable production and consumption patterns. SMEs can have considerable impact on the environment, not necessarily through individual pressure, but through their combined total impact across sectors. It is widely accepted that it would be too complex and burdensome for companies and public authorities to determine the detailed contribution made by SMEs to pollution (e.g. air pollution), in terms of the environmental load from different types of pollutants (e.g. CO2, SOx, NOx, etc.) in each Member State. Indeed, in many cases the data does not exist. Nevertheless, the often quoted rough figure of a contribution of 70% of industrial pollution in Europe seems reliable, and a number of studies attempt to provide 'insights' into particular environmental problems deriving from SMEs for specific countries. For example, a British report estimated that SMEs accounted for 60% of total carbon dioxide emissions from businesses in the UK and concluded that there was substantial room for improvement in energy efficiency and emissions reductions among SMEs. Again, estimates from the Netherlands and United Kingdom suggest that the commercial and industrial waste from SMEs represents on average 50% of the total. These studies further support the claim that SMEs can exert considerable pressures on the environment (SEC (2007) 907). Numerous regional and national studies show that the majority of SMEs have low awareness of their environmental impacts and how to manage them. Most SMEs are 'vulnerably compliant', since they do not always know enough about legislation to ensure that they are compliant. This is mostly due to lack of awareness of the environmental impacts of their own activities, ignorance of environmental legislation, inability to tackle their environmental impacts, and sometimes the excessive administrative and financial burden of compliance. Compliance is further hindered by the perception that environmental protection is costly and has little benefit for the business (SEC (2007) 907).

This scenario shows:

- the potential threat to the environment and to the effectiveness of many Community environmental protection measures due to lack of knowledge and awareness;
- the potential risk for the health and safety of SMEs employees
- the potential advantages that an enhanced environmental management could bring to SMEs in terms of economic and/or financial benefits (of which SMEs may be unaware). The complexity of the issues involving the SMEs' environmental compliance and their environmental performance, other than their capacity to fully respond in time to the new challenges placed by environmental issues (which allows them to catch the benefits in terms of competitiveness and innovation) needs a multiple approach, capable of putting into action a set of complementary measures. These would include education and awareness raising, on-the-spot assessment and identification of problems, targeted consultancy, exchange of information and best practices, removal of administrative burdens, availability of shared tools.

Generally, the issues are related to the environmental impact of the SMEs production activities and services, involving all the environmental aspects (air, water, soil and sub-soil, biodiversity, noise, land, etc.) normally/usually disciplined by the EU and the national and local legislation. They also include crucial new challenges that the EU is facing, regarding global warming, energy efficiency, renewable energy sources, sustainable use of resources, waste reduction, re-use and recycling.

All these considerations have stimulated the development of the so-called "cluster approach" to manage the environmental issues of a large number of SMEs located in limited territorial areas. In recent years, the "cluster approach" has been mentioned in some important official documents of the European Union, such as:

- EC COM (2007) 379 "Small, clean and competitive - A programme to help small and medium-sized enterprises comply with environmental legislation" (ECAP Programme) which encourages the use/implementation of environmental management in industrial clusters or districts of SMEs, using specific cluster or supply chain approaches;
- the new Regulation (EC) no. 1221/2009 (EMAS III), which in article 37 mentions SMEs clusters applying for EMAS registration.

Networking and cooperation among organisations emerge from several studies and empirical evidence as some of the most important factors fostering the dissemination of formal EMS (such as EMAS). Many authors (inter alia:Biondi et al. 2000, Hillary 2004) emphasise that working with groups of companies is an useful and efficient way of adopting EMAS, particularly for SMEs. Moreover, the European Commission has recently confirmed the key role of networking for overcoming the constraints and barriers for EMS adoption among SMEs (European Commission 2007). In fact, the Commission has highlighted its commitment to promote and encourage the use of EMAS in industrial clusters or districts of SMEs, using specific cluster- or supply chain- oriented approaches, since they can reduce consultancy and audit/verification costs for SMEs, and facilitate additional sharing of knowledge and exchange of experience among participants.

The effectiveness of the networking approach is clear among organisations operating in the same sector (such as the industrial sector, but also service sectors such as tourism or public institutions operating at different levels) and among organisations operating in the same region (or territorial area).

In the first case, enterprises can co-operate by identifying and assessing similar environmental aspects and by finding technological and operational solutions that can be applied to similar production processes and products, as well as by defining organisational structures suitable for the same kind of production cycles. In the second case, co-operation is facilitated by the 'physical contiguousness' and there are synergies both in improving the environmental impact on the same local eco-system, and in interacting and communicating with the same stakeholders (local population, authorities, etc..).

In some experiences, a network has been created among SMEs within a 'cluster', in order to foster information exchange and experience diffusion and to define and apply common solutions to similar environmental, technical and/or organisational problems, or to share environmental management resources (Iraldo & Frey, 2007). A specific kind of co-operation within a cluster of organisations takes place in the supply-chain: when a large customer is willing to support small suppliers in the EMS implementation process, then all the smaller organisations involved in the supply chain can benefit greatly from networking. This approach proved effective in some Member States such as Germany ("Konvoi" approach), Spain (co-operation in the tourism supply chain), Nordic Countries (Denmark and Sweden). In Italy it has shown a real effectiveness in promoting the environmental compliance of SMEs by means of the so-called APO "Ambiti Produttivi Omogenei".

This study aims at collecting and describing the most important international experiences concerning the cluster approach, with the following objectives:

- To analyse different types of inter-company dynamics in the cluster;
- To give information on the diffusion of the cluster approach at international level;
- To understand possible correlations between cluster approach and environmental issues management;
- To analyze some excellent experiences on shared and collective management of environmental aspects.

The structure of this chapter is organised in sections according to the different types of clusters considered. After a brief introduction on the methodological approach, each section analyses the various forms of cluster with the following structure.

The first section focuses on the identification and analysis of the definition of cluster to which we refer, outlining its key features and describing the cooperative dynamics that consolidate within it. The second section explains the presence of the abovementioned type of cluster in Italy, describing its prevalence and relevance for the national economy. The third section aims at outlining the European framework and offers a comparison with the information applied to the national context. The last section identifies the elements that characterise the management of environmental issues (in a cooperative logic) in the type of cluster analysed,

This study is based on the results emerged from a research carried out within the ECCELSA (Environmental Compliance based on Cluster Experiences and Local Sme-oriented Approaches) Life project. For this reason the authors acknowledge all project partners1 who contributed to the results presented in this chapter.

2. The methodological approach

The first references to the "Cluster organizational model", as an approach that can develop synergies resulting in more efficient production than would occur within a single large plant, are found in Marshall at the end of the XIX century. In the first half of the XX century the benefits of agglomeration of economic activities were also confirmed by Austrian economist Shumpeter, who stressed the importance of the cluster system in terms of business competitiveness. In 1991, Michael Porter in his *Competitive Advantage of Nations* (1991), stated the "cluster theory" in which he identified the most potential for growth and development for industrial clusters as opposed to the single enterprises, thanks to the presence of vertical relations *[customer/supplier*] and horizontal relations *[common customers, technology, channels*].

At European level, clusters have been formally recognized and defined in the Final Report of the 'European Commission Expert Group on Enterprise Clusters and Networks which offers a first "census" of the phenomenon, and in communication No. 652, October 2008[2].

In those documents, clusters are defined as "geographic concentrations of specialized companies that have workforce with advanced abilities and skills, and "support" institutions that make possible the spreading of knowledge and indirect positive effects as a result of their proximity".

According to this definition, the elements that characterize the cluster concept can be identified in *geographic proximity, specialization in production* and *interaction among different actors* in the cluster. Therefore, the definition of cluster is not restrictive; it covers a wide variety of approaches in which those elements are more or less relevant. In fact, the concept includes the classical configuration of the industrial district and the geographically confined

---

[1] The authors specifically acknowledge Sara Tessitore and Valentina Toschi (SSSUP); Besides Sant'Anna School of Advanced Studies (coordinator of the project) the other partners are: Ambiente Italia – Istituto di Ricerca, ERVET - Emilia Romagna Valorizzazione Economica del Territorio, IEFE Bocconi - Istituto di Economia e Politica dell'Energia e dell'Ambiente, SIGE - Servizi Industriali Genova, Gemini - Innovazione Sviluppo e Trasferimento Tecnologico.

[2] Communication from the commission to the Council, the European Parliament, the European Economic and Social Committee and the committees of the region 'Towards world-class clusters in the European Union: Implementing the broad-based innovation strategy ".

industrial areas such as the environmentally equipped production areas "APEA", as well as industrial parks up to the interactions along the supply chain. While being characterized by the presence of elements linked to the cluster, these network systems show major differences that allow to distinguish three main approaches within which to investigate.

For example, the Industrial District is a local system with the presence of a prevalent production carried out by a group of small independent firms highly specialized in different stages of the same production process. Production Areas instead represent an organizational model characterized by the territorial element but, unlike the District, they are concentrated in areas more easily defined and circumscribed geographically, and do not necessarily show the presence of one or more specialised production sectors.

Another organizational model that meets the definition of cluster is the supply chain: a network involving all stakeholders in the production chain, from the company that produces raw materials to the processing company, from carriers to distributors, from wholesale to retail. The feature of the production chain is the interaction between SMEs involved in a process of production, among which the exchange of information can flourish and the development of projects and a relationships based on trust may be encouraged. Compared to Districts and Production Areas the less important aspect in this context is the territorial delimitation and the concentration of enterprises.

The distinction between different aspects of the concept of clusters led to the identification of three inter-organizational interaction models on which to focus the analysis, particularly emphasizing the organizational arrangements implemented to manage common environmental problems. In detail, the types of clusters in this study are:

- Industrial districts, closely related to production;
- Ecologically equipped productive areas APEA and Industrial Parks;
- Supply chain management which aims to produce stable benefits for the companies that are part of the supply chain (buyers and suppliers) through process integration and long term relationships.

3. Industrial district cluster

3.1. "Industrial districts" cluster: Definition and boundaries

In literature, the industrial district is represented by a local system characterized by a main production activity performed by a group of small independent firms, highly specialized in different stages of the same production process. This peculiar entrepreneurial organizational model can develop synergies that result in a more efficient production than would occur within a single large plant. At the end of the XIX century Marshall (1890) had already highlighted the benefits coming from the agglomeration of economic activities in terms of availability of skilled labour and high level of specialization. Similarly, the Austrian economist Schumpeter in the first half of last century stated the existence of competitive advantages deriving from a business cluster. A significant contribution to the study of industrial districts and of internal relationship mechanisms able to generate competitive advantages for the cluster firms comes from G. Becattini who, in his article "From industrial sectors to industrial districts" introduced the concept of industrial district as a tool to support regional policies for territorial development.

Without any attempt to reorder the taxonomies in the field of industrial districts (which can be found in the literature according to the various configurations that such systems engage in), we mention some definitions and acronyms that have slightly different interpretation and, as a result, partially conflicting definitions of local systems with a high concentration of businesses: from industrial districts to system-sectors, from milieu to TPS (territorial production systems), to the RESS (regional economic and social systems). Connecting all the approaches underlying these concepts is the identification as the common element of the analysis of a system of usually small and medium enterprises operating in a homogeneous sector (or in sectors known as "auxiliary") and located in a limited socio-territorial area in which they have deep-seated social and economic relationships. The role which an industrial district can have in a competitive development of local production has stimulated in some national contexts the interest of policy makers. In Italy, Law 140, 1990, enacted to simplify and facilitate the set up of district areas, also fostered an institutional definition of the concept of Local Production System (LPS), which is an area characterized by:

- Homogeneous production contexts;
- High concentration of enterprises;
- Specific internal organization.

Based on that definition, the industrial district can be considered a specific LPS featuring:

- High concentration of industrial enterprises;
- Highly specialized production of business systems.

3.2. The "industrial districts" cluster in Italy

Law No 317 of 1991 "Action for innovation and development of small enterprises" introduced in Italy the concept of industrial district (art. 36) taken up and extended by Law No 140, 1999 with reference to the Local Production System. National legislation has given regional administrations the task to define criteria and procedures for the recognition of clusters and the legal form they should have once approved.

The Region of Lombardy under Regional Law (L.R.) 7 of 1993 and Regional Law 1 of 2000 regulates the procedures for the geographical boundaries of Industrial Districts, sets up the provision of development programmes in individual districts and the creation of innovative projects for the enterprises that belong to them. In 2000 Tuscany adopted Resolution No. 69 which defines districts as "monosectorial production systems with a high presence of small medium industrial manufacturers having a strong supply chain, social and institutional relationships also present in interprovincial areas". Italian Regions did not issue any rules to formally and rigidly acknowledge districts, so it remains a very flexible approach which is closely related to the characteristics of the area and primarily addressed to contexts consisting of SMEs.

The district approach has spread as a result of economic support programmes promoted at first at regional level (eg. I-C@AST programm - a textile project being reorganized - Lombardy 2006), and later by the Ministry of Economic Development (2003-2005 -2008)3. Art.3 of Law 266/1997 (so-called Bersani Law) for the first time establishes state funding for the Regions to be used in Industrial Districts identified on the basis of Law No. 317/1991, to fund programmes that improve service networks, in particular in the ITC sector.

Another important intervention of industrial policy in favour of Districts was carried out under the Decree for the liberalization of the electricity market, which envisaged the opportunity for those entities defined as "eligible customers" in Article 2 to sign supply contracts with any producer, distributor or wholesaler in Italy and abroad. Among those considered for this option are groups formed by companies whose total consumption reaches a value greater than 30 GWh, located exclusively in the same municipality or contiguous municipalities, or in areas identified with specific acts of regional planning. New policies to support technological development in the Districts were approved in 2001 and in 2005. The last supportive action took place under the 2008 Budget which included a call for bids to grant funding to districts also engaged in improving environmental performances at level of production area. Surveys by research bodies such as the "Districts Club" and the "IPI-Institute for Industrial Promotion" reported data showing a significant presence of the district system in Italy. Here is some information resulting from the last industry census carried out by ISTAT in 2000. In Italy there are 155 districts mostly located in the Northern regions. This model of production organization involves 4,929,721 employees and 1,180,042 businesses of the manufacturing, services and trade sectors[4]. The areas where this approach is prevailing are textiles with 45 districts, mechanics (38), household goods (35) and the tanning industry (20). Districts are less common in the production of paper and paperboard, in the food industry, chemicals and plastics.

The regions with greater presence of industrial districts (2001 data) are Lombardy (27), Marche (27), Veneto (22), Tuscany (15) and Emilia Romagna (13)[5]. Among the 27 districts of Lombardy the most important sectors are textiles and engineering, while in the Veneto and Marche regions districts are more present in the production of household goods. In Tuscany, along with textiles the most important districts are tanning and paper, while in Emilia Romagna the presence of districts in the mechanical industry is accompanied by the food industry and the production of household goods.

3.3. The "Industrial districts" in Europe

When comparing the characteristics of Industrial Districts as previously mentioned with the concept of cluster at European level, three important common characteristics emerge. Firstly, clusters are seen as *geographic concentrations* of specialized firms, of highly skilled

---

[3] Europe INNOVA Cluster Mapping Project: Report Italy Available in: http://www.clusterobservatory.eu/upload/Policy_Report_Italy_20080116.pdf>

[4] Club Distretti, Map of Italian districts, 2005.

[5] ISTAT, Industry Census 2000.

Region	No Districts	Workers
Piedmont	12	297,034
Lombardy	27	1,745,042
Trentino-Alto Adige	4	46,814
Veneto	22	861,546
Friuli-Venezia Giulia	3	123,244
Emilia-Romagna	13	574,432
Tuscany	15	466,494
Umbria	5	61,823
MarcheMarche	27	435,063
Lazio	2	31,542
Abruzzo	6	96,859
Molise	2	4307
Campania	6	26,177
Apulia	8	144,096
Basilicata	1	9,927
Sicily	2	3,236
Sardinia	1	2,085
ITALY	**156**	**4,929,721**

Source: ISTAT 2001

Table 1. Presence of manufacturing districts in Italy

Industry	Industrial Districts	Local manufacturing units	Workers
Textile and clothing	45	63,954	537,435
Mechanics	38	56,816	587,320
Household goods	32	42,287	382,332
tanning and footwear	20	23,441	186,680
Food	7	3,781	33,304
Jewelry and musical instruments	6	13,010	116,950
Paperrmaking and printing	4	4,342	35,996
Rubber and plastic	4	4,779	48,585
TOTAL	**156**	**212,410**	**1,928,602**

Source: ISTAT 2001

Table 2. Types of industrial districts in Italy

and capable workforce, and of supportive institutions that improve the flow and the spillover of knowledge. Secondly, the cluster is useful to reach the functional objective to *provide a range of specialized and customized services* to a specific group of firms. Finally, clusters are characterized by some social and organizational elements, called *"institutional social cohesion tools"*, which link the different and interconnected actors, thus facilitating a closer cooperation and interaction between them.

At European level, two different approaches are used to identify clusters. The best known is based on *"case studies"*, the gathering of *qualitative information* through interviews with local experts or research on documents and publications. The second approach regards the various *quantitative techniques* that rely on the most sophisticated economic models and are based on statistical methods which encourage to identify clusters indirectly, by measuring the effects that are supposed to be detectable in the presence of a cluster.

There are hundreds of case studies that document the history, activities and impact of clusters on regional development, on employment and innovation. The European Cluster Observatory[6] has published 25 case studies of European clusters, related to the areas and the sectors indicated in the picture below.

Through the collection of "case studies" each cluster tells "its own story" and sometimes it is difficult to compare different results. Furthermore, due to the rapid changes occurring in clusters, the results may arise from old data. Therefore, by describing relationships, processes and interactions among actors this methodology becomes an excellent tool that can be used to complement statistical analysis. With regard to *quantitative techniques*, the approach used by the European Cluster Observatory is based on indirect measurement of the effects revealed by coordinated localization of those elements that are assumed to be detectable in the presence of a cluster, such as the concentration of workers or high productivity. There are other techniques for quantitative mapping of clusters but, unlike the operational methodology[7] of the European Cluster Observatory, they are not constantly updated according to changes which take place in the countries analyzed.

The first results came in June 2007 with the establishment of a framework of regional clusters in 31 countries, divided into 38 areas. For the first time, the quantitative analysis performed is based on a fully comparable and consistent methodology in all European countries.This method identifies clusters based on regional employment data collected by EUROSTAT and national and regional statistical sources. The approach used is deliberately based on the measurement of the effects that relationships and spillover have on the companies' choice of location, and not on the direct measurement of the dynamic interactions among the forces driving the cluster.

The quality and quantity of the knowledge that circulate and the *spillover* among firms located in a cluster depend on the size of the cluster, on its degree of specialization and on how well the areas are equipped and focused on the production in the main industries that make up the cluster. Therefore, the three factors *size, specialization* and *focus* can be chosen to assess whether the cluster has reached a "specialized critical mass" likely to *spillover* and develop positive relationships. Statistical mapping of Clusters by the European Cluster Observatory identifies over 2,000 regional clusters in Europe, among which clusters

[6] The European Cluster Observatory was founded in September 2006 by Europe INNOVA.

[7] Cluster Mapping methodology developed by the Institute for Strategy and Competitiveness, Harvard Business School.

classified as *"industrial districts"* are 1380[8]. The following table shows the geographical distribution of "industrial districts" clusters.

European States	No Industrial Districts	Percentage of total
Austria	34	2.5%
Belgium	19	1.4%
Bulgaria	33	2.4%
Cyprus	2	0.1%
Denmark	18	1.3%
Estonia	6	0.4%
Finland	16	1.2%
France	103	7.5%
Germany	269	19.5%
Greece	18	1.3%
Ireland	7	0.5%
Iceland	2	0.1%
Italy	158	11.4%
Latvia	3	0.2%
Lithuania	9	0.7%
Luxembourg	1	0.1%
Malta	4	0.3%
Norway	12	0.9%
Netherlands	29	2.1%
Poland	97	7%
Portugal	27	2%
United Kingdom	58	4.2%
Czech Republic	61	4.4%
Romania	75	5.4%
Slovakia	23	1.7%
Slovenia	8	0.6%
Spain	104	7.5%
Sweden	27	2%
Switzerland	34	2.5%
Turkey	83	6%
Hungary	40	2.9%

Source: European Cluster Observatory

Table 3. Geographic Distribution of "Industrial Districts" clusters

[8] Clusters that do not comply with the definition of Industrial Districts were not included in the total number of clusters considered by the European Cluster Observatory. Specifically, they are: Agricultural, Business services, Distribution, Education, Entertainement, Finance, Fishing, Hospitality, Sporting, Transportation.

Even taking into account the different spatial dimensions of the European countries analyzed, the table and the graph above show a greater presence of clusters in Germany, Italy, Spain and France. Within the above mentioned countries, there is a predominance of the Construction and Food sectors with 15.5% and 10.7% respectively on the total number of districts. The Construction sector is more prevalent in Germany (27 out of 269 districts), in Italy (21 out of 158), and Spain (17 out of 104). In France, however, the greatest number of districts is in the Food sector, with 19 out of 103 districts.

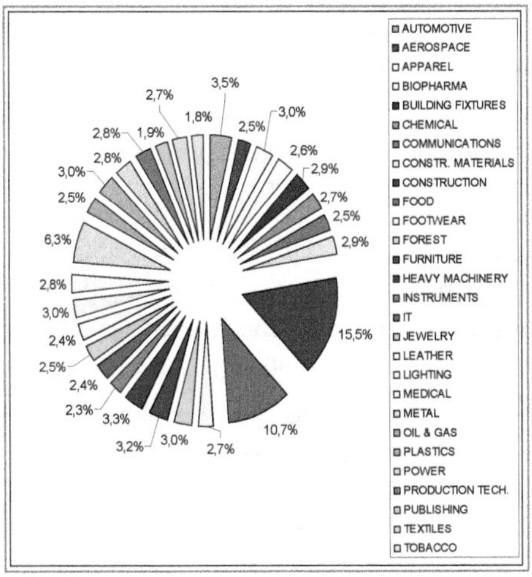

Source: European Cluster Observatory

Figure 1. Number of Districts by Industry

In recent years, many initiatives have been implemented in Europe in order to create favourable conditions for the establishment of new clusters and strengthen existing ones. To date, more than 130 specific national measures in support of clusters were identified in 31 European countries and registered by the INNO-Policy Trend Chart[9]. Nowadays, almost all European countries have specific measures for clusters or programmes developed at national and/or regional level, suggesting that they are a key element of the national and regional strategies in support of innovation

3.4. Cluster approach in the management of environmental issues

The size of clustering in a local context has critical relevance in the analysis of the environmental impact of industrial activities. When assessing the impacting factors related to a particular type of production, the characteristics of different local contexts in which that type of production produces its environmental effects have to be taken into account. Italy clearly shows how the environmental impact of some industrial sectors (textiles, tanning, ceramic) is localized around some areas where there is a high concentration of industries

[9] More detailed information in <http://www.proinno-europe.eu> and <http://cordis.europa.eu/erawatch>

from those sectors. In these cases local dimension becomes a key determinant of the significance of environmental issues for the entire industry sector and, at the same time, a key variable in coordinating an effective response by the companies.

There is no doubt that in terms of impacts on the environment, companies that operate in an industrial district have many elements in common.

First of all settlement, production and sales activities of these enterprises influence the same local ecosystem, characterized by specific and defined environmental aspects. Moreover, companies operating in one district often face similar environmental problems, because they dump the emissions from their production processes into the same receptacle: waste water that drains into the same river (eg the Bisenzio river that runs through the entire Prato textile area, or the Sarno in the Salerno tomato district) or solid waste that goes into the same landfill.

On the other hand, the high specialization of production and the usually very small size of enterprises (with all the implications in terms of limited availability of human, technical and financial resources) allows us to think of the district as an industrial area sufficiently homogeneous also in terms of production methods, degree of technology and organizational and managerial choices. The same technological and organizational matrix of the businesses in the district may show in common environmental problems that are related, for example, to the inefficiency and ineffectiveness of facilities to reduce pollution, to technology obsolescence, to inadequate structures for environmental management, cultural lag and so on.

Even relations with suppliers of equipment and components, according to the logic of "vertically integrated industry" that characterizes many districts, are often played at local level, thus also affecting the availability and appropriateness of the most innovative and advanced technological solutions for pollution prevention (think of the crucial role companies of the so-called mechano-ceramic play in the district of Sassuolo, as the almost exclusive repositories of technological know-how and, therefore, appointed to develop and propose new 'clean technology' to the ceramic businesses in the district).

A final aspect to highlight is the relationship with local stakeholders: for businesses in the district, interacting with the same community, the same institutions, the same local supervisory bodies means to deal with the same needs and requests concerning the quality of the environment. This is of fundamental importance if we consider that the significance of an environmental problem depends on the way in which it is perceived socially. The local dimension is a context where the relationship with company stakeholders is intensified, it becomes more straightforward (given the coexistence in the same area), more immediate (e.g. relationships with local institutions are more frequent than with national institutions), closer (just consider the number of residents employed by enterprises in the district). Besides, given the homogeneity of industrial activities, the physical proximity and frequent inability to attribute the environmental effects to any one production unit, enterprises in the district are considered by local partners almost as single entity.

The relational dynamics among companies and external stakeholders therefore become a crucial pressure factor to foster awareness on environmental issues within the district. By acting the same way and with the same incisiveness on a large number of similar businesses, it reinforces itself and strengthens its effects. For example, if the local population shows particular sensitivity to environmental issues, all enterprises in the district will undergo a high degree of examination from the public (which leads them to ensure continued compliance to regulations) and will be encouraged to use tools to enhance their environmental commitment to the local community.

Other important partners for companies in the district are local institutions. Sometimes companies interact with local authorities and supervisory bodies who are open to dialogue and willing to leave some room for negotiation, or with institutions that are particularly strict as regards law enforcement and extremely demanding on the compliance with obligations and deadlines. The different attitude of institutions can mitigate or amplify the context pressure, acting in the same direction for all firms in the district. Firms can be challenged with requests from local authorities that may focus on some environmental aspects (making them more problematic) or that may promote the application of certain environmental policy tools (e.g.: voluntary agreements at local level).

Local institutions may also prove to be particularly active in promoting common solutions (subsidiaries or consortium) to the most demanding and urgent environmental problems in the district, acting as a catalyst to encourage collaboration among businesses and promoting synergy in the commitment of human, technical and financial resources.

The local dimension represents an essential key in understanding environmental issues also because the same solution to environmental problems can be managed at district level. For example, the infrastructural equipment of a purification plant helps the industrial system in reducing the environmental impact. However, enterprises may find themselves having to directly invest in the installation of small treatment plants, which is known to result in a "scattered" distribution of facilities rather than in a systematic and consistent process.

Increasing awareness to environmental issues by the actors with whom the company interacts implies the need to meet certain "environmental questions".

This is especially significant for SMEs operating within an industrial district. In fact, efforts in the direction of environmental improvement by an individual company are here associated with new knowledge and with the onset of difficulties (the environment, as we have seen, is a challenge or new "turbulence") that once overcome constitute know-how that can be shared with other firms in the district. In this process of growth also appears/arises the need for support from (and relationships with) external actors, a need common to most SMEs, which fosters the development of new "answers" to the emerging needs/demands. In a territorial dimension the resulting "networking" takes peculiar forms, leading to the development of somewhat common solutions (i.e. based on sharing tangible or intangible resources) that are tied to the specific local environment in which businesses in the district interact.

Recent decades have shown the dynamic of those "common solutions" in industrial districts, connected to the different inputs and external forces that have enabled the development of strategies and tools to start up environmental management processes that could involve the whole district

4. Environmentally Equipped Industrial Areas (EEIA) Cluster

4.1. Cluster "Environmentally equipped production areas and eco-industrial parks: definition and boundaries"

"Industrial area" means an area with specific land uses, geographically limited, that is near or on the periphery of urban centers. The industrial area can affect one or more municipalities. "Environmentally equipped areas" were introduced in Italy by Legislative Decree no. 112/98 (the so-called "Bassanini Decree") that in Article 26 states: "The regions and the autonomous provinces of Trento and Bolzano govern, with their own laws, industrial areas and environmentally equipped areas, with infrastructures and systems necessary to guarantee the protection of health, safety and environment. The same laws govern the forms of single management of infrastructures and services in ecologically equipped areas by public or private actors... omission ... and procedures for land acquisition included in industrial areas,...omission .... The production plants located in ecologically equipped areas are exempt from the acquisition of permits concerning the use of services therein. The regions and autonomous provinces identify areas ... omission ...mainly choosing among already existing areas, zones or centres, even if partially or totally abandoned. The local authorities concerned participate in this identification procedure".

Hence, national legislation gives the individual regions the task to regulate the matter, giving some basic points of reference:

1. environmentally equipped areas have infrastructure and systems necessary to protect health, safety and environment;
2. environmentally equipped areas are characterized by forms of centralized management of infrastructure and services;
3. manufacturing plants located in environmentally equipped areas are exempt from the acquisition of permits concerning the use of services therein. [eliminato, era un copia e incolla che ripeteva la frase del punto 2).

A different approach must be used as regards the so-called "eco-industrial parks" (EIP), which are spread across Europe and the world and are in a way similar to Environmentally Equipped Areas, but are usually voluntary initiatives. Eco Industrial Parks as theorized by Lowe, Moran, and Holmes are communities of manufacturing and services firms linked by a common management, and they seek to improve their environmental, economic and social performances by collaborating when addressing environmental issues and using the resources (including energy, water and materials). This integrated approach aims to achieve collective benefits that exceed the sum of individual benefits each company would separately have from the optimization of its performances. The path to achieving this goal includes a new design or redevelopment of infrastructures, and planning of the production area, cleaner production, protection from pollution, energy efficiency and cooperation among enterprises.

The early theorists of the concept of industrial ecology as we consider it today were scientists Robert Frosch and Nicholas Gallopolus, who in an article published in Scientific American in September 1989 defined a new strategy for the manufacturing industry: "the traditional model of industrial activity, in which production processes generate products for sale and waste for disposal, must be transformed into a more integrated model: an industrial ecosystem. This system optimizes energy and raw materials consumption up to using residues from one process...to feed other processes".

Other important researchers (Tibbs, Allenby, Graedel, Lowe, Holmes, Moran) contributed to the development of the concept of industrial ecology. They developed industrial ecology into a discipline based on multidisciplinary contributions aiming at the improvement of the industry-environment relationship. In 1992 Tibbs, another important pioneer, held that "*in natural systems there is no waste, meaning something that can not be absorbed constructively elsewhere in the system" coming to a key concept which is closing cycles: "...making maximum use of recycled materials in new products, optimizing the use of integrated materials and energy, minimizing waste and recovering waste as raw materials for other processes*".

Thus industrial ecology considers the flow of matter and energy with the aim of significantly reducing the use of resources and pollution. It suggests the application to industrial systems and their processing/production cycles of the rules and principles that determine the functioning of non-human biological systems, of ecosystems that are characterized by symbiotic relationships and by the absence of the concept of waste. Every scrap is reintroduced into the cycle to generate energy or as raw material to start another process that is essential to maintaining the overall balance.

Both types of production area (APEA and EIP), therefore, aim at the so-called "closing cycles" of material, water and energy; they aim at sharing key environmental services (water, energy, waste) and at optimizing the organization of activities that have an impact on the environment.

The cooperative approach can be seen mainly in two basic aspects:

- The adoption of collective systems and infrastructures within the industrial area (e.g. purification plant, centralized area for storing waste, industrial water supply systems, power generators for the area);
- The identification of a single production manager that deals with common services within the production area (e.g. collective management of waste, energy, security).

4.2. Cluster "Environmentally equipped production areas and eco-industrial parks" in Italy

As mentioned above, Italian legislation lets the regions regulate the issue of environmentally equipped areas on their territory.

The spread of APEA in Italy depends on the choices promoted by each region. To date, the regions that have issued laws and regulations are:

- Abruzzo (R.G.D. October 10, 2003, No. 1122, "Leg. Decree March 31, 1998, number 112 - P.R.D October 20, 1998, No. 447 as amended by Presidential Decree March 31, 2000, No 440 Definition of the discipline of "Environmentally equipped areas").
- Calabria (Regional Law December 24, 2001. Number 38 "New legal system for Consortia for Areas, Centres and Industrial Development Zones);
- Emilia Romagna (Regional Law No 20/00 "General framework on the protection and use of the territory", Legislative Assembly Resolution No. 118/07 "Adoption of the Guidelines Act and technical coordination on the implementation of ecologically equipped areas in Emilia-Romagna).
- Liguria (Regional Council Resolution of December 28, 2000 No 1486 "Criteria, parameters and methods on the industrial areas and environmentally equipped areas)
- Marche (Regional Law n. 16/05 "Regulation of urban redevelopment and guidelines for the environmentally equipped production areas", and DGR n. 157 of 07/02/2005 "Guidelines for the ecologically equipped production areas (APEA) of Marche Region").
- Puglia (Regional Law January 31, 2003, No. 2 "Guidelines on actions for economic development, production activities, industrial areas and environmentally equipped areas).
- Tuscany (Tuscany Regional Law No. 87 of 22/12/2003 "Ecologically equipped production areas. Changes to the regional law December 1, 1998, No 87").

In other Italian regions in which there are no laws specifically dedicated to APEA, there are other standards that facilitate the environmental management of industrial areas or the creation of EIP. It is the case with regional laws governing consortia of industrial development which are suitable structures for the collective management of a number of issues in the area, including environmental issues (eg, Friuli Venezia Giulia, Sicily).

In other cases, although APEA characteristics are not regulated by regional standards, they are often referred to in regional planning documents (eg. DOCUP 2000-2006 Regione Piemonte).

An analysis of the related regional legislation shows that regulations concerning the management of industrial areas come from different disciplines, especially laws related to planning and environmental and production activities.

In some regions the choice to go towards an APEA is compulsory, since it was decided in urban planning (e.g. Emilia Romagna), while in others the choice is voluntary (e.g.Tuscany). Although the national law was enacted over 10 years ago, its application in different regions is not settled yet, but still under development.

APEAs could potentially become a very popular model of production area in Italy, but to date there are only few cases of full implementation.

In the various Regions there is the enactment of laws and technical regulations, the implementation of experimental projects and (albeit only recently) financing, but the reality of APEAs in Italy has not yet established itself in terms of actual implementation.

It is therefore difficult to identify APEAs in Italy. To date, the most advanced experiences are in Tuscany, where it is possible to identify 12 production areas involved in a qualification path towards APEA, and in Emilia Romagna, with a process of public funding for APEA which saw the application of 42 industrial areas.

This gradual development is also linked to the fact that the different regional regulations provide for environmentally equipped areas a very broad field of application, stating that issues such as waste management, water resources, transport and logistics, security, etc. should be dealt with, thus covering all environmental issues in a systematic way.

However, some environmentally equipped areas are precursors, i.e. industrial areas that implemented solutions fully in line with the contents of Legislative Decree No 112, 1998.

For these areas the definition 'eco-industrial parks' is more correct, since they are production areas that do not fully comply with regulatory requirements of APEAs, but where environmental management initiatives such as collective management of certain environmental problems (not all) were activated, or collective facilities were built.

These experiences represent case studies with established characteristics in the review of Italian good practices, while actual cases of APEA will be mostly developing initiatives.

4.3. Environmentally equipped production areas and eco-industrial parks Cluster in Europe

Eco industrial parks are not identical to the Italian environmentally equipped areas, but they certainly show strong similarities.

First, the reference cluster, identified in the industrial area intended as a geographically defined and limited area, with production activities; then the adoption of engineering and management solutions designed to reduce environmental impacts, in fact making the environment variable a lever for competitiveness.

Indeed, the main difference between APEAs and EIPs is that while the former are governed by technical and planning rules, often involving public entities, the latter are usually created on the basis of future economic gain.

The involvement of the public administration is also less frequent, although it still constitutes a major subject in many cases.

Eco-industrial parks are spread globally, and it is possible to find examples of success in Europe, North America and Asia. Although it is difficult to define the categories, it is interesting to highlight some elements that differentiate the European EIP from the American or Asian EIP.

Although there are exceptions, in general in European countries the initiator of the development process of the production area is a public actor which promotes the EIP as a solution to territorial problems. The involvement may be associated with a form of local governance. On the contrary, in Asia EIPs are usually linked to the economic value related to the application of the principles of industrial ecology, especially as concerns the issue of waste.

Another difference is related to the size of industrial areas (much bigger in Asia), which makes it easier to trade secondary raw materials. In Europe, collective solutions are usually linked to the sharing of financial and human resources (services, collective facilities).

The EIPs in North America are more similar to the European ones, although the public presence in developing the production area is lower.

A further difference between the eco-industrial parks in Europe and those outside Europe is that in Europe there are many technology parks that are classified as eco-industrial parks, i.e. areas that are not strictly involved in production, but which play a role in research, communication and promotion for the development of environmental technologies and sustainable management solutions in the production industry. It is difficult to gain a thorough knowledge of the distribution of eco-industrial parks in Europe, America and Asia since there are no institutional structures of reference.

The summary of the presence of eco industrial parks in the world was based on data collected from studies, networks and devoted sites.

A study by the University of Patras (Greece) in 2003 on the state of the art of eco-industrial parks in the world identified over 100 cases, of which 42% in the U.S., 36% in Europe, 11% in Asia and 6% in Canada. In Europe, eco-industrial parks are mainly concentrated in the North-West, especially in England, France and Scandinavia.

The geographic department of Hull University mapped a distribution of about 30 eco-industrial parks in Europe. It points out that in some cases they are not actual industrial parks but science and technology parks or initiatives and programmes aimed at sustainable management of production areas.

In a Finnish study on eco industrial parks conducted in 2006, data on the number of eco-industrial parks in Europe is lower and equal to 20, while North American EIPs are 36.

An investigation conducted by the Chinese 'National Commission for Development and Reform' (presented on the website of *Cleaner Production in China*) analyzed the Eco Industrial Parks initiatives undertaken in Asia. There are about 40 initiatives for environmental industrial networking or Environmental Management. These initiatives were launched in eastern Asia: China, India, Japan, Philippines, Malaysia, Taiwan, Vietnam, Thailand, Sri Lanka.

4.4. Features of the cluster approach in the management of environmental issues

The environmental performances of an APEA, both in consumption of non-renewable resources and in emission of pollutants in the air, water and soil, are based on three important aspects: urban planning, plant and infrastructure equipment, and management.

The search for performance excellence starts with the way in which the spaces within the industrial area are designed. The second aspect is based on using the best available techniques (eg. dual networks for the water cycle, own production of energy from cogeneration or from renewable sources) in line with national and EC legislation on Environmental Integrated Authorization (IPPC Integrated Pollution Prevention and Control) and on common spaces and facilities instead of individual ones (e.g. industrial sewage treatment plant for the entire area, common waste storage areas, centralized basins for the collection and treatment of stormwater).

The third aspect provides for the optimization of synergies already existing among the various businesses and the unified management of spaces and of centralized systems (e.g. provision of a grant for the recovery of waste among the firms located in the area, area mobility management, area energy management).

A collective management is the heart and engine of improving the environmental performance of the production area. Through the use of collective facilities and infrastructure it makes it possible to provide businesses with services that allow greater protection and environmental control and a reduction in costs.

The manager is generally an expression of the realities present on the territory, and it may be a public, private or mixed actor.

The single production manager is also responsible for the definition and implementation of an area environmental programme and for monitoring the environmental performance of the area. The Area Environmental Programme starts by analysing the existing issues and proposes solutions involving relevant local stakeholders (e.g. government, companies, trade associations, managers of public services).

Existing initiatives concern the activation of virtuous mechanisms among firms (sharing of human and technological resources, materials and energy flows in an industrial environmental perspective), the delivery of services to companies by the subject tha manages the area, the infrastructural and plant elements available to the production areas.

The following table shows the main solutions of good environmental management identified in the 32 analyzed cases (17 international cases[10] and 15 Italian cases[11])

---

[10] Eco-industrial park of Devens (USA); Burnside Industrial Park (Canada); Kokubo (Japan), Naroda Industrial Estate (India), LIK (Indonesia); Kalundborg (Denmark), Parc Industriel Plaine de l'Ain, Sphere EcoIndustrie D'Alsace, Syndival Lancadcres (France); Crewe Business Park, Sustainable Growth Park (U.K); Ecopark Hartberg (Austria); Vreten (Sweden); Value Park (Germany); S.Perpetua di Magoda, Parque Tecnologico de Reciclado Lopez Soriano.

[11] 1° Macrolotto Industriale di Prato, Z.I. Ponterosso - San Vito al Tagliamento (PN), Z.I.P. Padova, Z.I. Udine Sud, Z.I. Castello Lucento – Torino, Z.I. Valle del Biferno – Termoli, APO Ferrara, APO Ravenna, APEA Monte S. Vito (AN), APEA Pianvallico (FI), APEA Navicelli (PI), APEA Scandicci, APEA Ozzano (BO), APEA Ostellato (FE), APEA Colbordolo (PU).

RAW MATERIAL CONSUMPTION AND WASTE MATERIALS	
Good management of waste from construction activities	Recovery and safe disposal of waste generated during construction activities. The Italian experience demonstrates the success of the initiative when the monitoring activity carried out by the operator is accompanied by specific requirements in the Implementation Standards of Urban Planning.
Collective organization of waste disposal	It refers in particular to the special waste collection from businesses, which can be performed by the area operator or, more frequently, by third parties.
Promoting the principles of industrial ecology	Exchange activities among enterprises from the perspective of recovery and reuse of waste used as secondary raw materials. Usually there are specific agreements among a limited number of companies with synergic production processes.
SPREADING OF CLEAN TECHNOLOGIES	
Fostering of clean production technologies	Provision of research facilities; promotion of innovative solutions also through exhibition spaces; activation of pilot projects. Initiatives are generally aimed at creating of new jobs.
ADMINISTRATIVE SUPPORT FOR COMPANIES	
Interface between companies in the area and public administration	The area manager acts as mediator between businesses and the public administration with regard to administrative proceedings. The service can be simply informative (the manager contacts the local authority to obtain information and passes on said information to the interested companies) or it may concern the drafting of forms or the actual issuance of permits as delegated by the Municipality.
Environment info desk for businesses	Setting up an information service on environmental issues regarding regulatory obligations and as support for the processes of environmental certification.
Support to businesses for investment and funding	Activities carried out by the area manager for businesses as regards investments and research of grants and funding.
TRANSPORTATION	
Agreements with the provider of local public transportation	Study of traffic flows and routes from home to work, on the basis of which the single manager signs agreements with the provider of public transport to improve the service in terms of frequency and stops.
Car sharing and car pooling	Setting up a system of vehicles to stimulate the activation of a system of collective transport, as an alternative to the use of individual private vehicles. The initiative aims at reducing traffic in both commuting and missions.
TRAINING AND TECHNICAL SUPPORT	
Training on environmental issues	Courses for technical business personnel organized by the area manager, aimed at a proper application of environmental and

	safety regulations. Training and information activities for companies in the area offered directly or through third parties.
Eco – efficiency centers	Technical assistance and training to companies to improve their energy and water efficiency and in the production and disposal of waste.
SETTING UP OF BUSINESSES	
Environmental remarks in the selection of companies that wish to settle in the area	Generally carried out by drafting a questionnaire that the interested companies are required to fill in. The area manager is responsible for the selection as actor in charge of selling parcels of land or on behalf of the local authority.
Checking the companies status of compliance in terms of environmental requirements	Check up the relevant documentation prior to the settlement and visits to the site at the time of installation. This service is successfully carried out where companies have adopted an area Environmental Management System.
Envisaging environmental clauses in the contracts of sale for parcels	Provision of specific environmental clauses in contracts of sale for land parcels (e.g. when planning the areas for material handling or storage of waste) or request to adhere to the environmental policy.
Urban-environmental advantages	Removal of infrastructure costs, reduction of safety, health, and environmental taxes.
Landscape protection	Preserving the environment and integrating the business with the surrounding landscape through a centralized management of the development of the production area.
INNOVATIVE MANAGEMENT SOLUTION	
Implementation of environmental management system	Implementing an Environmental Management System complies with regulatory standards (ISO 14001 or EMAS) by the subject that manages the area. The EMS can provide a business involvement in varying degrees, for example in the adoption of common procedures or construction of a participatory environmental program. These elements also depend on the area of infrastructural facilities available
Creation of a territorial information system of the industrial	Development of a GIS on-line with all spatial information (geographical, cadastral and possibly environmental). The GIS is accessible by the companies settled and may represent a geo-referenced base that allows performance monitoring of the area. The management of the SIT is borne by the operator or company in charge of the area.
Agenda 21 meeting to define the action for cluster	Organise meetings within the local Agenda 21 to encourage the productive area to address environmental sustainability issues (energy supply, waste management, cleaning, transport and construction).

	Through working groups involving representatives of companies can address problems related to environmental issues and identify solutions that best suited for that territory.
Definition of an environmental action plan based on performance standards	Participation of local communities and businesses in programs of environmental action through the application of standards of environmental, social and economic performance and certification. The standards are defined by a committee made up of representatives of local production activities (trade associations) and citizens.

Table 4. Solutions for good environmental management

5. Cluster supply chain

5.1. Cluster "supply chain": definition and boundaries

The concept of this kind of cluster is not yet an 'autonomous' element in the economic literature, as is rather the case for other types of clusters (for example the large and consolidated strand of economic literature on *industrial districts,* or the most recent and innovative one on *environmentally equipped production areas*).

Although the meaning has not been conceptually developed yet, references to the 'supply chain' cluster are widespread and detectable in several contexts, both legal and academical.

Starting from Porter's remarks (1990), the *cluster theory* originally focused on identifying the key features of the 'cluster entity', with the aim of analyzing the ways in which its operation functioning mechanisms and its internal dynamics are able to determine a competitive advantage for businesses/industries belonging to the cluster itself. According to Porter, clusters are geographic concentrations of interconnected companies, specialized providers, service providers, and associated institutions in a particular field[12]. Firms located in the same area have the opportunity to operate with ease, in coordination along the value chain. The cluster is based not only on goods and *material* resources, but also on *intangible* resources such as development and exchange of knowledge, expertise, and relationships. These elements make a territorial area unique, an area in which human factor and knowledge make the difference when compared to any other area.

In other words, the supply chain is conceived as a geographically defined element within the cluster. This approach is confirmed by that part of literature aimed at investigating the connections between cluster theory, supply chain and supply chain management theory. This is a trend not developed theoretically and empirically yet, but which contributes to clarify some important aspects of the relationship between cluster and supply chain. Some authors observe that one of the key elements of Porter's cluster theory - the benefit deriving from knowledge exchange and cooperation between firms - is shared by the theorists of the

[12] *"A cluster is a geographically proximate group of interconnected companies and associated institutions in a particular field, linked by commonalities and complementarities"* (Porter, 1998).

supply management theory13 (De Witt et al., 2006; Mentzer et al., 2001). In fact, supply chain management aims at producing benefits for companies that are part of the chain through process integration and the set up of long term relationships (cooperation, trust) among companies. When companies belonging to the same supply chain operate in the same geographical context, they benefit from coordination and from the boost to competitiveness and innovation that comes from their geographical proximity (Mentzer et al., 2001).

To sum up, according to Porter the factors that characterize the cluster can be identified in geographical proximity, production specialization and in the interaction among the different actors in the cluster.

In this sense and for the purposes of our analysis it is relevant to note that proximity both upstream and downstream the supply chain is what facilitates interaction and promotes a continuous exchange of ideas and innovations.

Some studies aimed at providing empirical evidence of the existence of positive synergies, in terms of the impact on competitive performances that arises from belonging to a cluster. According to De Witt et al. (2001), the geographical proximity of firms belonging to a supply chain allows long-term competitive advantages that are more stable than those achieved by outsourcing to distant companies. Other studies highlight that the relational factors that characterize the relationship between buyers and suppliers belonging to the same cluster are able to enhance their performance in the long run (Noordewier *et al.* 1990; Corsten and Kumar, 2005). More specifically, the key elements that determine long-term competitive advantages are the *interdependence*, the mutual *trust* and *shared goals and commitments* that pervade the relationship among actors in the chain within a cluster, and the reduction of conflicts among the actors themselves (Ganeson, 1994, Doney and Cannon, 1997, Kumar et al., 1995).

A second important connecting element between cluster and supply chain refers to a particular connotation of the cluster, defined by the presence of a large multinational company surrounded by a 'halo' of suppliers ("A large demanding purchaser, such as a major multinational firm [...] surrounded by a 'halo' of suppliers", Johnston, 2003). In this respect, territorial location still plays a central role in the competitive dynamics of the cluster: the proximity of suppliers of a large enterprise allows for the development of agglomeration economies thanks to the direct or indirect links that are established between the economic activities upstream and downstream.

---

[13] In spite of the popularity of the *supply chain management* both in theory and in practice, there is no unanimous agreement on its meaning, also because its development in the managerial and academic fields is relatively recent. In his important work of 2001 ('*Defining Supply Chain Management*'), Mentzer defines the supply chain management, or management of the supply chain, *'a systemic, strategic coordination of the traditional business functions within a particular company and across businesses within the supply chain, for the purposes of improving the long-term performance of the individual companies and the supplu chain as a whole'* (Mentzer, 2001). With 'supply chain' he means *'a set of three or more companies directly linked by one or more of the upstream and downstream flows of products, services, finances, and information from the source to the ultimate customer'* (Mentzer, 2001).

More recently, the literature recognizes the need to introduce a new meaning of the word cluster alongside its more 'traditional' concept. Nowadays, the technological improvement of communication systems and of distribution networks at global level makes it possible to talk of 'virtual' clusters, where the element of geographical proximity is missing while emphasis is placed on features such as exchange and sharing of information and knowledge among the actors in the cluster (Johnston, 2003).

According to Rullani[14], this new type of cluster is characterized by an "evolved" form of proximity, not only physical but also virtual, among people and businesses that use technological mediation to develop 'close relationships' (easy, frequent, reliable, complex) even when they are physically distant from each other, thus replicating in a virtual space the benefits of proximity that in the past were a typical (and almost exclusive) characteristic of the territory.

The academic theory of virtual clusters is joined by their institutional recognition. In its working paper annexed to the EC Communication "Towards world-class clusters in the European Union: Implementing the broad-based innovation strategy" [COM(2008)652][15], the European Commission, while recognizing that most definitions of cluster focus on two factors - the concentration of one or more sectors in a given geographical area and the importance of networking and cooperation among enterprises and institutions - indicates that the spatial dimension of a cluster is variable and not necessarily limited to certain geographic boundaries, depending, among other things, on the ability and willingness of its actors to perform changes that are functional for the development and preservation of relationships that feed the cluster itself.

The definition of clusters developed by the OECD makes the "release" of the concept from the element of geographical proximity even more explicit, emphasizing once again the production and exchange of knowledge within the value chain as a key element of identity of the cluster "Clusters are characterised as networks of production of strongly interdependent firms, knowledge-producing agents and customers linked to each other in a value-adding production chain" (OECD, 1999).

A final element of analysis comes from recent literature aimed at defining the links between the concepts of cluster and supply chain in the context of today's 'knowledge-based economy". If it is true that the cluster theory in general takes a macroeconomic perspective in which the theories on supply chain refer to a purely microeconomic level, it is possible to recognize some key elements the two theories share and that can lead to the identification of significant similarities between the two concepts (Sureephong et al., 2008). In particular, the success of a cluster is often due to the presence of a Cluster Development Agent (CDA), an organizational entity that represents the different socio-economic actors in a cluster, acts as coordinator and facilitator of cooperative dynamics among the actors, promoting knowledge sharing, innovation, the ability to communicate and the mutual recognition and trust[16].

---

[14] Enzo Rullani, *Cluster: tendenze e scenari nell'economia globalizzata,* "Pattern of clusters evolutions" conference proceedings, Venezia.

[15] European Commission (2008b).

[16] Notice how this role and functions correspond to what in the context of E.U. policies are the tasks of the governing body of a cluster that is established when a cluster is 'institutionalized' (Cfr.: CE, 2002).

The literature recognizes a strong similarity between this role and the one that in the context of the supply chain management theory is played by the Supply Chain Facilitator (SCF), the person that guides the relationships among actors in a supply chain, and encourages the development of relationships and cooperation among them. If this role was originally played by the large multinational company, in the context of the current "'knowledge-based economy"' its function can be successfully carried out by other bodies - such as universities, associations, local institutions, *ad hoc* coordinating organizations - that belong to the same cluster as the socio-economic actors in the supply chain (Sureephong *et al.* 2008).

5.2. Supply chain cluster in Italy

As abovementioned, for the purposes of this analysis we should highlight how within this theoretical model it is possible to trace a first description of cluster as a type of "supply chain", relating to a local chain dimension, coinciding or included in the 'classic' industrial district, physically located in a given territorial area. The cluster supply chain is especially notable for the presence of a primary industry supported by several companies specializing in various phases of the industry. Thanks to territorial proximity and the presence of a socially cohesive community, these companies have the ability to easily operate by coordinating along the value chain (Sacco, Ferilli, 2006).

With respect to this first description, the literature emphasizes how clusters linked to local industries represent an area of strong identification and specialization in the economy of many Italian regions, although generally the cluster does not reach the regional dimension (Bardi, Bertini, 2005). In many contexts, some leading companies' ability to grow and export can foster local induced activities and is an example increasingly imitated by dynamic new small businesses, a catalyst of growth processes and territorial specialization. Examples of chains related to this meaning may be found in many traditional Italian industrial districts. Consider, for example, the area between Carpi and Reggio Emilia for knitwear and clothing, the area of Forlì and Cesena for the food industry, the Bassa Bresciana for shoe manufacturers, the furniture chain located between Matera and Puglia, etc.

More recently, the literature shows how in many cases this type of articulation of the cluster supply chain reached maturity during the 90s, and how the highest degree of complexity of the industry and the increased competition on international markets led to an expansion of the classical concept of district, which meant an area in which a single homogeneous production chain was focused. To respond to these changes, the new articulation of a cluster is as metadistrict, that is a territorial chain not concentrated locally, but that creates a widespread, sectorially specialized network - on a regional, multi-regional or national scale - with strong interaction/interdependence/competition among the business realities belonging to it. In many cases this connotation can be traced back to the development of products that identify the so-called 'Made in Italy' in some specific areas (typically the North East of the country and Tuscany), mainly focusing on sectors such as fashion, home, typical food products (e.g, buffalo mozzarella in Campania, bresaola of the Valtellina), light engineering (e.g. the biomedical field in Emilia-Romagna) (Rullani, 2002).

In this respect, the cluster supply chain is characterized by the different role of territorial contiguity of businesses with regard to cooperative and competitive dynamics that characterize it. Studies and remarks note how the metadistrict was established to overcome the old links among companies - where proximity is a prerequisite for the existence of relationships that can trigger processes of technological exchange and learning – by creating new links related to the development of new technologies and services necessary to keep those companies competitive. Globalization and new information technologies do not "disengage" companies from the territory, but help to provide the cluster with another operational dimension, in addition to and not as a replacement of the local dimension (Zucchetti, 2003).

The change of competitive scenarios also influence the *nature of relationships among companies within the supply chains and production chains*. While within the traditional districts of the '70s and '80s the relationships among companies along the supply chain are basically egalitarian, mostly informal and usually direct, beginning from the '90s these characteristics deeply change. The stability of the suppliers' relationship with a limited number of clients, typically located within the district, is no longer a 'dogma', i.e. the degree of 'mobility' of the system is increasing.

Furthermore, to support the internationalization and export activity aimed at enhancing the 'Made in Italy', territorial production chains sometimes adopt a form of 'twinning' among the original industrial districts. Aggregation, in these cases, also aims at improving the awareness of belonging to a quality industry, promoting high visibility of the territories, increasing transparency and credibility with all stakeholders - investors, tour operators, consumers, etc. - and especially at increasing cooperation among industrial districts within a sector at national level[17] (Fontana, 2007). Consider, for example, the chain of leather products for shoes and clothing in Tuscany, or the textile chain in Lombardy.

Finally, the literature recognizes that even in our country the previously defined *virtual clusters* are taking shape. According to Rullani[18], a virtual cluster may arise from very different evolutionary paths:

1. From *previously existent clusters* that learn to master new technologies, combining the advantages of the existing local network with the advantages deriving from long distance relationships and multi-territorial sharing;
2. From *medium enterprises*, once "plunged" in a regional system, that use new information technologies to extend their supply and distribution networks in the global circuit (without losing their roots in the area of origin);

- From *multinational companies,* that discover the importance of differences and specificities of each territory they have access to, eventually anchoring themselves permanently to certain territorial specializations. These specializations are enhanced

[17] An example is the 'agro-ichthyc-food industry' twinning between the districts of *San Daniele del Friuli* (famous for the prosciutto *Dop*), *Nocera Inferiore-Gragnano* (famous for its tomatoes and pasta), *Mazara del Vallo Co.S.Va.P.* (fisheries) and *Vulture* (specialized in wine, fruit and vegetables, olive oil, cheeses and dairy products) (Fontana, 2007).
[18] *Ibidem.*

and expanded through access to that global network of virtual proximity and coexistence made possible by the multinational company itself.

If, in general, our country sees setbacks in the rise of virtual clusters (mainly due to the difficulty for the traditional districts/clusters of SMEs to evolve rapidly[19]), the third path here outlined leads to the identification of a third type of supply chain cluster in Italy, defined in terms of **'trade mark' (or 'brand') industry.** The connotation, though not prevalent in the Italian production reality (characterized by a structure of small and medium enterprises), can be recognized in all those cases in which large companies and/or multinational companies or large chains (whether the property be domestic or foreign) operate on a national scale through a widespread and 'loyal' network of suppliers and subcontractors. The following section presents in detail this specific type of supply chain cluster, widely present in the European and international context.

5.3. Supply chain cluster in Europe

While in Italy the patterns adopted by the supply chain cluster are significantly influenced by the specificity of the economic and social fabric, in the rest of Europe chain clusters are generally configured as business combinations, often multi-sectoral, linked by supply relationships (of goods and services) at different levels with large leading companies and/or multinational corporations.

Due to globalization, in recent decades the European context has wirnessed the emergence of the concept of cluster supply chain in terms of virtual aggregation. As anticipated, this type developes around large companies that become a drive of the territorial development of some industrial areas at regional level, through processes of "virtualization" of the flow of information and of internationalization of the supply and distribution networks. These companies are able to stimulate and support the growth of competitiveness poles of global significance, within which operate production chains belonging to one or more related manufacturing sectors.

Many corporations belonging to different sectors, such as IKEA, Ericsson, ABB, H&M, Volvo, belong to this type of cluster (Sölvell, 2006).

Therefore, in this type of cluster the globalization of the supply chain and of innovation goes hand in hand with the strategic importance of the so-called *'local environments'*. While local markets seem to have exhausted their function of driving force for most goods and services, the local concentration of activities related to various production sectors (old and new) continues to act as a driver of (multinational) business innovation. In other words, the multinational companies' ability to compete is linked to their ability to act as *"insiders"* (through their branches or subsidiaries) within the most dynamic regional clusters, and to

[19] Among the main factors that hinder the 'virtual' development of Italian districts and clusters, the literature identifies chronic saturation of work environment (full employment), of infrastructures (congestion), of free space (crowding), and of environmental tolerances (with balanced under constant stress) (Rullani, ib.).

globally coordinate activities in order to connect in large-scale networks global markets and the innovative thrust deriving from local clusters[20] (Sölvell, 2002).

In summary, the literature recognizes in this type of cluster the persistence of some distinctive features of the supply chain cluster, except for the element of proximity, now redefined on a virtual basis. The specialization of production and the interaction among actors in the production chain continue to classify the cluster, developing through a virtualization of the flows of knowledge that stimulates the constant change and "upgrading" of products and services, thus creating the bases for advanced supply chains that are diversified geographically as well.

5.4. Features of the cluster approach in the management of environmental issues

This paragraph analyzes the characteristics of the "supply chain" cluster with reference to the elements that characterize its management from an environmental point of view. To this end we gathered a review of existing theoretical and empirical studies, oriented along the three trends of literature identified as priority areas of investigation about the connections between *cluster approach*, *environmental management* and *supply chain management*:

- The trend of the so-called **Green Supply Chain Management,** or the development of supply chain management practices in environmental terms and their connections with the application of Environmental Management Systems (EMS);
- Studies and experience related to the adoption and implementation of **strongly product-oriented EMS**, so-called **POEMS** (*'Product-Oriented Environmental Management Systems'*), mainly based on the need to identify, assess and manage the so-called "indirect environmental aspects" (introduced by EMAS II Regulation) related to the product;
- The literature on the application of the concept and methodologies of **Life Cycle Assessment (LCA) in line with the chain/supply chain.**

These three trends have in common (albeit with different meanings at times) the acknowledgement of the need to adopt an inter-organizational approach to environmental management based on coordination and co-operation among different actors in the sector and not necessarily linked the territory. This need is the culmination of an evolution in the application of environmental management tools, characterized by the progressive recognition and assimilation by the businesses of a management approach inspired by the so-called life cycle thinking[21], marked by a greater sense of awareness and responsibility towards the environmental impacts that their activities have outside the confines of the production site (Carnimeo et al., 2002).

---

[20] In literature such clusters are sometimes called *"Hollywoods"* with reference to the californian cluster, world leader in the *entertainment* industry.

[21] In summary, *Life Cycle Thinking* can be defined as a 'cultural' approach that aims at focusing all management aspects of a product through a single 'magnifying glass': its life cycle. Under this approach, the environmental impacts (actual or potential) generated during the life cycle should be considered in an integrated manner when designing, developing and managing a product. For further reading see: Carnimeo, Frey, Iraldo (2002).

As previously stated, the first trend refers to the so-called Green Supply Chain Management, or that part of the literature born by integrating and "contaminating" environmental management and supply chain management studies, and targeted at investigating supply chain approaches and management logic from an environmental point of view (Srivastava, 2007)[22]. This is a trend literature regards as underdeveloped although potentially fertile for investigation, and so far studied in depth only in relation to certain phases or activities of the supply chain *viewed individually* (eg. green design, green procurement, reverse logistics, etc.) (Srivastava, 2007; Sharfman *et al.*, 2009). Conversely, the same literature agrees in giving particular importance to studies and contributions that on the one hand account for the reasons for adopting environmental supply chain management practices, and on the other hand account for the benefits and difficulties related to the implementation of the same practices.

At the empirical level, studies have shown the existence of a wide range of internal and external factors that may push companies to expand the environmental management to upstream and downstream supply chain activities: from the need to respond to increasing *pressures from external stakeholders* (e.g. consumers or institutions), to the need to *ensure compliance to more stringent environmental regulatory requirements*, to reasons of a *strategic nature*, or related to the opportunity to gain a competitive advantage (Sharfman *et al.* 2009; Darnall *et al.* 2008; Nawrocka, 2008). Corbett and Decroix (2001) affirm that nowadays the need to extend environmental management practices to the supply chain to improve performances is so recurrent in studies it has become a 'mantra' ("We have heard [...] various versions of the 'mantra': 'the next step forward in environmental improvement lies in supply-chain coordination', Corbett and Decroix, 2001).

Important for the purposes of this analysis are those studies that investigate the reasons that push companies to extend the environmental management to the supply chain under a cooperative approach. In this context, the empirical contributions emphasize how - beyond the reasons more closely tied to a proactive and '"value-driven"' environmental management - the key element that pushes companies towards a logic of cooperation with suppliers is mainly due to the uncertainty of information that governs the nature and extent of environmental impacts associated with the production process upstream and downstream activities, and the complexity and difficulty of decision-making processes frequently created by this uncertainty (Vermeulen and Ras, 2006 ; Sharfman et al., 2009).

In the context of a relatively scarce literature it is significant to note that some recent studies focus on the relationship linking the adoption and implementation of an Environmental Management System and the development of supply chain management practices, with the main objective to investigate (i) if and what kind of correlation exists in the adoption of these management practices, that is (ii) if and to what extent the adoption of one practice influences the adoption of the other (Darnall et al. 2008; Nawrocka 2008).

The process of gradual 'opening' of EMS and their redefinition in inter-organizational terms is recognized by the literature as a significant innovation in environmental management

[22] Srivastava's work (2007) offers a very detailed review of GSCM literature.

practices, characterized by the gradual assimilation by enterprises of the management logic inspired by the so.called life cycle thinking. This trend is identified by some authors with the gradual extension of the objectives and scope of management systems to environmental issues related to the lifecycle of the product (or service). The result of this "integration" is called POEMS - Product Oriented Environmental Management Systems (Klinkers et al., 1999).

The links between environmental impacts related to production and product make the environmental management tools within the company ineffective by shifting the emphasis on the relationships the company itself has with the actors that influence those impacts. By this logic, the goal of the management system is not only to ensure the implementation of the company's environmental policy by governing the processes and internal resources, but also to manage relations with the outside world, to foster and promote dissemination of that policy to other actors that share it. In other words, the scope of implementation of the management system becomes that of the actions and interactions through which many actors manage the impacts related to the different stages of the product's lifecycle (Sharfman et al., 1997).

The literature on POEMS highlights the benefits and limitations of these tools. In fact, although taking into account the entire lifecycle can provide more opportunities to reduce the environmental impact associated with the products - either through actions on specific issues and joint efforts involving different stakeholders along the chain - (Sharfman et al., 2009), thus contributing to the achievement of tangible improvements in performance (van Berkel et al. 1999; Charter and Belmane, 1999; Brezet and Rocha, 2001), the process of 'opening up' the environmental management systems is not immediate or 'painless' for businesses.

At European level, some experiences of POEMS developed since the '90s. In the Netherlands the signing of voluntary agreements between various industries and local authorities has allowed the development and funding of some pilot projects for product-oriented EMS. A study of sixty Dutch companies shows that the most sensitive element in the development of such tools is the difficulty for organizations to obtain the necessary information from their suppliers, because of the other companies' unfamiliarity with POEMS and the little influence of single organizations, especially small ones, on the whole product chain. Other experiences have demonstrated the 'compatibility' of the approaches and the requirements of international and European standards on environmental management systems (ISO 14001 and EMAS) and POEMS.

In France, a project to develop product environmental management in the automotive sector involved 250 companies, mostly SMEs, with the aim of honing tools and methodologies for the development of POEMS in organizations belonging to the sector. In this case the main project stimulus was of a regulatory nature, in connection with the requirements of the European Directive End of Life Vehicle, which demands specific knowledge of the supply chain[23] (Andriola et al., 2003a).

---

[23] Directive 2000/53/CE.

In Italy as well several POEMS experiences were carried out in medium-large businesses, or under programme agreements with public institutions (Ardente et al., 2006; Andriola et al., 2003a). Overall, the Italian experiences and the experimental applications in Europe (mainly in the Netherlands, France and Denmark) have not led to a 'coding' of POEMS by any international or Community standard. However, it must be noted how these experiences have contributed significantly to the development of methodological and management tools aimed not only at accounting for the impacts associated with the lifecycle of the product, but also at involving the suppliers in the assessment and quantification dynamics of these impacts. Among these tools, the methodologies of Life Cycle Assessment (LCA) are predominant[24].

A vast literature deals with the technical and methodological aspects as well as with numerous case studies and methods of dissemination of the LCA in Italy and in other advanced countries. For the purposes of this analysis, it is particularly important to observe the studies that investigate the role and mode of application of the LCA in the context of the management of supply chain relationships and management. In this respect, the literature recognizes that the potential of a LCA is revealed mainly in the form of a methodology capable of triggering an inter-organizational and networking approach within the chain. In fact, the steps that make up the tool[25] imply the need to activate along the entire supply chain the information and communicational channels necessary to collect and process data and information that allow to quantify the interactions between the 'product system' analized and the environment (Pesonen, 2001, Lefebvre 2000, Krikke et al. 2004; Sarkis, 2001; Sroufe et al., 2000).

Finally, it is important to note that some studies go further in analyzing the relationship between LCA and supply chain. They apply the methodology of Life Cycle Assessment as an opportunity to "restructure" the supply chain, with the aim of improving the environmental performance associated with a specific "product system". It is essential to take into account a number of factors for the LCA to be able to effectively contribute to

---

[24] The most common definition of this method comes from the Society of Environmental Toxicology and Chemistry (SETAC): *"a LCA is a process to evaluate the environmental burdens associated with a product [...] by identifying and quantifying energy and materials used and wastes released to the environment to assess the impact of those energy and materials used and releases to the environment; and to identify and evaluate opportunities to affect environmental improvements. The assessment includes the entire life cycle of the product [...] encompassing, extracting and processing raw materials; manufacturing, transportation and distribution; use, re-use, recycling, and final disposal"* (SETAC, 1993). If the inter-organizational approach of the POEMS consists in considering the product-system as *"a network of operations linked together by flows of materials and energy [...] that ties activities and processes in different organizational contexts"* (Heiskanen, 2000), the use of a LCA proves itself extremely useful, since it can be aimed at finding the optimal solutions for the entire product-system, regardless of what would be preferable from an environmental point of view for each single process (or single organization) (Carnimeo *et al.*, 2000).

[25] The structure of the LCA derives from the scheme set up by SETAC in the early '90s with the aim of proposing a common approach for all the analyses carried out until then. This format is still the basic structure from which the subsequent changes and amendments derived. Internationally, the scheme proposed by SETAC was complemented by the ISO Regulations 14040 series, which rule the drafting of LCA studies. According to the ISO general regulation, the assessment of the life cycle must include the following steps: 1) definition of the objective and scope of the study, 2) inventory analysis, 3) impact assessment; 4) interpretation of results.

improve the environmental performances associated with the implementation of a product or service which combine multiple actors along the value chain (Hagelaar G. and van der Vorst J., 2001):

- The results of the LCA methodology are closely tied to the definition of the objectives and scope of this tool that are set at the startup of the evaluation process;
- The application of LCA in the perspective of the supply chain involves strong cooperation among all actors involved, not only in terms of trust and openness, but also of *transparency in the shared data and information* and of *consistency in their policies*.

In conclusion, it is possible to account for some empirical evidence related to the implementation of tools linked to *Life Cycle Thinking* in the logic of supply chain management. When LCA is integrated into the environmental management system in a logic of supply chain dynamics management, it is to be assumed that companies that use and promote this approach are able to affect the environmental impacts or influence the behaviour of actors that are external to the "boundaries" of the companies' organization. This is a pre-condition for the integration of product logic to be effective. If this pre-condition occurs, the use of LCA within the supply chain relationships provides important contact points, synergies and complementarity with an "extended" environmental management system. The above mentioned approach of "indirect environmental aspects" introduced by the EMAS II Regulation is well suited to be the "leverage" through which to introduce these tools.

The experiences in the analyzed scenario refer to the implementation of a LCA aimed at including *product-oriented* logic in an environmental management system. A first empirical evidence refers to the transition from a single company to a supply chain perspective, in the implementation of the Initial Environmental Analysis (as provided by ISO 14001 and EMAS) that crosses the company's boundaries and allows to identify and properly assess the indirect environmental aspects thanks to the adoption of the LCA. Although in practice the EMAS Regulation does not require the company to conduct a thorough LCA, it is clear that knowledge of these impacts is necessary, especially when it is functional to the identification of the most significant indirect environmental aspects. Understanding the impacts of product disposal, or of the product's packaging, may be decisive for a company that uses large distribution channels and targets/is addressed to the final market.

The same applies when a company uses a transport network for the delivery of its products to intermediate customers "spread" on the territory. This activity can be carried out as part of a chain. In the case of the local supply chain that drove the PIONEER project26, for example, a simplified LCA of paper products was implemented (streamlined or screening LCA, to use English terminology), useful to companies in the supply chain to identify

[26] The Life PIONEER Project (2003-2006) had the objective to define and experimentally implement a methodology based on the EMAS Regulation to the paper industry district of Lucca. The methodology has promoted a cooperative and integrated approach to the environmental management at local level, aimed at involving all stakeholders in actions to improve the environmental performance of the territory. For further reading, see: Frey and Iraldo, 2008 and the website www.life-pioneer.info

particularly relevant indirect aspects and to determine the connections among the various business activities of companies operating at different stages of the supply chain, in order to focus on critical points on which it was then possible to work with joint programmes and improvements.

LCA can be applied effectively in the supply chain environmental management also in the pursuit of ways to improve eco-efficiency. The so-called Life Cycle Costing (LCC), for example, provides guidance on how to integrate "conventional" accounting with an approach that allows to identify longer-term strategic opportunities and efficiency margins. An interesting example of application of LCC is that of the logistics of the large company Xerox (Bennet, James, 1999). Careful analysis of logistics costs by the company (related to product distribution and the recovery of remains to be reused) showed efficiency margins in its supply chain. However, in order to seize these opportunities some interventions were needed to design, manage and reorganize logistics flows which would mean a different allocation of costs and benefits among the different stakeholders (Xerox suppliers and customers).

Examples of these solutions were the internalization of the costs of packaging (including disposal) in accordance with the suppliers, and the standardization of packages so that they were adaptable to every product and, above all, reusable by customers to pack the product being replaced (and returned) at end of life. With an initial investment of 4-5 million dollars, Xerox estimated annual savings of $1.2, to which some “intangible” benefits (e.g. in the management and organizatin of logistics: handling of homogeneous packaging, reduction of operations’ time, etc.) added up. Although the overall comparison gave a positive outcome, some phases of the chain showed an increase in costs. Had the company not analyzed costs and potential benefits in the different stages of the lifecycle using the LCC method and reasoning in terms of actions promoted and coordinated by Xerox in an integrated logic with the other industry players, they would have never independently decided to engage in the improvement programme.

There are other examples that show how attention to the product under a “supply chain” environmental management system driven by the logic of the life cycle can constitute a solid basis on which to build a strategy oriented to an “environmental” customer satisfaction27. An interesting case is that of Baxter International. After stating in its policy that "we will work with our clients to help them tackle their environmental problems", the company managed to translate this principle at the operational level by applying an approach heavily oriented towards the green supply chain management. Resolutely going beyond the boundaries of its organization, management at Baxter decided to take charge of the issues related to the disposal of waste that results from the use of intermediate products in the "downstream" activities. By this logic, a waste auditing service was activated with the aim to

[27] Although the concept of environmental *customer satisfaction* is not totally alien to the logic of a management system, it is usually related to social actors, since the main “customer” in environmental terms is the one that suffers the externalities of production. When this customer coincides with the traditional customer, however, the dynamics with which relationships can be managed differ from those that govern relationships with social *stakeholders*.

verify the needs in the waste management of their intermediate customers. The activity developed by this service has produced a set of suggestions/recommendations, then implemented through a thorough and extensive redesign of products and services in order to minimize their impact downstream of the production process (Fuller, 1999).

Another interesting case is that of the initiatives undertaken by a group of U.S. chemical companies in a "horizontal supply chain" (Elwood and Case, 2000) for the management of orders from customers in a logic of eco-efficiency. In this case, the needs of production efficiency go together with the attempt by the manufacturers to reduce the environmental impacts of their products in all stages of the lifecycle.

The examples here reported show some approaches that were made possible only through the cooperative relationship among the companies in the supply chain and, although not directly related to the application of LCA, they clarify that a product-oriented logic can effectively engage on the environmental management system of a company that operates as a producer or as a customer in any type of supply chain. This logic provides crucial support for the management of customer relationships, and bridges the gaps of the system when it comes to identifying the customers' needs, defining interaction modes, handling complaints and returns (for environmental reasons), reviewing the contract (which can include requirements concerning the product's impact) and measuring customers' satisfaction, which is essential in order to assess the environmental competitiveness and "green" marketing strategies implemented by the company itself.

In terms of marketing and environmental communication, there is another interesting example of application of product logic within a supply chain. It is the opportunity, now available and potentially very effective, to certify the environmental impact of a product of a "local" chain or a group of producers on the basis of an international scheme based on the ISO standard. The EPD international system (Environmental Product Declaration), currently managed by a body comprising representatives of some EU countries (including Italy) was born with the objective to certify the environmental performances of a product or service of a single firm.

Because of the evolution of the system and its gradual spreading (currently it counts more than 100 companies) there is the the need to promote products and services that come from an entire production system (i.e. a "cluster" or chain of companies) precisely because of environmental marketing objectives of a typical product or a product tied to a "brand". Thanks to the recent amendment of the EPD certification system groups of producers (from a district, a chain, a geographical area) were offered the opportunity to develop an *Environmental Product Declaration* that can enhance the excellent environmental performances of their "average" product.

6. Conclusion

SMEs are to be considered a crucial target if policy makers really want to pursue sustainable development. The environmental problem does not fully emerge if one considers individual firms (although in some cases there can be serious impacts on local environments and communities exerted by a single SME), but it pertains to their combined and cumulative impact across sectors. Therefore. these companies are responsible for a large share of business environmental impacts.

These observations have stimulated the development of the so-called "cluster approach" to manage the environmental issues of a large number of SMEs located in limited territorial areas. As above mentioned, this concept is not limited to the classical configuration of the industrial district (geographically confined industrial areas) but it also encompasses environmentally equipped production areas ("APEA"), or industrial parks up to the interactions along the supply chain.

For example, the Industrial District is a local system with the presence of a prevalent production activity carried out by a group of small independent firms highly specialized in different stages of the same production process. On the contrary, Production Areas represent an organizational model characterized by the territorial element but, unlike the District, they are concentrated in areas that are easier to define and circumscribe geographically, and do not necessarily show the presence of one or more specialised production sectors.

The networking approach allows enterprises to co-operate by identifying and assessing similar environmental aspects and by finding technological and operational solutions that can be applied to similar production processes and products, as well as by defining organisational structures suitable for the same kind of production cycles

In the APEA case, co-operation is facilitated by the 'physical contiguousness' and there are synergies both in improving the environmental impact on the same local eco-system, and in interacting and communicating with the same stakeholders (local population, authorities, etc.).

In some cases, a network was created among SMEs within a 'cluster' in order to foster information exchange and experience dissemination and to define and apply common solutions to similar environmental, technical and/or organisational problems, or to share environmental management resources (Iraldo & Frey, 2007). A specific kind of co-operation within a cluster of organisations takes place in the supply-chain: when a large customer is willing to support small suppliers in the EMS implementation process, then all the smaller organisations involved in the supply chain can greatly benefit from networking. This approach proved to be effective in some Member States such as Germany ("Konvoi" approach), Spain (co-operation in the tourism supply chain), Nordic Countries (Denmark and Sweden) but in particular in Italy, where by means of the so-called APO "Ambiti Produttivi Omogenei", it has shown its effectiveness in promoting the environmental compliance of SMEs.

Therefore, we collected some empirical cases to demonstrate how the innovative approach to environmental management called "cluster approach" can be an effective tool available to SMEs in order to improve their environmental performance and find innovative management solutions.

The goal for future research and experimental initiatives should be the development of the cluster approach and its structural inclusion in policy-making.

Author details

Francesco Testa[1,*], Tiberio Daddi[1], Fabio Iraldo[1,2] and Marco Frey[1,2]
[1]*Sant'Anna School of Advanced Studies, Pisa, Italy*
[2]*IEFE – Institute for Environmental and Energy Policy and Economics, Milano*

7. References

ADEME (2007). L'environnement et la maîtrise de l'énergie dans les PME, Paris.
Andriola L., Luciani R., Sibilio S. (2003b), 'I sistemi di gestione ambientale orientati al prodotto: POEMS, un nuovo strumento', Ambiente, n. 8
Bardo A., Bertini S. (a cura di) (2005), Dinamiche territoriali e nuova industria, Dai distretti alle filiere, V Rapporto della Fondazione istituto per il Lavoro, Maggioli Editore, Milano.
Becattini G. (1979). "Dal 'settore' industriale al 'distretto' industriale. Alcune considerazioni sull'unità di indagine dell'economia industriale", Rivista di economia e politica industriale, n. 1,
Calcagno D., Del Borghi A., Gaggero P.L., Sacerdote I. (2006), The contribution of LCA methodology to the environmental sustainability of cement production, 15th International Symposium on Mine Planning & Equipment Selection MPES, Torino - Italy, September 20 - 22, 2006
Cancila E., Bosso A. e Ottolenghi M. (a cura di), (2006), La gestione sostenibile delle aree produttive. Una scelta possibile per il governo del territorio e per il rilancio delle politiche industriali, Ervet, Bologna.
Carnimeo G., Frey M., Iraldo F. (2002), Gestione del Prodotto e Sostenibilità, Le imprese di fronte alle nuove prospettive delle politiche ambientali comunitarie e della IPP (Integrated Product Policy), Franco Angeli, Milano
Charter M., Belmane I., (1999), Integrated Product Policy (IPP) and Eco-Product Development (EPD), Journal of Sustainable Product Design.
Communication from the European Commission to the Council, the European Parliament, the european economic and social committee and the committee of the region "Towards world-class clusters in the European Union:Implementing the broad-based innovation strategy" (COM (2008) 652).
Corbett C. J., DeCroix G. A., (2001), Shared-Savings Contracts for Indirect Materials in Supply Chains: Channel profits and Environmental Impacts, Management Science, Vol. 47, n. 7

---

* Corresponding Author

Corsten D., Kumar N. (2005), Do suppliers benefit from collaborative relationships with large retailers? An empirical investigation of efficient consumer response adoption, Journal of Marketing, Vol. 69, n.3

Darnall N., Jolley G. J., Handfield R. (2008), Environmental Management Systems and Green Supply Chain Management: Complements for Sustainability?, Business Strategy and the Environment, n.18

De Witt T., Giunipero L. C., Melton H. L. (2006), Clusters and supply chain management: the Amish experience, International Journal of Physical Distribution & Logistics Management, Vol. 36, n 4.

Doney P.M., Cannon, J.P. (1997), An examination of the nature of trust in buyer-seller relationships, Journal of Marketing, Vol. 61, 1997.

ENEA - Progetto Life-Siam (2007), Linee Guida per l'insediamento e la gestione di aree produttive sostenibili [on line], disponibile su <http://www.life-siam.bologna.enea.it/> [data di accesso: 23 marzo 2009]

Fontana D. (2007), Distretti Italiani in quindici anni di vita di strada ne ha fatta, Amministrazione & Finanza, n. 22/2007, Inserto

Frosch R. e Gallopolus N., (1989), Strategies for manufacturing, Scientific American, 261: 94

Fuller D.A. (1999), New decision boundaries: the product system life cycle, in Sustainable Marketing, SAGE Publications, Thousand Oaks, London, New Delhi.

Ganeson S. (1994), Determinants of long-term orientation in buyer-seller relationships, Journal of Marketing, Vol. 58

Graedel T.E. and Allenby B.R., (2002), "Industrial Ecology", Prentice Hall, 363 http://www.chinacp.org.cn, Cleaner Production in China, [data dell'ultima consultazione: 25 marzo 2009]

Hagelaar G., van der Vorst J. (2001), Environmental Supply Chain Management: using Life Cycle Assessment to structure supply chains, Paper IAMA 2001, Sydney, Australia.

Heiksanen E, (2000), Managers'interpretation of LCA: enlightment and responsibility or conflusion and denial, Business Strategy and the Environment, Vol. 9

Johnston R. (2003), Clusters: A Review, The Australian Centre for Innovation Limited.

Lowe, E.A., Moran S.R. and Holmes D.B. (1996). Fieldbook for the development of Eco – Industrial Park, final report, Washington

Marceau J., Dodgson M., (1999) Systems of Innovation, Paper N. 1, Innovation Summit, Department of Industry, Science and Resources, Canberra

Marshall A., (1980). Principles of Economics: an Introductory Volume. London.

Marshall C. (1998). Report for HM Treasury.Economic Instruments and the Business Use of Energy. The Stationary Office, London.

Mentzer J.T., DeWitt W., Keebler J.S., Nix N.W., Smith C.D., Zacharia, Z.G. (2001), "Defining supply chain management", Journal of Business Logistics, Vol. 22 No. 2.

Morikawa M., (2000), Eco–industrial development in Japan, Indigo Development Center

Nawrocka D. (2008), Inter-Organizational Use of EMSs in Supply Chain Management: Some Experiences from Poland and Sweden, Corporate Social Responsibility and Environmental Management, n. 15

Noordewier T.G., John G., Nevin J.R. (1990), Performance outcomes of purchasing arrangements in industrial buyer-vendor relationships, Journal of Marketing, Vol. 54

Pesonen H-L. (2006), Multinational Corporations and Clusters, Vinnova Conference Proceedings "Challenges to National Innovation Systems in a Globalizing World", Stockholm, 19th January 2006.

Porter M. E., (1998). On Competition, Harvard Business School Press, Boston.

Porter, M.E. (1990). The Competitive Advantage of Nations, Macmillan, London.

Regional Council of Etela – Savo, (2006), Eco Industrial Parks, background report for the eco – industrial park project at Ratasalmi – Interreg IIIB.

Rullani E. (2002), Dallo sviluppo per accumulazione allo sviluppo per propagazione: piccole imprese, clusters e capitale sociale nella nuova Europa in formazione, East West Cluster Conference Proceedings, 28-31 Ottobre 2002, OECD, LEED, Udine.

Sacco P., Ferilli G. (2006), Il distretto culturale evoluto nell'economia post industriale, Università Iuav di Venezia, DADI Dipartimento delle Arti e del Disegno Industriale, WP WORKING PAPERS DADI/ WP_4/06.

Schumpeter J.A. (1971), Teoria dello Sviluppo economico, Sansoni, Firenze.

Sharfman M.P., Ellington R.T., Meo M. (1997), The next step in becoming "green": life-cycle oriented environmental management, Business Horizons, vol. 40

Sharfman M.P., Shaft T. M., Anex R. P. Jr (2009), The Road to Cooperative Supply-Chain Environmental Management: Trust and Uncertainty Among Pro-Active Firms, Business Strategy and the Environment, n. 18.

Sinding K. (2000), Environmental management beyond the boundaries of the firm: definitions and costraints, Business Strategy and the Environment, Vol. 9, n. 2

Sölvell Ö. (2002), The Multi-Home Based Multinational – Combining Global Compentencies and Local Innovativeness, Institute of International Business – Stockolm School of Economics

Srivastava S. K. (2007), Green supply-chain management: A state-of-the-art literature review, International Journal of Management Reviews, Vol. n. 9, Issue 1

Sureephong P., Chakpitak N., Buzon L., Bouras A. (2008), Cluster Development and Knowledge Exchange in Supply Chain, The Proceeding of International Conference on Software Knowledge Information Management and Applications (SKIMA 2008), Katmandu, Nepal.

Tibbs B.C, (1992), Industrial Ecology: an Environmental Agenda for industry, Whole Earth Review, 4 - 19

Van Berkel R., van Kampen M., Kortman J. (1999), Opportunities and constraints for Product-oriented Environmental Management Systems (P-EMS), Journal of Cleaner Production, n.7(6)

Vermeulen W. J. V., Ras P. J.,(2006),The Challenge of Greening Global Product Chains: Meeting Both End, Sustainable Development, vol. 14.

Zucchetti S. (2003), Una nuova generazione di distretti industriali, Impresa & Stato.

Chapter 2

The Economic and Financial Feasibility of a Biodigester: A Sound Alternative for Reducing the Environmental Impact of Swine Production

Antonio Zanin and Fabiano Marcos Bagatini

Additional information is available at the end of the chapter

1. Introduction

The modern landscape is one where environmental impact encroaches upon our quality of life. The search for viable technologies which both alleviate and lessen environmental pollution has become a priority, especially in the arena of production. Business, as with society, is now focused on minimizing environmental degradation, reviewing its strategies, structures and responsibilities. Tinoco (2001) explains that this modern responsibility is defined by the environmental and social demands of our day, where it is not merely a question of profit but also social conscientiousness. Business is now geared towards the interests of society, where environmental policy features high on its agenda.

It seems that human being has the tendency to risk his existence and wellbeing when the environmental impact of business proves a constant disruption to our natural world. Rural swine production and the environmental dangers it poses, is one example: waste, in remaining exposed releases methane gas into the atmosphere. As Brilhante and Caldas (1999) on this point, over the last few decades the dissipation of gasses has been affected by such practices, resulting in an increased concentration of carbonic gas (CO^2); methane (CH4); chlorofluorocarbons (CFCs), nitrous oxide and atmospheric ozone. As we now know, these gasses disrupt the energetic equilibrium of the Earth's atmosphere, and by consequence, our climate system.

Swine production has contributed substantially to Brazilian trade; it has received large investment incentives from genetics and other technologies, in order to provide and ensure a quality product. Panty (2008) highlights, that Brazil and its State of Santa Catarina have a model integrating both industry and producer, where specialists and rural farmers are well attuned to the competitive advantages of swine production.

Rural producers of swine do however face a lack of financial assistance in their search for an environmentally friendly solution. But an alternative is available, one able to reduce the endangerment of even more natural resources. Various "techniques" bringing to light and lessening the environmental effects of swine waste have been developed and put into practice. For instance, the process of biodigestion transforms methane gas into carbon dioxide, lessening environmental impact. The implementation of a biodigester means that waste can be reused and transformed into a renewable resource, proving an important mechanism for both business (finance) and society (the environment) alike. In other words, the biodigester is today's alternative solution for the rural farming industry, capable of minimizing the environmental effects of swine production. Waste on each farm, can be ably "re-directed" benefiting financial return; and where the farmer's quality of life is enhanced economically and financially, so too is that of the population at large who rely on the natural world for sustainability and survival. Indeed, we not only have the importance of increased productivity and the ensured success of new markets, but of being alert to a challenging future of sustainability and social responsibility.

Our research assesses the financially and environmental viability of installing a biodegster on a farming property. We divide our present study into 7 subsections, not counting the introduction. Beginning with *Environmental Management* we seek to contextualize the discourse and importance of our research by drawing upon recent literature in order to underline the contemporary shift in practice – namely the move towards more environmentally sensitive and socially sympathetic business strategies. Environmental strategy is vast becoming the defining characteristic of market competitiveness, and the success of a business to be environmental conscious and aware in its strategy and decision making, will prove the measure of its market edge and value. Our third section, *Swine Production and the Environment,* introduces the readership to the unique global positioning of Brazilian swine production and the very real possibilities of introducing Biodigestion as a viable and important measure ensuring both environmental integrity as well as cost effectiveness for the farming business. Section four, *Biodigesters,* introduces anaerobic biodigestion as the natural mechanism for both farming and environmental integrity. We seek to detail this anaerobic process and set out the processes and procedures of installing a functional biodigester on farming properties. Our sections on *Methodology* and the *Interpretation of Data* introduce the reader to the sampling method and data specifics of the herd studied. We then follow on with a section dealing with initial investment into the biodigester project, detailing the projected return and revenue, this, qualifying such investment as both environmentally timely and financially cost-effective for the business.

We conclude our study by underlining the necessity and urgency of biodigestion for swine producers, this, supported by the modern context of social conscientiousness and the benefit of sound financial return in line with our research projections.

2. Environmental management

The more our environment is damaged, the more our planet earth is compromised. We are witnessing, and experiencing, the progressive extinction of fauna and floral species, the pollution of groundwater and global warming.

In the face of these threats, is environmental management which aims to minimize environmental impact and maintain the wellbeing of people by redefining practices, processes and procedures for private, public and rural life. Good, proper working conditions and environmentally sound products complying with environmental laws and regulations fall to the responsibility of environmental management, as does the proper handling of waste produced in rural areas and the legal remit within which organizations can operate. This said, environmental management seeks to strike the legal and ethical balance between quality, productivity and competitiveness with the minimization of environmental degradation. Thus, for Moura (2000), the workings of environmental management involve putting specialized concepts and management techniques into environmental practice.

An environmental management system also signals greater competitiveness for the business which can equally retain and attract modern, learned consumers, whilst meeting the growing demands of external markets. Barbieri (2007) tells us that environmental management is defined by its administrative guidelines and key activities such as the planning, direction, control, and allocation of resources. Its main objective, furthermore, is to achieve positive, environmental results by reducing and eliminating the damage caused by human practices, or indeed to prevent such damage from even arising now and in the future.

This is why companies are now looking to develop and implement environmental management at the core of their operations in order gain an environmental advantage over market competitors, in strict accordance with the principles of sustainable development. Thus in retaking the concept of Barbieri, Tinoco and Kraemer (2004), we can assert that the role of environmental management is to effectively minimize and eliminate the environmental risks of private, public and rural businesses.

Companies can therefore meet the demands of the environment (and likewise of society), by tallying the expenditure of resources with legislation, restoring the natural resources extracted from the environment.

In this context, environmental management underlines the importance of environmental certification and accreditation which aims to help companies engage with, and commit to, the environment. Ribeiro (2006), for example, states that it is necessary to determine the particular strategy of implementing guidelines so as to more broadly define a company's environmental status and profile. Here, tools such as economic planning can be implemented into environmental programs that seek to change the current management system. Such programs, moreover, must be constantly checked through environmental audits. Yet aside from the virtues of the environmental management model, there is a constant need for businesses to be aware of the rapid changes which are occurring in information technology and strategic cost initiatives.

Indeed, models are constructed using the concepts defined by companies to guide and achieve goals. Barbieri's findings (2007) underline that the adoption of a model is critical because activities can be developed by different people at different times, in different places and through different ways of perceiving and positioning crucial issues. Companies can create their own environmental management models or take advantage of the various generic models that have been with us since the mid-1980s.

We see then that companies are finally coming into the age of environmental awareness, where standards and environmental legislation feature high on the agenda of those wanting to be maintain a market edge, nationally and internationally. We have a new series of standards defining our environmentally sensitive and fragile modernity, where the environmental status of a particular company has become the internationally accepted standard, seal and guarantee.

3. Swine production and the environment

For Marion (2002 p. 24) rural business, "explores the productive capacity of soil through the cultivation of land, as well as breeding and processing of certain agricultural products". The author classifies such rural activity as (1) agricultural; (2) zootechnical; and (3) agro-industrial. Similarly, Araújo (2003, p. 31) points out that, "agro-industries are businesses defined by the processing, handling and transformation of natural agricultural products into commercially packaged goods." Such businesses – deemed "agro-industries" by Marion (2002) and Araújo (2003) – are those which transform the agricultural/zootechnical product: the process of breeding, raising and slaughtering pigs for example, is for the purpose of transforming and commercializing derivatives. In a strictly economic context, the swine industry plays an important role in the movement of the food and supply chain. Sobestiansky (1998) shows then, that the modern swine industry is primarily focused on the production of pigs for slaughter and / or the breeding of livestock.

One of the most evident changes occurring both worldwide and in the Brazilian pig industry is the linear trend where a decrease in the number of production systems runs parallel to an increase in the number of system matrices. In terms of international agriculture, Brazil has the specialized workforce capable of producing technology that ensures a competitive advantage – it is a country fit to compete on equal terms with any other in the agricultural business, heavily investing in research and production strategies. The work of Gonçalves and Palmeiras (2006) shows us that the Brazilian swine industry has received greater international attention for its advantageously competitive edge: swine production in Brazil has lower costs than its major worldwide competitors; its system of production is vertically integrated (meeting agro-industrial demand); foodstuffs and basic grains such as soybeans and corn are plentiful, and there is technological investment.

In retrospect, market realities and the accelerated growth of a global economy meant that agro-industrial businesses had to seek improvements and overhaul their organizational structures in order to guarantee market presence and competitiveness. The swine industry was by no means an exception to this trend, requiring both injections of investment into its processes and new facilities. As Leite (2008) makes clear, swine production underwent change because technological innovation became the rule-of-thumb for a new generation of engineers focusing on the economic and environmental viability of such practices. Zuin and Alliprandini (2006, p. 255) similarly assert that, "over the years innovation and invention proved powerful tools in achieving better efficiency in farming systems." An equally important factor to this is the partnerships established between pig farmers and agro-industrial businesses, which have ensured the commercialization of production and added market value to the final product. Yet Mior (2005) equally points out that at the close of the 1990s in the western region of Santa Catarina, partnerships between agro-industries and swine producers heralded a new epoch in pig production, by reducing the number of contracts at the same time as increasing production.

According to findings from the Associação Catarinense de Criadores de Suínos (2011), Brazil has 2,460 matrices for the housing of swine, where production sprang from 2,708 million tons in 2003 to 3,240 million in 2010 – this, an increase of 19.65% in merely five years, signaling market growth and the importance of swine in the supply chain. Yet the concentration of pig herds on small rural farms nevertheless carries environmental impacts owing to the vertical model of production adopted by Brazil, which is characterized by partnerships with industry.

Generally, animal waste is treated in liquid form: water runs into deposits which are stored and the soil then used as organic fertilizer (EMBRAPA, 2008). This model coupled with the growth of swine production in Brazil heightens the risks of environmental degradation whereby the measures for treating swine waste are not only costly but require constant precision – mishandled waste can lead to water, soil and air pollution with both an unwelcome stench and mosquitoes.

In addition to industry and urbanization (domestic sewage), swine production is monitored by regulatory bodies and environmental agencies, so much so that the law clearly identifies the environmental dangers of such practices (GUIVANT e MIRANDA 2004). This is due to the large number of contaminants found in effluents which represent a potential source of air, water and soil contamination. Indeed, due to the high concentration of livestock at these rural sites, swine waste can easily exceed the capacity of local ecosystems, potentially disrupting the natural environment and human health through organic matter; nutrients; pathogens; odors and microorganisms generated in the atmosphere (PEREIRA, DEMARCHI e BUDINO, 2009).

A series of requirements can therefore be provisionally laid out, aimed at preventing and correcting increased environmental degradation. Among these we can list:

1. The need to maintain a permanent boundary of preservation set at a distance of 30 meters, with distances between dwellings and settlements of at least 300 meters, and distances to roads at least 50 meters (Bezerra, 2005).
2. The need to prohibit and monitor the dumping of waste and/or effluents from any polluting source, including waste from livestock, into Class I rivers intended for domestic supply. Such material may be released directly or indirectly into Class II and III rivers only after appropriate treatment and having satisfied the conditions, standards and requirements set forth by government Decree (RESOLUÇÃO CONAMA nº 357, from 17th March, 2005).
3. The need to research ways of combining the use of waste for crops (fertilizers), or for the production of energy. This would reduce the degree of environmental pollution in line with the realities faced by rural farmers. Given this context, some environmental problems could be solved if environmental measures for swine production were effectively researched and put into practice. However, not all swine producers have the awareness or financial resource to treat waste correctly.

4. Biodigesters

Seganfredo (1999) notes that the continuous use of large quantities of swine waste as fertilizer has proven environmentally detrimental, not only in terms of air pollution but in terms of the progressive accumulation of nutrients in the soil and the presence of excess nitrates in water. Likewise Sampaio et al. (2010) show that the mismanagement of remaining waste water can lead to an excess of pig manure in the soil (depending on the capacity for absorption some of these nutrients may lead to water contamination).

Nogueira (PALHARES, 2008) therefore does well to remind us that in 1086 the Englishman Humphrey Davy identified a gas rich in carbon and carbon dioxide resulting from the decomposition of manure in humid places. Released into the atmosphere, the gas attacks the ozone layer and causes global warming. Jordan (2005) states, moreover, that methane - one of the gases produced by the degradation of waste – is twenty-one times more volatile than carbon dioxide (CO2).

It is precisely here that the process of anaerobic biodigestion can prove the necessary environmental solution in that it destroys pathogenic organisms and parasites. In this way the treatment of waste by such means carries great advantages, transforming harmful gasses into a source of energy (bio-gas) *for the better*. In addition to this, solid matter decanting in the bottom biodigesting tank acts as biofertilizer, with liquid matter the (treated) mineralized effluent. Nogueira (1986) further points out that such a process offers multiple advantages. The production of fuel gas; the control of water pollution and odor; the elimination of pathogens from organic matter and the preservation of fertilizer are the immediate benefits of such waste removal. We can likewise emphasize that anaerobic digestion helps to minimize negative environmental impact, at once reducing relative risks and improving quality of life issues.

A digester is essentially built from a tank sealed with canvas suitable for storing waste. The process of biodigestion occurs in rotating fashion where organic matter enters through the side and anaerobic fermentation taking place without the presence of air. The result of this process is the transformation of methane into carbon dioxide.

The biogas storage balloon is used for the intake and utilization of highly corrosive gases which require beneficiation treatment. Such processing is essential and increases their efficiency, for without it the use of biogas is not recommended. The process has specific stages of washing; cooling and compression. Finally, there is the intake of biogas by the storage (balloon), as evidenced in Figure 2.

The liquid effluent resulting from this process then exits the digester for maturation ponds wherein the release of gases is completed: even after waste digestion in the machine, some substances still remain and must be released into the atmosphere so that the liquid can be used for fertilization.

As Nogueira explains (1986) waste is transformed by agitation which distributes the substrate and bacteria, efficiently using the volume of the biodigester and reducing/eliminating supernatant scum matter. In order to have a guaranteed and precise process, agitators must be inserted into the biodigester in order to correctly agitate the substances necessary for transforming waste. It is also important to maintain the biomass at a heated temperature within the biodigester, for as Nogueira (1986) further points out, the biodigester has to be constructed and set up below ground-level, since this depth serves as a thermal insulator. Temperature plays an important part and it is advisable to ensure internal or external heat (of course depending on the agricultural needs of the producer) because bacteria can reproduce in this way, thereby transforming waste into biogas.

Figure 1. The biodigester set up on the property where research findings were collated.

Figure 2. Biogas storage balloon

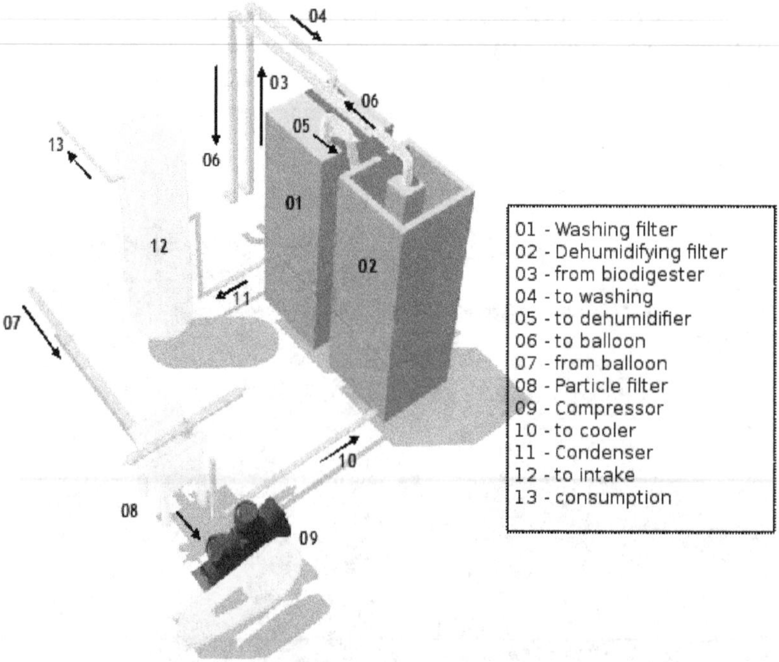

Source: Gter Energias Renováveis.

Figure 3. The biogas treatment process prior to intake.

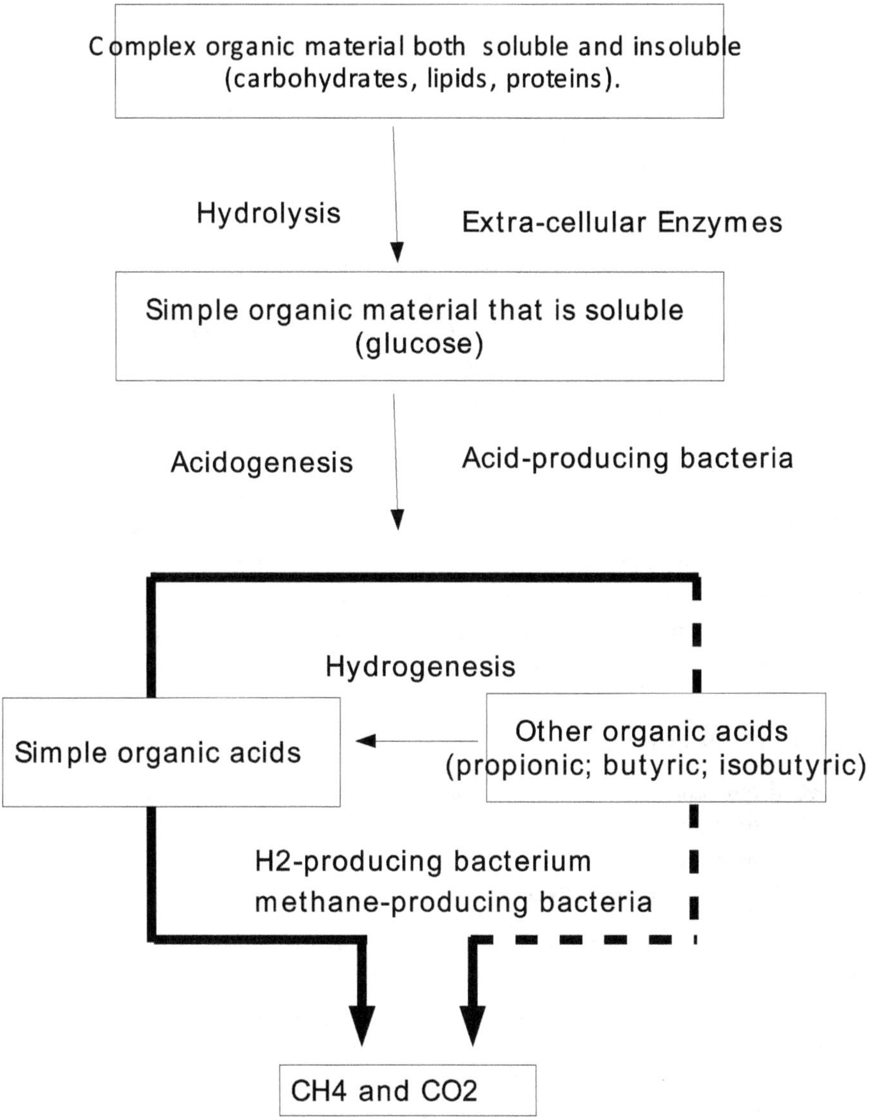

Source:http://www.biodieselbr.com/i/energia/renovavel/biogas/processo-biodigestor.gif.
Accessed: 18/5/08.

Figure 4. The metabolic stages of anaerobic digestion occurring in the biodigester: the transformation of organic materials into gases used by farms, means that the environment is benefitted through the reduction of harmful methane gas emission.

The introduction of biodigesters to rural swine production can thus equally benefit producers and the environment, reducing and possibly eliminating risks of environmental degradation whilst ensuring the quality of life required by human being.

5. Methodology

Our present study focuses on a rural farm chiefly producing swine. Research was carried out on a farming property with a herd of 5, 362 pigs. All processes and procedures undertaken at the farm are officially approved and licensed by the governing environmental agency. We also investigated whether the implementation of a biodigester is an economically and financially viable alternative: where there is the appropriate treatment of swine waste from production and the financial return of investment for the farmer.Data was solicited by semi-structured interview, with the additional analysis of documentation. The methodology of our study consisted of exploratory (qualitative) research, case studies of site procedures as well as quantitative research. Collated data was then tabled, and calculations / financial projections were made for the economic viability of investment.

For the sake of anonymity, "Business A" is used to designate the slaughterhouse and "Business B" the company which invested funds into the biodigester.

Calculations for the economic and financial viability of investment drew upon data collected in the first half of 2009. These data were tabled and are the basis for projections. In January 2012, we revisited the businesses which had participated in the research study in order to measure the strength and validity of the 2009 projections. We found that the generation of biogas and the financial return were in line with our 2009 forecasting/calculations. Adjustments were made in order to ensure a greater reliability for the return of investment.

6. Interpretation of data

The object of our study is Farm III and its 2 nuclei. The farm provides raw material to Business A. Our study seeks to measure the economic and financial feasibility of installing a digester and the proper treatment of swine waste this promises. All necessary data was therefore collected from Farm III/Nuclei I and II.

Farm III currently has a herd number of 5,362 pigs between nuclei I and II. It has a legally authorized waste treatment system. Each nucleus on the site has six pools and a Geomembrane Biodigestor of HDP (High Density Polyethylene) and LLDP (Linear Low Density Polyethylene) of 1.00 mm with a volume capacity of 800m ^{3} in nucleus I and 1400m ^{3} in nucleus II. Each nucleus has a homogenization tank which heats waste prior to it entering the biodigester. This tank has a volume capacity of 75m ^{3} in nucleus I and 150m ^{3} in nucleus II. There is also a homogenization pump in each nucleus with a power of 5 hp.

7. Initial investment in the project

Installing a biodigester requires building a workable structure. Costing included reservoirs of clean biogas; biogas cleaning kits and labor. It was also necessary to purchase equipment such as biogas dryers; piping; compressors. Table 1 presents the total costing of initial investment necessary for installing a biodigester.

Description	Qty	Value (R$)
Biodigester 1.400 m³	1 und	131.700,00
Biodigester 900 m³	1 und	102.426,00
Earthwork/excavation	N/D	37.000,00
Machinery room	2 und	26.000,00
Biogas cleaning kit	2 und	49.640,00
Biogas dryer set	N/D	38.960,00
Reservoirs of clean biogas	3 und	54.000,00
Pipe connections	N/D	12.000,00
Compressor	2 und	24.000,00
High pressure network	N/D	9.600,00
Pipeline	4 km	41.735,00
Pipeline excavation	4 km	22.000,00
Control fittings	N/D	28.716,00
Command table	2 und	24.800,00
Burners / heating equipment	N/D	32.321,00
Unforeseen costs	N/D	15.873,00
Total		**650.771,00**

Source: data collected from Business B.

Table 1. Initial investment.

7.1. Revenue generated by investment

Business B is responsible for funding the biodigester. It receives a return through the use of biogas consumed in the slaughterhouse and the singeing of swine. Values are invoiced using the following calculation:

$$CBG(R\$) = \frac{m^3 \text{biogas} \times \text{cost per kg of GLP} \times 0.80}{2.3}$$

KEY: CBG = value of biogas consumed
m^3 biogas = quantity of biogas consumed;
cost per kg of GLP (liquefied petroleum gas);
0.80 = eighty percent of the value of LPG (as per the binding contractual clause between investor and slaughterhouse);
2.3 = conversion factor for energy equivalence GLP/biogas.

Table 2 shows the net income accrued over 2006; 2007; 2008; 2009; 2010 and 2011, and the projected income for 2012; 2013; 2014 and 2015. The projections were made by calculating the average between 2009, 2010, 2011 and applying a 5% growth estimate (subsequent years are projected to grow by 5% when compared to previous years).

Year	Revenue
2006	94.922,09
2007	115.993,24
2008	107.239,10
2009	111.354,05
2010	116.921,75
2011	122.767,84
2012	122.865,27
2013	129.008,54
2014	135.458,96
2015	142.231,91

Source: Data collected from Business B.

Table 2. Generated and projected revenue.

In 2006 we see that the investing company had revenues of R $ 94,922.09 (ninety four thousand, nine hundred and twenty-two Reals, nine cents). This, representing a nominal return of 14.95% from the value invested. The rate of inflation for 2006, using the IPCA index (adopted by the Brazilian government as the official measure of inflation) was 3.14% (according to the Central Bank of Brazil) and stands at 11.45% as the effective rate of return for that year.

In 2007, the investing company achieved a total revenue of R $ 115,993.24 (one hundred and fifteen thousand nine hundred and ninety-three Reals, twenty-four cents), representing an 18.27% nominal return on the value invested. The IPCA inflation rate for 2007 stood at 4.46% (according to the Central Bank of Brazil), with 13.22% as the effective rate of return on investment for the period.

In 2008, total revenue accrued by the company amounted to R $ 107,239.10 (one hundred and seven thousand, two hundred and thirty-nine Reals, ten cents), reaching a nominal return of 16.89% on invested capital. IPCA (according to Central Bank of Brazil) registered inflation at 5.90%, thus qualifying 10.38% as the effective return of investment.

In 2009 the total net revenue obtained by the company amounted to R $ 111,354.05 (one hundred and eleven thousand, three-hundred and fifty-four Reals, five cents), reaching a nominal return of 17.54% on invested capital. Using IPCA as an index (and according to the Central Bank of Brazil) inflation stood at 4.31%, thus qualifying a 12.68% effective return of investment.

2010 had total revenues of R $ 116,921.75 (one hundred and sixteen thousand, nine hundred and twenty-one Reals, seventy-five cents), reaching a nominal return of 18.42% on invested capital. In 2010 IPCA (according to the Central Bank of Brazil) pegged inflation at 5.91%, this putting the effective return of investment at 11.81%.

In 2011 net revenue was R $ 122,767.84 (one hundred and twenty-two thousand, seven hundred and sixty-seven Reals, eighty-four cents), reaching a nominal return of 19.34% on invested capital. Inflation stood at 6.50% according to IPCA (and the Central Bank of Brazil), with the effective return of investment at 12.06%.

It appears that the annual return of savings (investment considered low risk and yield equal in all financial institutions in the country) for the years 2006; 2007; 2008; 2009; 2010 and 2011 was 8.41%; 7.80%; 7.74%; 7.09%; 6.81% and 7.50% respectively. The rate of cumulative nominal return on investment in question was higher in all the periods analyzed when compared to the accumulated rate of return of savings. From a purely financial viewpoint, such investment is attractive in terms of yields.

Year	Initial Investment	Net revenue	Return
2006	-650.771,00		-650.771,00
2006		94.922,09	-555.848,91
2007		115.993,24	-439.855,67
2008		107.239,10	-332.616,57
2009		111.354,05	-221.262,52
2010		116.921,75	-104.340,77
2011		122.767,84	18.427,07
2012		122.865,27	141.292,34
2013		129.008,54	270.300,88
2014		135.458,96	405.759,84
2015		142.231,91	547.991,75

Table 3. *Payback and return of investment.* The payback method of analysis presents to shareholders the necessary time to recover the financial resources invested in the project. Source: authors.

One of the weaknesses of payback strategy is the depreciation of monetary value over time. This said, a reversal seems to have occurred between 2010 (negative) and 2011 (positive). In terms of our research and its specific scenario, a recovery of initial investment kicks in at about five years and 11 months. The projection of discount rates for the following years was calculated by using the simple average of yields recorded over the previous three years, this being approximately 7.13%.

Year	Initial investment	Net revenue	Tax deduction	Net income	Deducted return
2006	-650.771,00				-650.771,00
2006		94.922,09	8,41%	87.558,43	-563.212,57
2007		115.993,24	7,80%	99.814,85	-463.397,72
2008		107.239,10	7,74%	85.747,65	-377.650,07
2009		111.354,05	7,09%	84.666,25	-292.983,82
2010		116.921,75	6,81%	84.107,70	-208.876,12
2011		122.767,84	7,50%	79.548,84	-129.327,29
2012		122.865,27	7,13%	75.866,74	-53.460,55
2013		129.008,54	7,13%	74.358,33	20.897,78
2014		135.458,96	7,13%	72.879,90	93.777,68
2015		142.231,91	7,13%	71.430,88	165.208,56

Source: authors.

Table 4. *Deducted* return: this table shows the calculation of discounted payback. The discount rates applied to net annual income were from savings accounts checked for the years 2006, 2007, 2008, 2009, 2010 and 2011.

Considering this data, a sign inversion occurs between the projected years 2012 (negative) and 2013 (positive), pitching the recovery of initial investment at roughly 7 years and 9 months. Using the information in Table 4 as a reference point, that is, the initial investment and net income both generated and projected, it is also possible to calculate the internal rate of return for the project – this being 12.16% per year.

Financial analysis reveals the benefits of investing in a biodigester. The farm reported a total revenue of R $ 650,771.00, with a net income of R$ 94.922,09 for 2006; R$ 115.993,24 for 2007; R$ 107.239,10 for 2008; R$ 111.354,05 for 2009; R$ 116.921,75 for 2010 and R$ 122.767,84 for 2011. For 2012 net income is projected to be R $ 122,865.27 with 5% real annual growth applied.

The payback period of investment recovery stands at 5 years and 11 months, and the discounted payback at 7 years and 9 months. Comparing the annual return of investment to the cost of savings, the rate of return on investment yielded higher gain over all periods analyzed, this, confirming the economic and financial viability of such a project. Financial and economic optic thus evidences the viability of financial investment into the project. This, combined with the environmental benefits, marks the entrepreneurial quality of such an enterprise.

8. Conclusion

The goal of our study was to demonstrate the relationship between swine production and the environment – where the former can be developed through new, innovative technologies, the latter is not impervious and remains vulnerable. There is a need to consider the negative environmental effects of such practices, highlighting the urgency for due care and attention in waste management and business strategy.

The current market calls for fundamental changes in the economic situation and nature of organizations as well as flagging demands for a new benchmark in business management. An innovative approach to the way new realities are both understood and dealt with is a modern requirement.

There is nevertheless a lack of accurate information about the chemical concentration of swine waste. Alternatives remain limited and the rationale underdeveloped. Ongoing research is required into the suitability of different soil types and crops receiving fertilizer. Likewise, both the short and long-term environmental effects of swine waste need to be studied and known. What we do know is that the indiscriminate disposal of swine waste in natural environments, runs the elevated risk of contaminating soil; water supplies; rivers; effluents and the air itself – this, a condonable practice directly affecting the health of rural and urban communities.

Any waste distribution system then, must take into account the cultural and economic realities of farmers and local producers of swine. It is essential to raise the awareness of farmers and society alike, bringing to their attention issues of waste pollutants, as well as the benefits of implementing a technology that combines the agronomic use of manure as fertilizer, providing the economy with greater input and systems which minimize the effects of pollution.

As we previously asserted, our modern scenario is one requiring a social conscientiousness for the ecosystem is on the verge of total collapse. It is where modern man is faced with the urgency to change his world-view through sustainable business practices, igniting a change of values and a new direction in operating systems which engage with sustainable development and environmental preservation.

According to Freitas (2008), a great many experts fear that if emissions of greenhouse gases (mainly carbon dioxide, methane and nitrous oxide) continue to increase, then the planet's temperature will rise and the results will be drastic if not unimaginable. Formidable changes in our climate with extreme cold and elevations of wetland will lead us to experiencing periods of drought; fertile farmland will not too far into the future succumb to desertification; incidences of severely destructive storms, tornadoes, hurricanes or typhoons will be frequent; the dwindling and complete loss of floral species and fauna in different parts of the natural world will be commonplace, as will the melting of ice caps and the consequent increase in global sea levels. With the unbridled increase in global warming, mainly due to high levels of carbon dioxide and methane, serious consequences can already be felt on a global scale, threatening the survival of the earth's inhabitants.

The aim of our study is to demonstrate the benefits generated by installing a biodigester as a financially and economically viable alternative in the management of swine waste. Given this, biodigester technology and its environmental value, is more than a mere hypothesis but a tried and tested means of treating and reusing detrimental waste material – this, important for both swine producer and society. Besides being an alternative source of renewable gas, such a technology reuses and recycles, improving and maintaining soil without jeopardizing environmental standards and human wellbeing.

Author details

Antonio Zanin and Fabiano Marcos Bagatini
Universidade Comunitária da Região de Chapecó – UNOCHAPECÒ, Brazil

9. References

Associação Catarinense De Criadores De Suínos. Available at: http://www.accs.org.br/dados_ver.php?id=4 Accessed on 10/7/2011.

Araújo, Massilon J. Fundamentos do agronegócio. São Paulo: Atlas, 2003.

Barbieri, José Carlos. Gestão ambiental empresarial: conceitos, modelos e instrumentos. 2. ed. Atual. Ampl. São Paulo: Saraiva, 2007.

Bezerra, Severino Antunes. Gestão ambiental da propriedade suinícola: Um modelo baseado em um biossistema integrado. Revista Ciências Empresariais da UNIPAR, Toledo, v.6, n.2, jul./dec., 2005.

Brasil. Resolução CONAMA nº 357, 17th March, 2005. Ministério do Meio Ambiente. Available at: http://www.mma.gov.br/port/Conama. Accessed on 17/2/ 2012.

Brilhante, Ogenis Magno; CALDAS, Luiz Querino de A. Gestão e avaliação de risco em saúde ambiental. Rio de Janeiro: Fiocruz, 1999.

Embrapa. Licenciamento ambiental para criação de suínos. Available at: http://hotsites.sct.embrapa.br/diacampo/programacao/2008/licenciamento-ambiental-para-criacao-de-suinos. Accessed on: 30/1/2012

Gonçalves, Rafael Garcia; PALMEIRA, Eduardo Match. Suinocultura Brasileira. Observatório de Economia Latino americana, Número 71, 2006. Available at: http://www.eumed.net/cursecon/ecolat/br/06/rgg.htm. Accessed on 5/4/2011.

Guivant, Julia S.; MIRANDA, Cláudio R. de. Desafios para o desenvolvimento sustentável da suinocultura: uma abordagem multidisciplinar. Chapecó: Argos, 2004.

Jordan, Danielle. Dejetos de suínos e outros animais podem ser convertidos em créditos de carbono. Homologa.ambiente.sp.gov.br/proclima, September, 2005. Available at: //http://noticias.ambientebrasil.com.br/exclusivas/2005/12/09/22129-exclusivo-dejetos-de-suinos-e-outros-animais-podem-ser-convertidos-em-creditos-de-carbono.html. Accessed on: 2/3/2012.

Leite, João Guilherme Dal Belo. Inovações Tecnológicas na Suinocultura Sul – Brasileira sobre a Ótica da Sustentabilidade. ENGEMA- Encontro Nacional de Gestáo Empresarial e MeioAmbiente (National Congress for Business and Environmanetal Management), 9th- 12th November, 2008. Porto Alegre, Brazil.

Marion, José Carlos. Contabilidade Rural: Contabilidade Agrícola; Contabilidade Da pecuária; imposto de renda pessoa jurídica. 7. ed. São Paulo: Atlas, 2002.

Mior, Luiz Carlos. Agricultores Familiares, Agroindústrias e redes de desenvolvimento rural. Chapecó: Argos, 2005.

Moura, Luiz Antonio de. Qualidade e gestão ambiental: sugestões para implantação das Normas ISSO 14.000 nas empresas. 2ª ed. São Paulo: Editora Juarez de Oliveira, 2000.

Nogueira, Luiz A. Horta. Biodigestão: a alternativa energética. São Paulo: Nobel, 1986.

Palhares, Julio César Pascale. Biodigestão anaeróbia de dejetos de suínos: aprendendo com o passado para entender o presente e garantir o futuro. Available at: http://www.infobibos.com/Artigos/2008_1/Biodigestao/Index/.htm. Accessed on: 9/1/2012.

Panty, Eliana. I simpósio Brasil Sul de Suinocultura. Agromais, Chapecó, n. 9, p. 66, April/May, 2008.

Pereira, E.R.; Demarchi, J.J.A.A.; Budiño, F.E.L. A questão ambiental e os impactos causados pelos efluentes da suinoculturaa. 2009. Article in Hypertext format. Available at: http://www.infobibos.com/Artigos/2009_3/QAmbiental/index.htm. Accessed on: 30/1/2012

Queiroz, Timóteo Ramos (Org.). Agronegócios: gestão e inovação. São Paulo: Saraiva, 2006. p. 251-280.

Ribeiro, Maisa de Souza. Contabilidade ambiental. São Paulo: Saraiva, 2006.

Sampaio, Silvio C; Fiori, Marciane G.; Opazo, Miguel A. U.; Nóbrega, Lúcia H. P. Comportamento Das Formas De Nitrogênio Em Solo Cultivado Com Milho Irrigado Com Água Residuária Da Suinocultura. Eng. Agríc., Jaboticabal, v. 30, n1, jan-feb. 2010.

Sasaki, Yosuke; Saito, Hikari; Shimomura, Ai; Koketsu. Consecutive reproductive performance after parity 2 and lifetime performance in sows that had reduced pigs born alive from parity 1 to 2 in Japanese commercial herds. Livestock Science. School of Agriculture, Meiji University, Kanagawa 214–8571, Japan, 2011.

Seganfredo, Milton Antonio. Os dejetos de suínos são um fertilizante ou um poluente do solo? Cadernos de Ciência e Tecnologia, Brasília, v. 16, n. 3, p. 129-141, Sept./Dec. 1999. Available at: http://www.uov.com.biblioteca_arquivos/Curso55-3.pdf. Accessed on: 4/2/2012.

Silva, Benedito Albuquerque da; Robles Junior, Antonio. Modelagem de custos das atividades integradas de suinocultura: cria, recria e engorda. XVI Congresso Brasileiro de Custos – Fortaleza, Ceará, Brasil, 3rd -5th November, 2009. Annals.

Sobestiansky, Jurÿ; et al. Suinocultura intensiva: produção, manejo e saúde do rebanho. Brasília: Embrapa, 1998.

Tinoco, João Eduardo Prudência. Balanço Social: uma abordagem da transparência e da responsabilidade pública das organizações. São Paulo: Atlas, 2001.

Tinoco, João Eduardo Prudêncio; KRAEMER, Maria Elisabeth Pereira. Contabilidade e Gestão Ambiental. São Paulo: Atlas, 2004.

Zanin, Antonio; Bagatini, Fabiano Marcos; Pessato, Camila Batista. Viabilidade econômico-financeira de implantação de biodigestor: uma alternativa para reduzir os impactos ambientais causados pela suinocultura. Custos e agronegócio on line, ISSN 1808-2882, Recife, volume 6, número 1, p.1-161, Jan/April, 2010. Available at: http:// www.custoseagronegocioonline.com.br. Accessed on: 12/3/2012.

Zuin, Luíz Fernando Soares; AlliprandinI, Dário Henrique. Gestão de inovação na produção agropecuária. In: Zuin, Luiz Fernando Soares.

Chapter 3

Beach Erosion Management with the Application of Soft Countermeasure in Taiwan

Ray-Yeng Yang, Ying-Chih Wu and Hwung-Hweng Hwung

Additional information is available at the end of the chapter

1. Introduction

The total length of Taiwan's shoreline is approximately 1,100 kilometers including sand, rock, cliff, gravel and reef coasts (see Figure 1). Almost half of the shoreline has been protected by seawalls. From the viewpoint of shore protection in coastal area, these seawalls actually play an important role of costal protection that prevents people and infrastructure from coastal disasters. Furthermore, detached breakwater and groyne are built to protect the coastal area with serious erosion problem. These efforts made our land safe over the last fifty years to some extent. Due to the martial law, it was not so easy for people to walk or visit near the coastal area in Taiwan during 1949 to 1987. However, after 1987 people gradually valued the coastline for environmental protection and, recreational use as well as the economic activity. The purpose of the coastal protection is diversified by these new demands. In this study, we will introduce environmentally and user-oriented coastal protection works as well as technically sound creditable coastal protection works to meet these new trends. Therefore, the purpose of this study is to evaluate on how to join soft solution strategies into current shore protection system throughout Taiwan's coast. Moreover, a feasible application for hard solutions complemented by soft issues for beach erosion management has also been evaluated. In Taiwan, beach erosion has become more serious in the recent past. The time for this beach erosion to become apparent chiefly depends on how fast the rate of longshore sediment transport decreased from the up-coast area and on the river sediment supply. Many industrial, commercial and fishery harbor construction projects were also observed to have disturbed the continuity of littoral sediment transport, and lead to the retreat of the shoreline in the downcoast area. However, sufficient knowledge on nearshore hydrodynamic forcing (incoming wave energy, wave-induced currents in the surf zone and tidal range), sediment transport processes and morphological features along coasts, will be helpful to the improvement of shore protection work. Countermeasures for beach erosion control should depend on the local conditions of hydrodynamic forcing characteristics, littoral sediment transport and various morphologies. Therefore better applications of the various soft methodologies available for beach erosion management will be proposed in this study.

Figure 1. Different types of coasts in Taiwan

2. Evaluation of soft engineering structures

The protection of beaches against erosion has always been an important aspect of coastal engineering works in Taiwan. History is replete with the loss of valuable coastal lands such as beaches, reclamation areas, harbors, and other valuable coastal property to erosion induced by sea encroaching. On the other hand, there are also some cases of harbors being abandoned because of infilling by sediments, which is quite a different coastal engineering problem but a significant one. Erosion control measures should incorporate a reduction in the cause of the beach erosion where possible, when, for example, the erosion is caused by human activities along the coastal areas such as hard engineering structures or harbors. For each coastal erosion mitigation measure, it is important to know how they work. In fact, some of the mitigation schemes are able to reduce the wave energy at the shoreline, or simply provide a sacrificial beach, whereas others try to impede the long shore transport of sand. In a particular situation, one mitigation scheme will work better than some others owing to the difference in operation. In fact, some methods will fail in one situation and do very well in others. Therefore, in order to mitigate the erosion problems due to the current hard shore protection system around Taiwan coast, detailed evaluation of various feasible soft engineering structures should be conducted in advance. The evaluation of soft countermeasure includes beach nourishment, near shore disposal berms, geosystems, artificial oyster reef, fluid-elastic sheet and aquatic vegetation.

2.1. Beach nourishment

Beach nourishment is the mechanical or hydraulic placement of sand on the beach and/or shoreface to advance the shoreline or to maintain the volume of sand in the littoral. It is a soft protective and remedial measure that leaves a beach in a more natural state than hard structures and preserves its recreational value. Of the many remedial measures for beach erosion, beach nourishment is the only approach that introduces additional sand sources into the coastal system. Without the construction of coastal structures, beach nourishment seldom causes damage to the landscape, and can flexibly responds to changes of the littoral environment. Beach nourishment, with its expected widening of beach, is used to accomplish several goals as follows: formation of additional recreational area; land reclamation; maintenance of shoreline; reinforcement of dunes against breaching; protection of coastal structures; reduction of the wave energy near shore and creation of a sacrificial beach to be eroded during a storm; provide, in some cases, environmental habitat for endangered species.

Sand nourishment can be carried out at various locations in the profile or along the shoreline. The options of nourishments in cross-shore profile are shoreface (underwater nourishment or profile nourishment), dune zone (landward and seaward above dune toe), beach and swash zone. Leonard et al., 1990 evaluated 155 beach nourishment projects in the U.S.A. In all, about 300 million m^3 of sand was placed along 700km of shoreline (470km along Atlantic coasts, 180km along Gulf coasts and 50km along Pacific coasts). In 1996, Rijkswaterstaat also evaluated nine nourishment projects (volumes between 50 and 100 $m^3 / m / yr$; sand size between 0.15 and 0.3mm) carried out along the coasts (tidal range of about 2m) of the Netherlands in the period 1975-1994. Several characteristics, including ratio of design nourishment volume and required volume to compensate annual erosion volume in active zone before nourishment and after nourishment, were analyzed in his evaluation. Leo C. van Rijn, 1998 summarized the sand nourishment characteristics under micro and meso-tidal conditions in great detail from five projects (Delft Hydr., 1987; Dette & Raudkivi, 1994; Mϕller, 1990; Rijksw, 1996; Work & Dean, 1991). From his result, it shows that beach nourishment can be mostly utilized on coastal areas of low or moderate wave energy with micro-tidal condition. Meanwhile, the three basic elements including the eroded area, the borrow area and the transportation/ dumping methods should be investigated in detail when sand nourishment is applied to beach erosion control. Dean, 1986 recommended a mitigative approach for armoring on an eroding coastline that calls for the placement of sand annually in the amount that has been prevented from entering the system by the armoring structure. This approach maintains a more natural littoral system. Often, the nourishment scheme is remedial rather than preventive (Hamm et al., 1998). In summary, beach nourishment is the approach that directly addresses the deficit of sand in the coastal system without at least the potential of causing adverse effects on adjacent property. Bridges & Dean, 1996 concluded that beach nourishment is the most benign and acceptable approach to beach erosion mitigation. However from the new demands of shore protection now in Taiwan, beach nourishment can not be the only option of beach erosion control.

2.2. Near shore disposal berms

Open-water disposal of dredged material has been practiced worldwide for over sixty years. The initial attempts have arisen from the search for a beneficial use of the large amounts of dredged material obtained from navigation channel maintenance operations. The removed material which varies in size and quantity has been placed in nearshore disposal sites seaward of the surf zone. A major cost savings often accrues if beach fill material can be placed offshore rather than on the beach in the expectation that natural processes will move the material to the beach. The performance of underwater berms has been investigated both in the laboratory and through field monitoring programs. Hands, 1991 provided a thorough review of the behavior of 11 berms and their performance. Furthermore, Otay, 1994 presented a summary of submerged berms and their characteristics, including whether they were judged to be stable or migrated. Of the berms placed to benefit the landward beaches, possible designs could be a feeder berm, in which sand would be transported to the beach from an active berm or as a stable berm that causes damping of the waves and thus sheltering of the landward beach. In his research, Otay also described the monitoring results of an underwater berm placed off Perdido Key, Florida. Monitoring included repetitive beach profiles and wave measurements. His result showed that the berm had exerted a stabilizing effect on the beach leeward of the berm.

2.3. Geosystems

Geosystems (tubes, containers) have already found various applications in coastal engineering. The tubes and containers are mainly applicable for construction of groynes, perched beaches, and offshore breakwaters, and as bunds for reclamation works. Application of these systems has executed by a number of projects in the Netherland, Germany, Japan and U.S. Some information on U.S. experience with geotubes can be found in Fowler et al., 1995 including the application of geotubes for dewatering of contaminated maintenance dredged material. Geosystems have much applicability in erosion control, water control (small weirs and reservoirs), flood control, etc. For example, breakwaters made of sandbags, geotubes, etc, have been used successfully in the United States of America under conditions for low tidal range and low wave activity (Krystian, 2000). Under gentle wave climates such structures may not only attenuate waves, but can also encourage the accretion of sediment between them and the shore. Geotube can also be used to assist in dike, groin and breakwater construction. Krystian summarized the examples of application of geotube, as dune reinforcement, core of breakwater, and bunds for dike construction, from a number of projects executed in the Netherlands and Germany. The main advantages of geosystems in comparison with more traditional methods (rock, concrete armor units, block mats, asphalt, etc.) are: a reduction in work volume, execution time and cost, the use of local materials, low-skilled labour and locally available equipment. However, until now, geosystems were mostly applied as temporary structures. The reason for that was their relatively low resistance to the loading of waves and currents, the lack of proper design criteria, and a low durability in respect to UV-radiation and vandalism.

2.4. Artificial oyster reef

There is increasing interest in oyster reefs used to restore eroding coastlines. Occasionally, subtidal oyster reefs can be found offshore. These immense natural submerged breakwaters protect the beaches from storms and wave erosion by dissipating wave energy. The study of how artificial and natural reefs have protected shorelines has been conducted by Hamaguchi et al., 1991. They investigated the effects of an artificial reef on the Niigata coast in Japan. It was found that a significant amount of sand was deposited landward of this artificial reef. This reef was developed to mimic the effect of the natural coral reefs in the area. There has been an effort to find different methods of restoring oyster reefs in various estuaries around the world. O'Beirn et al., 2000 conducted the experiments by using oyster shell, concrete, and rubber tire chips as oystercultch material. A structure termed an "oysterbreak" was designed to stimulate the growth of biological structures in an optimal shape to serve as submerged breakwaters (Foret, 2002). Oysterbreak can form immense structures that can protect shorelines and coastal communities by reducing wave energy.

Currently, mineral accretion amelioration on gabion that was filled with oyster cultch & rock to form a new biological unit has been investigated in field experiment by Hwung et al., 2008. It is hoped that this combination of oyster cultch, mineral accretion and cage meshed into berm breakwater can improve the toe revetment and berm advance, and simultaneously enrich the local environment to a higher level.

2.5. Fluid-elastic sheet

Fluid-plate hydro-elastic interaction problems have been of common interest for a long time because of their engineering applications. During the past decades, for instance, there has been a gradual increase in interest in the use of flexible plates or membranes as alternative effective inexpensive wave barriers in a beach zone. Currently, developing of the new design of floating wave breakers in a beach zone using coating of the sea surface by an elastic plate, which absorbing the energy of sea waves, is investigated by Hwung et al., 2008. A properly designed horizontal flexible membrane can be a very effective wave barrier and its optimal design can be found through a comprehensive parametric study using the experiments, theory and computer programs developed. In particular, the membrane is light and rapidly deployable; thus, it may be an ideal candidate as a portable temporary breakwater. Since a horizontal membrane does not directly block incoming waves, the transmitted and motion-induced waves need to be properly cancelled for it to be an effective wave barrier.

2.6. Aquatic vegetation

Aquatic vegetation provides important ecosystem services to coastal marine systems. They influence their environments through wave attenuation, the stabilization of sediments, increased setting of the suspended particulate particle and nutrient cycling. For environmental and esthetic purposes, projects on natural development of wetlands and restoration of river basins toward natural development have been promoted recently. The growth of vegetation in these areas is favored. Such vegetation increases the resistance of the watercourse, leading to an increase in water depth and reduction of flow velocity. In estuarine and coastal areas with vegetation, in addition to freshwater flow upstream, waves and tidal currents exist and will play a significant role in the hydrodynamics and mixing processes. Waves over vegetation will be attenuated due to the resistance offered by the vegetation. The bidirectional nature of wave motion will increase the mixing between the water column and that within the vegetation (Li & Yan, 2007). Wave motion tends to be highest in the shallow waters where, in combination with tidal currents, water movement imposes a shear stress on bottom sediments. If bottom shear stress exceeds a critical value, sediment will be resuspended, increasing turbidity and light attenuation (Wright, 1995).

For waves propagating over vegetation, Kobayashi et al., 1993 developed an analytical model to predict wave attenuation over vegetation by assuming an exponential decay of incoming regular waves. Vegetation meadows can reduce suspended sediment concentrations; friction from vegetation leaves reduces current velocity and attenuates waves, thus reducing the stress on bottom sediments, decreasing resuspension, and promoting sediment settling within the vegetation bed (Fonseca & Cahalan, 1992; Rybicki et al., 1997). Vegetation beds may also increase particle settling shoreward of the bed (Chen et al., 2007).

3. Hydrodynamic energy and morphology classification of Taiwan coast

The hydrodynamic and morphological processes in the coastal zone are governed by two primary phenomena, namely, winds and tides. The winds are directly responsible for the transport of sand on the dry beach and for the generation of waves, currents and water-level fluctuations, while the tides express themselves in a periodic rising and falling of the water and in tidal currents. Therefore, coastal classification based on hydrodynamic energy was presented by Davis & Hayes, 1984. The classification is shown in Table 1. The wave climate is generally characterized, as: low wave energy, if annual mean significant wave height at edge of surf zone (say, depth of 6m) is $H_{s,am}<0.6$m; moderate wave energy, if $H_{s,am}$ between 0.6m and 1.5m; high wave energy, if $H_{s,am}>1.5$m.

However, tides are classified as micro-tidal, if the tidal range (TR) <2m, meso-tidal for TR between 2m and 4m and macro-tidal for TR>4m. Furthermore, the relative strength of tide-induced (tidal range TR) and wave-induced forces (mean annual nearshore wave height H) acting in coastal system, the following classification may also be given as: wave energy-dominated coasts (TR/H=0.5 to 1.0), tide energy-dominated coasts (TR/H>3); mixed energy coasts (TR/H=1 to 3).

Regarding the long-term marine observation data, we refer to the research reports analyzed by Tainan Hydraulics Laboratory [THL], 2002 and then summarize the hydrodynamic energy classification of Taiwan coast shown as Figure 2 and Figure 3. Therefore based on Davis and Hayes's classification, the results show that northeastern coast, east coast and south coast of Taiwan belong to micro-tidal coast. However, Yun-Lin, Changhua (mid-western coast) and north coast are meso-tidal coast. As mentioned about classification of wave energy coast, the coast from Taipei county to Hsinchu county (northwestern coast), and coast between Yun-Lin county and Tainan county (west coast) are moderate wave energy coast.

Wave Energy	low	moderate	high
	Hs,am<0.6m	0.6m <Hs,am<1.5m	Hs,am>1.5m
Tidal Energy	micro-tidal	meso- tidal	Macro- tidal
	TR<2m	2m<TR<4m	TR>4m
Coastal classification based on hydrodynamic energy			
Wave Energy-dominated			TR/ Hs,am=0.5~1
Tide Energy-dominated			TR/ Hs,am>3
Mixed Energy			1<TR/ Hs,am<3

Table 1. Hydrodynamic forcing in the coastal zone

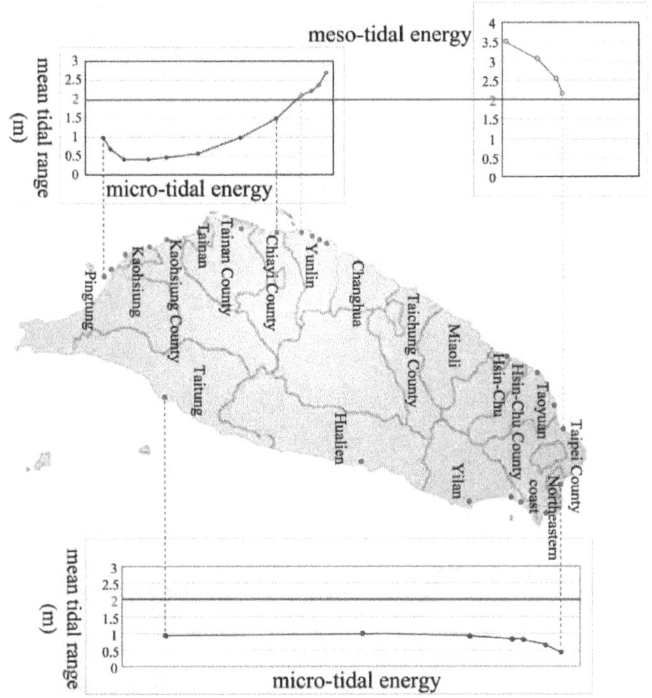

Figure 2. Tidal energy classification of Taiwan coast

However, southwestern coast from Tainan city, Kaohsiung to Ping-Tung county belong to low or moderate wave energy coast. Regarding to the east coast of Taiwan, besides partial coast of Taitung county is moderate wave energy coast, most parts of the east coast, (Hualien, Yi-Lan) and northeastern coast, are high wave energy coasts. For morphology classification, the slopes of beach profiles around Taiwan coast are shown in Figure 4. Due to these hydrodynamic energy and morphology classification of Taiwan coast, a strategic management proposal can be made to integrate soft countermeasure into the current hard shore protection system around Taiwan coast.

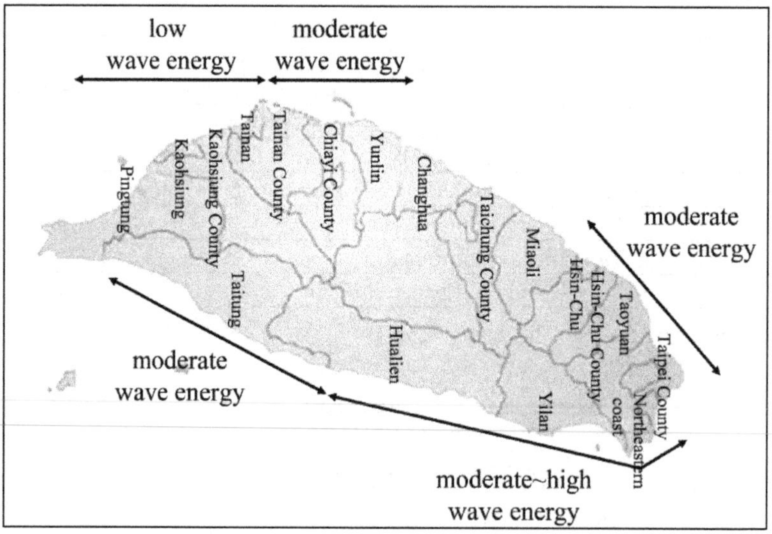

Figure 3. Wave energy classification of Taiwan coast

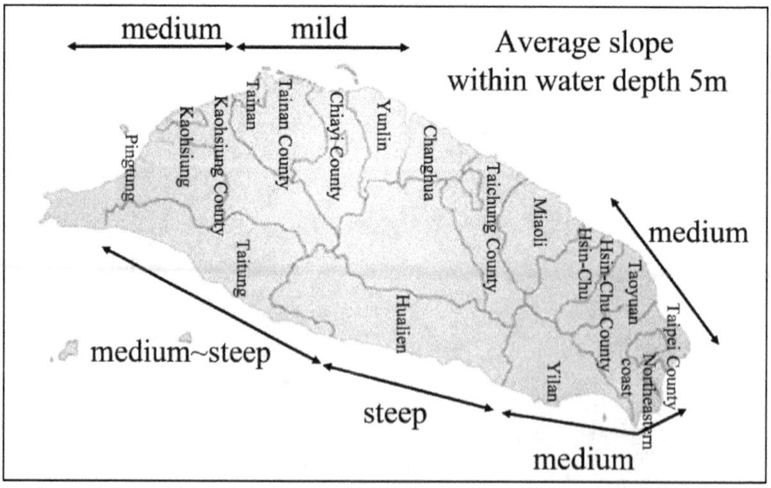

Figure 4. Morphology classification, the beach slopes around Taiwan

4. Discussion on application of combining hard and soft solutions for beach protection in Taiwan

Water Resource Agency, Taiwan, has claimed nearly 95 percent of fulfillment on coastal protection after more than 30 years' efforts. However, under the disaster prevention demand in earlier days, coastal protection has been accomplished mostly be lining up the hard engineering structures such as seawalls, groyne and armour units along the coastlines. Until now, the constriction of the seawalls for shore protection does not always work well on each coast around Taiwan. Some shores still get eroded seriously with seawalls being damaged partially. Meanwhile, the coastal engineering development in Taiwan currently has changed the previous strategy, only focused on shore protection, to a new one taking into consideration several aspects like safety, economy, construction, recreation, landscape and ecology. According to this new trend, the current shore protection system around Taiwan can be properly mended by soft countermeasures, however, the specific characteristics and requisite at each local site should be taken into consideration. Therefore, we have divided Taiwan coast into several categories based on the collected long-term observation information such as geological characteristics, hydrodynamic forcing, the intensity of beach erosion and shore protection. Then, the national beach management and protection problems were evaluated and the solutions for shore protections and further improvements were proposed.

After evaluation of a number of soft solution results, it is indicated that beach nourishment is a natural and popular soft shore protection technique that has been applied worldwide recently. This method can be utilized on the coastal areas under low or moderate wave energies with mild or moderate bottom slope for engineering purpose. However from the new demands of shore protection now in Taiwan, beach nourishment can not be the only option of beach erosion control. Another soft solution should be also taken into consideration for integration. Based on these criteria and the categories of Taiwan coast, the countermeasure of integrating soft solution into current hard shore protection system around Taiwan coast will be proposed as follows:

Because the north coasts in Taiwan are meso-tidal and moderate wave energy coasts, and their beach profiles are of mild slope, the eligible improvement criteria for current shore protection scheme are based on headland control strategy plus sand nourishment. However, volume of sand, sand size, beach and swash zone placement should be taken into account in the works of beach nourishment.

On the other hand, many barrier islands in the offshore of Yun-Lin, Chia-Yi and Tainan coast, can be treated as natural offshore breakwaters. In fact, those barrier islands can form a defense line of low-lying coastal plains and back-barrier basin against storms attacking these areas. Therefore, the shore protection strategy for these areas should be focused on how to protect these barrier islands. However, most important for the formation and maintenance of barrier islands and inlets is the relative strength of the wave processes and of the tidal processes. In order to protect these barrier islands, detailed field investigations on sediment supply (sources and sinks), hydrodynamic forces (waves, tides and rate of sea level change) and geomorphic setting (shoreface profile shape, sub-strata composition) should be carried out in advance. Meanwhile, oyster cultivation is an important fishery industry in the coast of these areas. Thus, there is large volume of oyster cultch in these coasts. Therefore, one new shore protection technology using mineral accretion technique is proposed by Hwung et al., 2008. Regarding to this new technology, mineral accretion (an advance on cathodic protection) amelioration combined with oyster cultch and rock is used to form the new biological unit in order to enhance the efficiency of anti-rusting and function of shore protection. Several field experiments of this new shore protection technology have been done on Chigu Lagoon in Tainan County. It is hoped that this new shore protection technology can be successfully applied to coastal engineering in the near future.

Offshore sills or breakwaters have proved to be much effective when used in combination with beach nourishment schemes. The retention capacity of a perched beach not only helps to reduce wave attack but is clearly beneficial from a recreational point of view. Since many offshore breakwaters already exist in the coasts of Kao-Shung and Pin-Tung counties, sand sources can be filled in the region between offshore breakwaters and sea dikes. The expanding beach faces will be helpful for wave damping and sightseeing. However, as many successful fisheries exist in the coastal regions of Kao-Shung and Pin-Tung Counties, when the beach nourishment is taken as the shore protection method in these areas, the influence of beach nourishment to coastal fisheries should be taken into considerations.

As for steep beaches and high wave energy, such as those of the east coast region, more specific parameters should be taken into account. For example, the erosion problem at the Tou-Chen beach in Yi-Lan County can be remedied by headland control strategy plus sand nourishment. The Tou-Chen beach now is defended only by a seawall and some short groynes. The headland control strategy can be based on reconstructing two long arc-shaped groynes with a submerged breakwater to support the recreational beach fill in front of seawall. Meanwhile, for the purpose of recreational activity, the fluid-elastic shirt can be applied as a portable temporary breakwater in this beach zone for wave damping. However in order to mitigate the erosion of gravel beaches in Hualien and Tai-Tung counties, the beach nourishment of mixed grain sizes is to become an alternative solution. Since there is little experience on the movement of gravel on steep coast of Hualien and Tai-Tung counties, a comprehensive field investigation should be done on the mechanism of sediment movement especially for action of typhoon wave. Moreover, a detailed physical model study about the effect of dynamic nourishment as a countermeasure against erosion should also be conducted before gravel nourishment can be carried out on the Taiwan eastern coast.

5. Experimental application study on integration of soft solution into current hard shore protection system

Two local sites in the southwestern Taiwan coast and one biggest offshore barrier island in Taiwan are selected for the experimental application study on integration of soft solution into current hard shore protection system. These in-situ experimental studies are therefore designed to improve the security as well as to involve ecological and scenic remediation for the beach erosion problem.

5.1. Ching-Tsao-Lun coastal area

With the length of 5 kilometer, Ching-Tsao-Lun coastal area is located between Zeng-Wun River mouth and Lu-Erh-Men River mouth at the middle section of Tainan coast. Without manmade intrusion, Zeng-Wun River mouth was once a natural estuary for more over three decades ago. Unfortunately, nowadays Ching-Tsao-Lun coastal area was invaded by constructing concrete and pebble dikes or other artificial protection. Therefore, the beach in front of Ching-Tsao-Lun concrete dike had been eroded ten years ago (Figure 5 and Figure 6). In order to have an overall study on rebirthing Ching-Tsao-Lun coastline, the in-situ study (Figure 7) focuses not only on safety evaluation of remedial Ching-Tsao-Lun sea dike but also on low to reform the beach back with ecology evaluation (Liou et al., 2007). Furthermore, by joining government resource and local manpower together, advance coastal management will enlighten the environment and landscape of Ching-Tsao-Lun coast again. The overall coastal protections and environment rebirths for Ching-Tsao-Lun coastal are listed as follow:

The coastal area around Ching-Tsao-Lun concrete dike:

1. Short-term goal: If rebuild is required, dike section adjustment could be put into consideration.
2. Long-term goal: Three schemes with beach nourishment and offshore breakwater are proposed for beach rebirth. Physical remediation will be half the way whiles the beach reversible.

Figure 5. The erosion area in the Ching-Tsao-Lun coastline

The coastal area around Ching-Tsao-Lun pebble dike:

1. Apply "Vegetation Evolution Method" to enhance vegetation diversity and landscape.
2. Advance coastal management and use floating logs to enlighten the environment and coastal protection.

The natural beach zone:

1. Short-term goal: Build wind fence with planting to enhance the environment and landscape.
2. Long-term goal: Build artificial sand dune and increase dune elevation by using local floating logs, oyster cultch or dredged sedimentation form Lu-Erh-Men River.

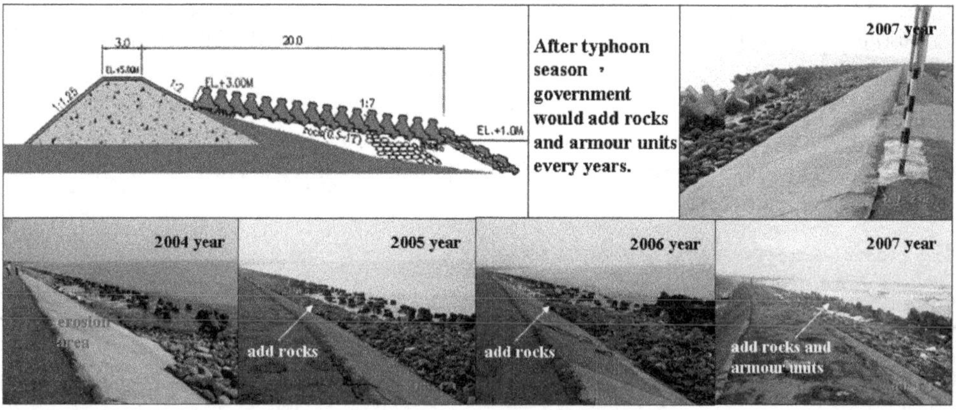

Figure 6. The erosion beach in front of Ching-Tsao-Lun concrete dike

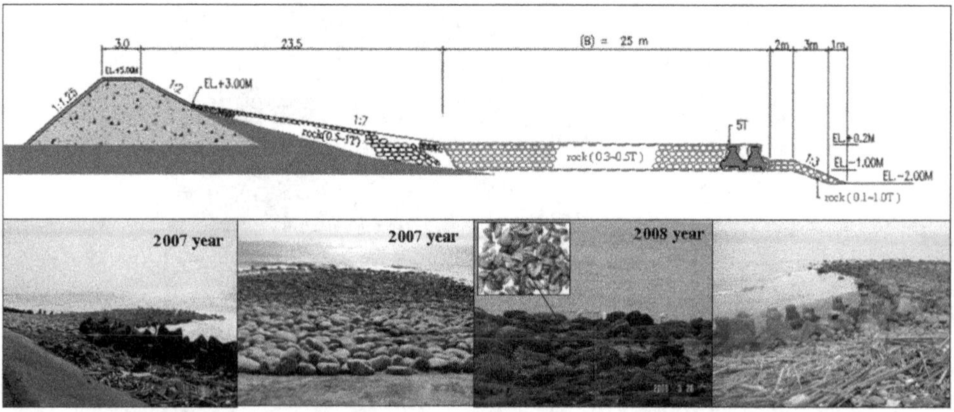

Figure 7. In-situ study on beach erosion control in the Ching-Tsao-Lun area

5.2. Shuang-Chun coastal area

With a beautiful sight and a mangrove ecosystem preservation area, Shuang-Chun coastal area (Figure 8) locates between Ba-Jhang River mouth and Ji-Shuei River mouth at the northern section of Tainan coast. Historical evolution of the coastal morphology shows that one third of the Shuang-Chun coastline at northern section has revealed shoreline retreat problem in the recent years. It can be shown by the evolution of coastline, satellite photos and aerial photographs from 1993~2002 (Figure 9, Figure 10 and research reports by THL, 2004).

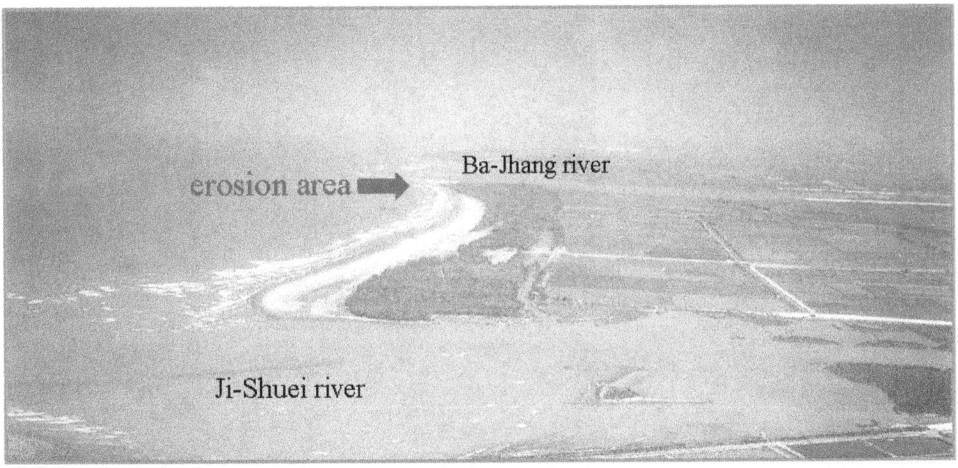

Figure 8. The location of Shuang-Chun coast between two river mouths

After analyzing the long term field investigation data, the hydrodynamic characteristics of Shuang-Chun coast are moderate wave energy coast and micro-tide condition. The dominant hydrodynamic characteristics (incoming wave climate, tidal range) and local morphological information are considered for the proposed countermeasure to control beach erosion. Based on parabolic bay orientation on equilibrium shape (Hsu & Evan, 1989), the final select countermeasure is show in Figure 11 and Figure 12. The design is to establish three offshore breakwaters with a southwest stretched breakwater as a down-coast artificial headland on the existing seawall of Ba-Jhang River mouth (Yang et al., 2004). All the new breakwaters are set up by geobags filled with sand from the northern deposited area of Ba-Jhang River mouth. Behind the shelter areas of three offshore breakwaters, oyster booth is used as wave energy dissipation and sand trapping. Meanwhile, the existing dune is reinforced by dune zone nourishment. Monitoring of this experimental application study has still been conducted from 2008~2012. The final results of this proposed countermeasure will be verified by measurements of hydrodynamics and topography, sand sampling and monitoring of geosystem and oyster booth for their function of anti-damage and suitably applied environment.

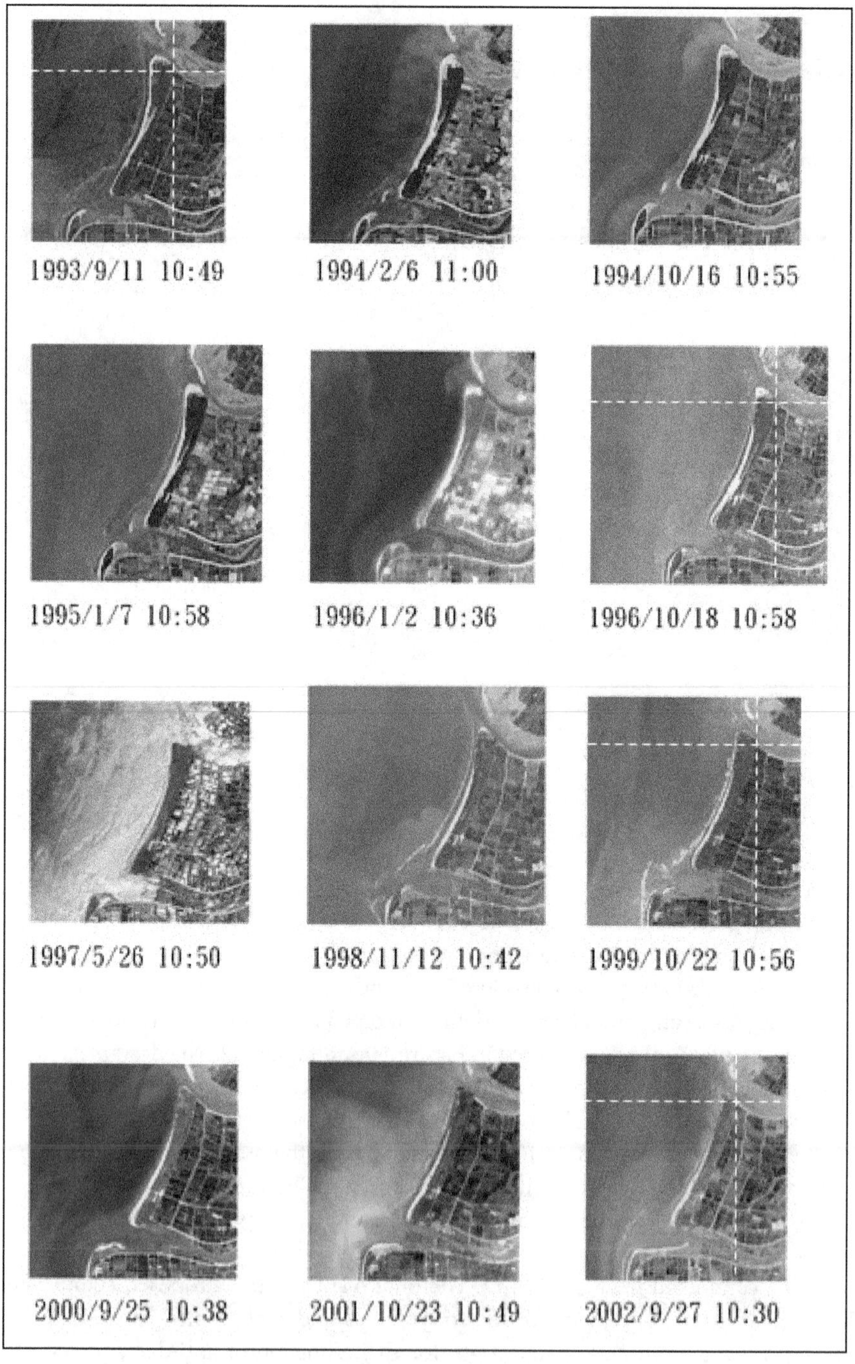

Figure 9. The satellite photos of Shuang-Chun coast from 1993~2002

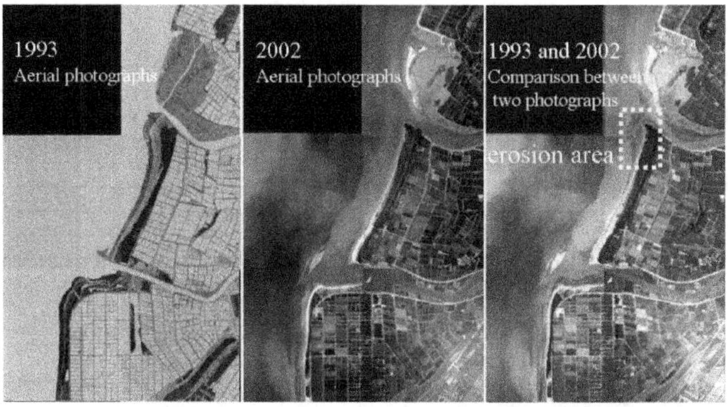

Figure 10. The comparision of Shuang-Chun coastal morphology between 1993 and 2002

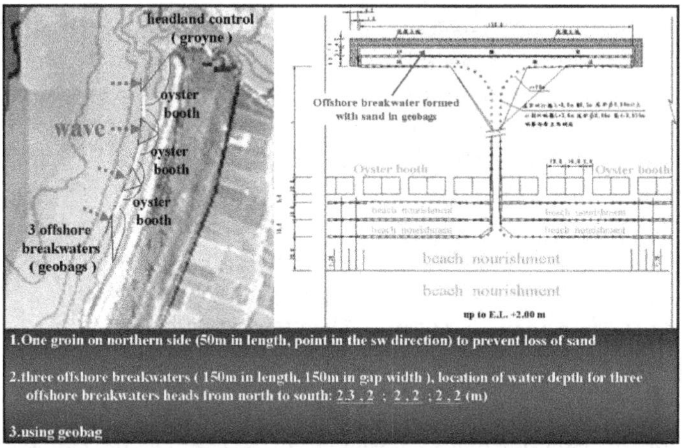

Figure 11. The final sketch of shore protection countermeasure for Shuang-Chun coastal area

Figure 12. Beach nourishment and geobag application in experimental study of the Shuang-Chun

5.3. Wai-San-Ding barrier island

Barrier islands as their name implies, they form a protective barrier between coastal shorelines and wave action that originates offshore. Barrier islands are also ecosystems that border coastal shorelines and physically separate the offshore oceanic province from inshore wetlands, bays and estuaries. Coastlines fronted by barrier islands also include some of the greatest concentrations of human populations and accompanying anthropogenic development in the word. The native vegetation and geological stability of these ecosystems are coupled and vulnerable to erosion events, particularly when also disturbed by development.

The Wai-San-Ding barrier island (Figure 13), protruding at about a forty-five-degree angle from the natural trend of the mainland shoreline at the mouth of the Peigang Shi River, is the largest remaining barrier island off the Taiwan coast. The overall length of this barrier is twenty kilometers, and her area is around two-thousand hectares during the Mean Water Level (M.W.L.). Much of the island shoreline is investigated to have been eroding at a rate of 50m~60m per year in recent years. Furthermore, this island holds some sort of "land speed" with continuing 0.2 degree/year counter-clockwise rotation to migrate southeastware to the mainland shoreline and gradual submerging into the sea. The Wai-San-Ding barrier island located on the southwestern Taiwan, is normally treated as natural offshore breakwater. In fact this biggest barrier island can form a defense line of low-lying coastal plains and back-barrier basin against storms attacking the southwestern coastal area in Taiwan. However, the erosion problem of the Wai-San-Ding barrier island has become more serious in the recent past. Therefore, how to protect this barrier island is always an important issue both from the consideration of coastal hazard and sustainable environment in Taiwan.

The objective of this experimental application study is to find the suitable measure for mitigating the existing erosion problem of the Wai-San-Ding barrier island. After collecting enough hydrodynamic and morphodynamic data from the long-term field investigation, the erosion mechanisms of the barrier island were analyzed in detail. Figure 14 shows that the time for this beach erosion to become apparently chiefly depends on how fast the rate of longshore sediment transport decreased from the up-coast area and on the river sediment supply. Meanwhile, run up mechanism under various waves, storm surge and overwash threshold on sand barrier during typhoon are also the important factors to be investigated. However, sufficient knowledge on nearshore hydrodynamic forcing, sediment transport processes and morphological features along this offshore barrier island, is helpful to the countermeasure control work. Based on the analysis of the erosion mechanisms, consideration of some measure options were proposed and firstly simulated by numerical model to find the two better solutions. Then two better applications (Figure 15) of the various soft methodologies available for the beach erosion control were proposed after numerical model analysis and further investigated by physical model test in the Near-shore Wave Basin (NSWB, 150x60x1.5m) at the Tainan

Hydraulics Laboratory (THL), National Cheng Kung University (NCKU), Tainan, Taiwan to validate their effect. The results showed that soft groins in the downstream and submerged artificial berms in the midstream are the effectively integrated measure to mitigate the continuing erosion problem of the Wai-San-Ding barrier island. Meanwhile, the plant evolution method and oyster cultch with aquatic vegetation were also proposed to apply in mitigation of wind sand transport and stabilization of sand dune. In order to protect this offshore barrier island, the more detailed field investigations on sediment supply (source and sinks), hydrodynamic forces (waves, tides and rate of sea level change) and geomorphic setting (shoreface profile shape, sub-strata composition) should be continuously conducted. Furthermore, the in-situ experimental study based on two proposed countermeasures is suggested to apply in improving the security as well as to involve ecological and scenic remediation for the erosion problem of Wai-San-Ding barrier island.

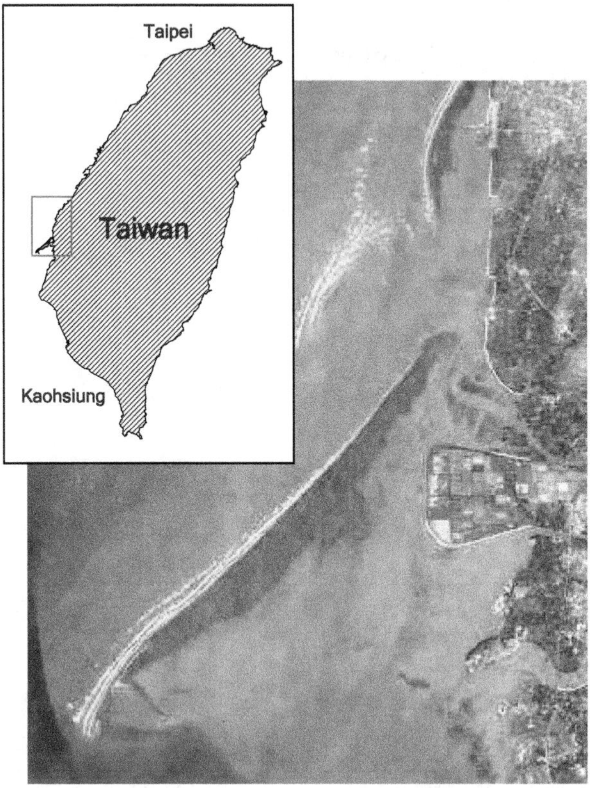

Figure 13. Studied area Wai-San-Ding Barrier Island image from satellite SPOT(2001)

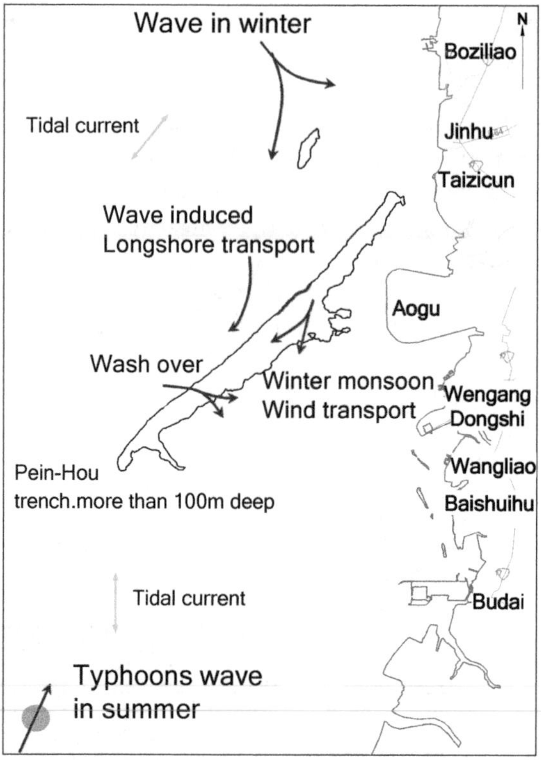

Figure 14. Hydrodynamic forcing characteristics, littoral sediment transport and morphology dynamics of the offshore barrier island

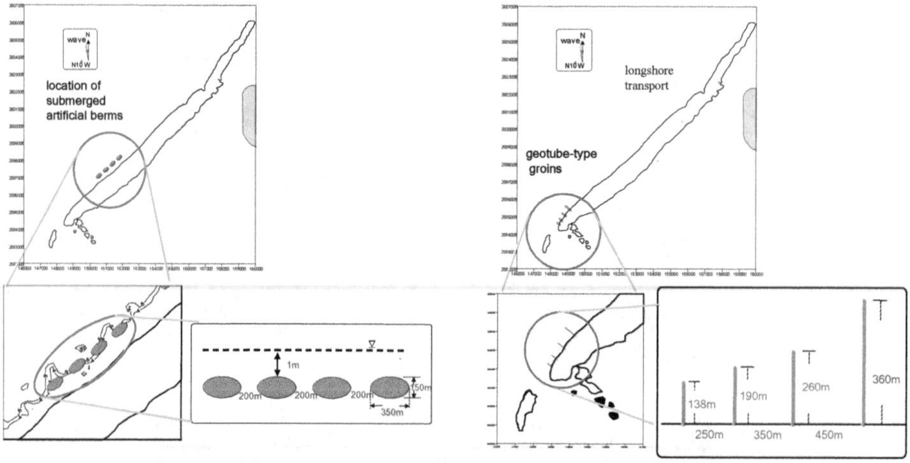

(a) submerged artificial berms (b) soft groins used by geotube

Figure 15. Two better countermeasures for mitigation the erosion problem of Wai-San-Ding Barrier Island

6. Conclusion

The efficiency and productivity of the infrastructure facilities is strictly required nowadays owing to escalating fiscal deficit of the government. We have questioned ourselves about what the people in Taiwan expect on coastal protection work for the next decade; moreover, is it worth applying soft solution instead of keeping the previous reinforced concrete revetment under this tight budget condition? The objective of this study is therefore to present various soft solution strategies available for beach erosion control in the hope of providing better efficiency and cost-effectiveness as well. The results also reveal that the current shore protection system around Taiwan can be properly controlled by beach nourishment. However, the specific characteristics at each local site should be taken into consideration. Accordingly, we divide Taiwan coast into categories based on the collected information such as geological characteristics, hydrodynamics, and the intensity of beach erosion. The national beach management and protection problems will therefore be evaluated followed by the offering of resolutions for shore protection and further improvements. For the purpose of beach erosion management, we also have completed collecting and analyzing coastal data around Taiwan and constructed a database as well a geographic information system (see Figure 16) as reference. Related units of coastal management agency, in Taiwan, are permitted to log on to and use the system via the Worldwide Web with an authorized username and password. The actual locations and related information of the current shore protection constructions with suitable principle and countermeasure of the future beach erosion control around Taiwan can be obtained via this geographic information. It is helpful for future reference of beach erosion management for the governmental agency in charge of shoreline policies.

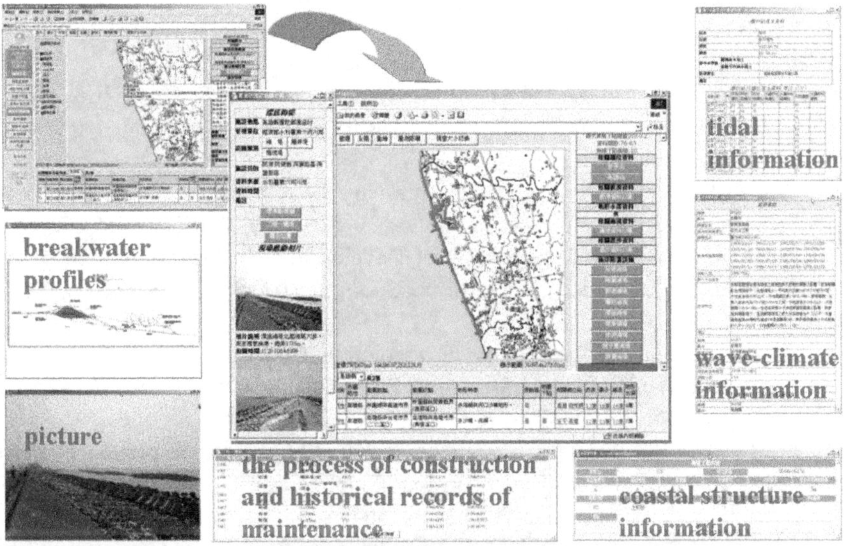

Figure 16. GIS information system of Taiwan coast

Author details

Ray-Yeng Yang and Ying-Chih Wu
Tainan Hydraulics Laboratory, National Cheng Kung University, Taiwan
Department of Hydraulic and Ocean Engineering, National Cheng Kung University, Taiwan

Hwung-Hweng Hwung
Department of Hydraulic and Ocean Engineering, National Cheng Kung University, Taiwan

Acknowledgement

The authors wish to acknowledge the generous support from Water Resources Agency, Ministry of Economic Affairs and National Science Council through the project "KUN–SHEN" of ANR (France) and NSC (Taiwan) joint funding initiative -- grant no. NSC-100-2923-E-006-001-MY3.

7. References

Bridges, M. & Dean, R.G. (1996). Erosional Hot Spots: Characteristics and Causes, *Proceedings 10th National Conference on Beach Preservation Technology*, Florida Shore and Beach Preservation Assoc.

Chen, S.N.; Sanford, L.P.; Koch, E.W.; Shi, F. & North, E.W. (2007). A Nearshore Model to Investigate the Effects of Seagrass Bed Geometry on Wave Attenuation and Suspended Sediment Transport, *Estuaries and Coasts*, Vol.20(2), pp.296-310.

Davis, R.A. & Hayes, M.O. (1984). What Is a Wave-Dominated Coast ? , *Marine Geology*, Vol. 60, pp. 313-329.

Dean, R.G. (1986). Coastal Armoring: Effects, Principles and Mitigation, *Proceedings 20th Intl. Conference Coastal Engineering*, pp. 1943-1857, ASCE, Taipei, Taiwan.

Delft Hydraulics (1987). *Manual on Artificial Beach Nourishment*, Delft, The Netherlands.

Dette, H.H. & Raudkivi, A.J. (1994). Beach Nourishment and Dune Protection, *24th ICCE*, pp. 1007-1022, Kobe, Japan.

Fonseca, M.S. & Cahalan, J.A. (1992). A Preliminary Evaluation of Wave Attenuation by Four Species of Seagrass, *Estuarine, Coastal and Shelf Science*, Vol.35, pp.565-576.

Foret, J. (2002). Role of Artificial Oyster Reef Development in the Restoration of Coastal Louisiana, *6th International Conference on Shellfish Restoration*, Charleston, SC, USA, NOAA/Sea Grant.

Fowler, J.D.; Toups, Ch. & Gilbert, P. (1995). Geotextile Contained Contaminated Dredged Material, Marina del Ray, Los Angeles and Port of Oakland, California, *Proceedings 14th World Dredging Congress (WODA)*, Amsterdam.

Hamgauchi, T.; Uda, T.; Inoue, C. & Igarashi, A. (1991). Field Experiment on Wave-Dissipating Effect of Artificial Reefs on the Niigata Coast, *Coastal Engineering in Japan*, Japan Society for Civil Engineers, Vol. 34, pp.50-65.

Hamm, L. et al. (1998). *Beach Fills in Europe; Projects, Practices and Objectives*, Book of Abstracts, 26th ICCE, Copenhagen, Denmark.

Hands, E.B. (1991). Unprecedented Migration of a Submerged Mound off the Alabama Coast, *Proceedings 12th Ann. Conference Western Dredging Assoc.*, pp.1-25.

Hwung, H.H.; Huang, H.Y.; Wu, Y.C.; Liou, J.Y. & Liu L.L. (2008). Mineral Accretion Technique during Biological Attachment In-Situ, *30th Ocean Engineering Conference in Taiwan*, pp.553-558, National Chiao Tung University, *Taiwan*. (in Chinese).

Hwung, H.H.; Yang, R.Y. & Igor V.S. (2008). Sea Wave Adaptation by an Elastic Plate, *Proceedings 18th International Offshore (Ocean) and Polor Engineering Conference*, pp.296-302, Vancouver, Canada.

Hsu, J.R.C. & Evans, C. (1989). Parabolic Bay Shapes and Applications, *Proceedings Instn. Civil Engers.*, Part 2. London: Thomas Telford, Vol.87, pp.557-570.

Kobayashi, N.; Raichle, A.W. & Asano, T. (1993). Wave Attenuation by Vegetation, *J. Waterway, Port, Coastal, Ocean Engineering*, Vol.119(1), pp.30-48.

Krystian, W.P. (2000). Geosynthetics and Geosystems in Hydraulic and Coastal Engineering, A. A. Balkema, Rotterdam, Netherlands.

Leo C. van Rijn (1998). *Principles of Coastal Morphology*, Aqua Publications.

Leonard, L.A. et al. (1990). A Comparison of Beach Replenishment on the U.S. Atlantic, Pacific and Gulf Coasts, *Journal of Coastal Research*, SI6, pp.127-140.

Li, C.W. & Yan, K. (2007). Numerical Investigation of Wave-Current-Vegetation Interaction, *Journal of Hydraulic Engineering*, Vol.133, No.7, pp.794-803.

Wright, L.D. (1995). Morphodynamics of Inner Continental Shelves, Boca Raton, Florida :CRC.

Liou, J.Y.; Huang, H.Y.; Kuo, C.H.; Shieh, C.T. & Chiang, W.P. (2007). An Amelioration Study upon Ching-Tsao-Lun Dike in Tainan Coast, *Proceeding of the 29th Ocean Engineering Conference in Taiwan* (in Chinese).

Mφller, J. T. (1990). Artificial Beach Nourishment on the Danish North Sea Coast, *Journal of Coastal Research*, SI6, pp. 1-10.

O'Beirn, F.; Luckenbach, M.; Nestlerode, J. & Coates, G. (2000). Toward Design Criteria in Constructed Oyster Reefs: Oyster Recruitment as a Function of Substrate Type and Tidal Height, *Journal of Shellfish Research*, Vol.19, No.1, pp. 387-395.

Otay, E.N. (1994). *Long-Term Evolution of Nearshore Disposal Berms*, Ph.D. dissertation, Dept. of Coastal and Oceanographic Engineering, University of Florida.

Rijkswaterstaat (1996). *Evaluation of Sand Nourishment Projects along the Dutch Coast 1975-1994(in Dutch)*, Report RIKZ 96.028, The Hague, The Netherlands.

Rybicki, N.B.; Jenter, H.L.; Carter, V.; Baltzer, R.A. & Turtora, M. (1997). Observations of Tidal Flux between a Submersed Aquatic Plant Stand and the Adjacent Channel in the Potomac River near Washington, D. C., *Limnology and Oceanography*, Vol.42(2), pp.307-317.

Tainan Hydraulics Laboratory Technical Report. (2002). *The Research of New Shore Protection Technology (3/4)*, National Cheng Kung University, Taiwan, Bulletin No.285. (in Chinese).

Tainan Hydraulics Laboratory Technical Report. (2004). *The Research of New Shore Protection Technology (4/4)*, National Cheng Kung University, Taiwan, Bulletin No.312. (in Chinese).

Work, P.A. & Dean, R.G. (1991). Effect of Varying Sediment Size on Equilibrium Beach Profile, *Coastal sediments*, Seattle, USA.

Yang, R.Y.; Wu, Y.C.; Liou, J.Y. & Hong, C.S. (2004). A Research of Field Case Protection on Shuang-Chun Coastline, *Proceeding of the 26th Ocean Engineering Conference in Taiwan*, pp. 691~698 (in Chinese).

Chapter 4

Overview of Environmental Management by Drill Cutting Re-Injection Through Hydraulic Fracturing in Upstream Oil and Gas Industry

Mansoor Zoveidavianpoor, Ariffin Samsuri and Seyed Reza Shadizadeh

Additional information is available at the end of the chapter

1. Introduction

For the reason of worldwide increased activities of upstream oil and gas industry for future energy demands which will be associated with more waste generation, zero discharge is considered an environmentally friendly approach of complying with environmental legislations. Drilling is one of the major operations in upstream oil and gas industry that can potentially impact the environment through generation of different types of wastes. The drilling process generates millions of barrels of drilling waste each year; primarily used drilling fluids and drill cuttings especially oil-contaminated drill cuttings. In the early years of the oil industry, little attention was given to environmental management of drilling wastes. The rapid development of drilling operation in order to fulfill the global energy demands and so the drilling environmental regulatory requirements have become stricter, drilling and mud system technologies have advanced, and many companies have voluntarily adopted waste management options with more benign environmental impacts that those used in the past. Moreover, it is crucial to find out why drilling wastes are important nowadays, how they generated and by which means those waste could be disposed off with higher efficiency and acceptable HSE and economically concerns. Drill Cutting Re-Injection (DCRI) is one of the processes that developed as an environmentally friendly and zero discharge technology in upstream oil and gas industry.

A variety of oil field wastes are disposed of through injection, such as produced water that re-injected through tens of thousands of wells for enhanced recovery or disposal. Other oil field wastes that are injected at some sites include work over and completion fluids, sludge, sand, scale, contaminated soils, and storm water, among others. The focus of this chapter is injection of wastes related to the drilling process, which involve processing cuttings into small particles, mixing them with water and other additives to make slurry, and injecting it into a subsurface geological formation at pressure high enough to fracture the rock. DCRI has been given other terms by different authors such as fracture slurry injection, grind and inject, and drill cuttings injection.

The most critical aspect in waste injection through hydraulic fracturing (HF) in upstream oil and gas industry, which is DCRI, will be reviewed in this chapter. The subject of this chapter, DCRI, is a specialized area in upstream petroleum industry; even though many brilliant papers presented on various environmental areas, overview papers that present a context for those more specific studies are needed. This chapter will presents in an effort to review the environmental management of DCRI in upstream petroleum industry. The aims are firstly, to review the drilling process and different types of drilling fluid. Afterwards, because it's considered as a key in identifying containment formations to prevent waste migration to water resources and environment in DCRI operations, HF technology will be introduced in the second part of this chapter. Finally, after reviewing the essential parts of DCRI, drilling wastes and HF, the nature of DCRI and its role in environmental management will be presented in details.

2. Overview of drilling operation

Oil and gas wells are drilled to depths of several hundred to more than 5,000 meters. Figure 1 shows a schematic of typical drilling rig, which uses a rotating drill bit attached to the end of a drill pipe. Drilling fluids (muds) are pumped down through the hollow drill pipe, through the drill bit nozzles and up the annular space between the drill pipe and the hole. Drilling mud mixture is particularly related to site and hole condition; it used to lubricate and cool the drill bit, maintains pressure control of the well as it is being drilled, and helps to removes the cuttings from the hole to the surface, among other functions. In fact, the technology of mud mixing and treatment has been recognized as a source of pollutants.

Mud and drill cuttings are separated by circulating the mixture over vibrating screens called shale shakers. As the bit turns, it generates fragments of rock (cuttings), which will be separated from the mud by shale shakers that will moves the accumulated cuttings over the screen to a point for further treatment or management. Consequently, additional lengths of pipe are added to the drill string as necessary. As a common practice in drilling of oil and gas wells, when a target depth has been reached according to the drilling plan, the drill string is removed and the exposed section of the borehole is permanently stabilized and lined with casing that is slightly smaller than the diameter of the hole. The main function is to maintain well-bore stability and pressure integrity. (Three sizes of casing depicted in Figure 1). Cement is then is pumped into the space between the wall of the drilled hole and the outside of the casing to secure the casing and seal off the upper part of the borehole. Each new portion of casing is smaller in diameter than the previous portion through which it is installed. The final number of casing strings depends on the total depth of the well and the sensitivity of the formations through which the well passes. The process of drilling and adding sections of casing continues until final well depth is reached.

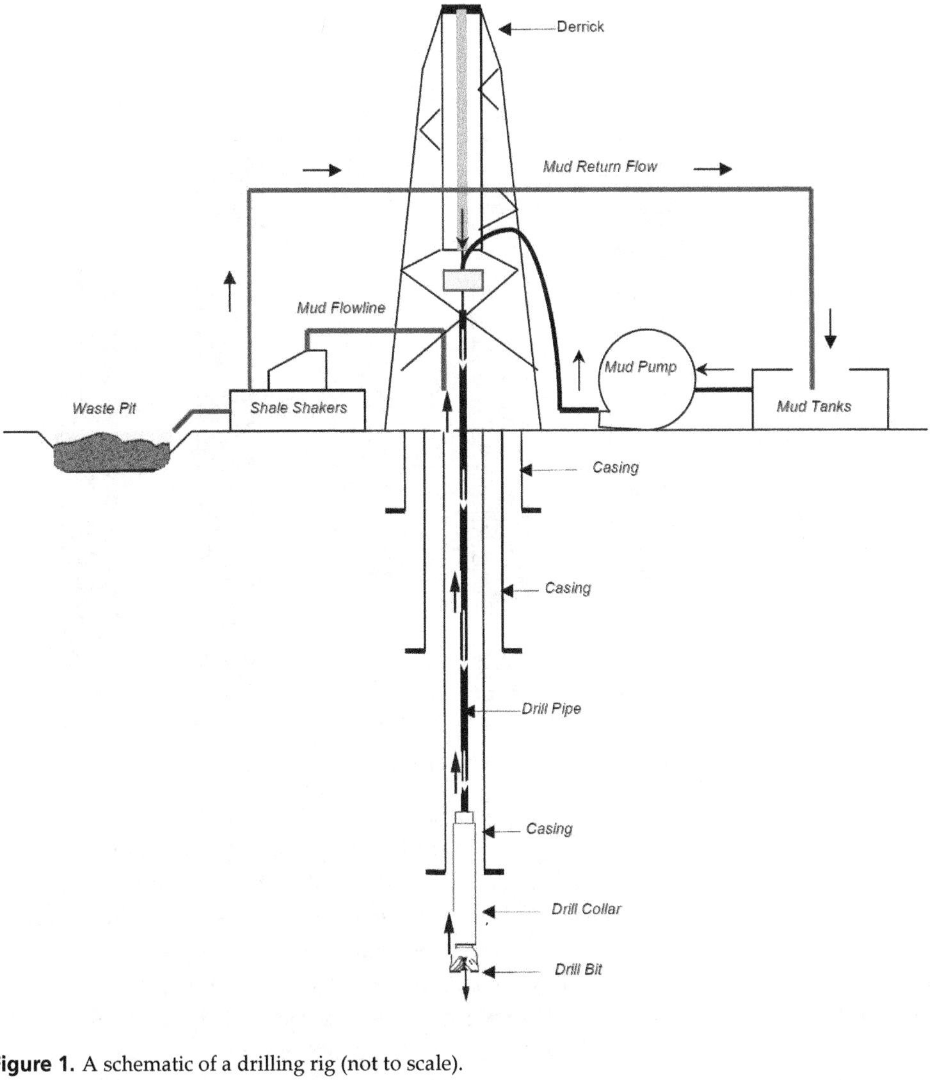

Figure 1. A schematic of a drilling rig (not to scale).

Two primary types of wastes are generated in drilling of oil and gas wells; drill cuttings and drilling fluids. Most drilling fluids contain bentonite clay, water, barite, specialized additives, and some types of muds also contain hydrocarbons. Large volumes of drilling muds are stored in aboveground tanks or pits. The liquid muds pass through the screen and are recycled into the mud system, which is continuously treated to maintain the desired properties for a successful drilling operation. Depending on the depth and diameter of the well bore, the volume of drilling wastes generated from each well varies; typically, several thousand barrels of drilling waste are generated per well. Figure 2 is a demonstration of the generated drilling waste from a 2400 meters well depth that comprises of four different borehole sizes.

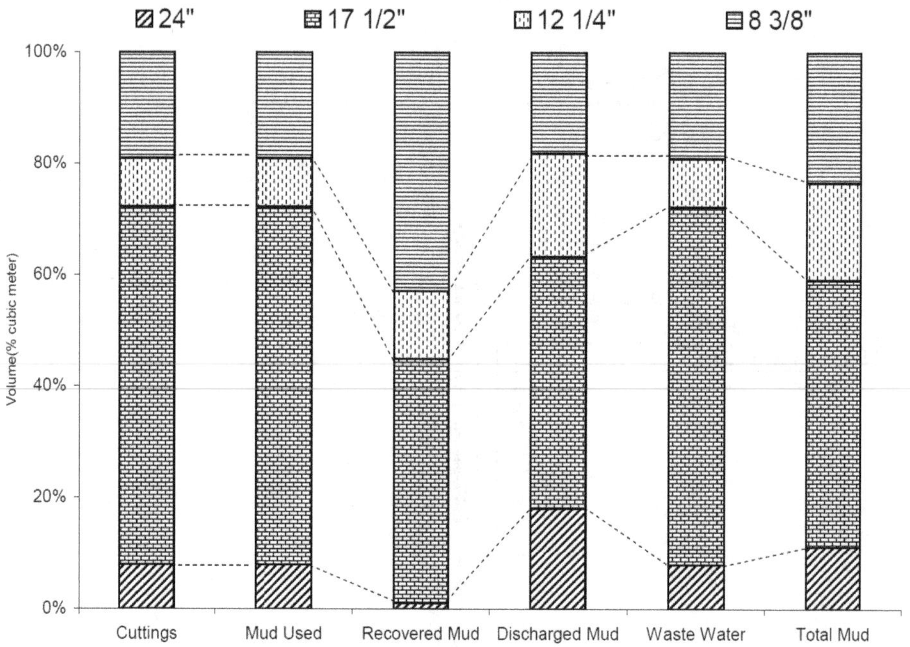

Figure 2. A typical drill cutting and mud volumes for a 2400 meters well depth.

The generation of wastes from drilling fluid and drill cuttings could be recognized at different stages of the drilling operation. When drilling at the first few hundred meters to run conductor casing or surface casing, higher quantities of cuttings are produced; that's because borehole diameter is the largest during this stage. Substantial waste fluid must be handled when drilling deep wells that encountered shale's and/or unstable formations. So, oil based muds (OBMs) is utilized to overcome those problems which will be mixed with other drilling fluids in waste pit and disposed to the environment. Furthermore, higher volume of wastes must be displaced in the completion phase of drilling operation which is replaced by completion fluids and equipment. Physical condition of a waste pit during and after drilling operation is illustrated in Figures 3 and 4, respectively. More details could be found by Shadizadeh and Zoveidavianpoor, (2008).

Figure 3. Mud pit condition during drilling operation.

Figure 4. Mud pit condition after drilling operation.

3. Environmental impacts of drilling muds

In upstream petroleum industry, drilling is the major operation that can potentially impact the environment. Drilling operation generates a significant volume of wastes. The composition of drilling fluid constituents is depicted in Table 1. Environmentally responsible actions require an understanding of the characteristics of these wastes and how they are generated in order to minimize their environmental impacts by known environmental protection methods. In this section, environmental impacts of a drilling mud will be presented along with a case study on mud pit samples for heavy metals (Cd, Cr, Ni, and Al) concentrations during and after the drilling operation. For more details please consult Shadizadeh and Zoveidavianpoor, 2008 and 2010.

Elements	Water	Cuttings	Barite	Clay	Chrome-lignosulfonate	Lignite	Caustic
Aluminum	0.3	40,400	40,400	88,600	6,700	6,700	0.013
Arsenic	0.0005	3.9	34	3.9	10.1	10.1	0.039
Barium	0.01	158	590,000	640	230	230	0.26
Cadmium	0.0001	0.08	6	0.5	0.2	0.2	0.0013
Chromium	0.001	183	183	8.02	40,030	65.3	0.00066
Cobalt	0.001	183	183	8.02	40,030	65.3	0.00066
Copper	0.0002	2.9	3.8	2.9	5	5	0.00053
Iron	0.003	22	49	8.18	22.9	22.9	0.039
Lead	0.5	21,900	12,950	37,500	7,220	7,220	0.04
Magnesium	0.003	37	685	27.1	5.4	5.4	0.004
Mercury	4	23,300	3,900	69,800	5,040	5,040	17,800
Nickel	0.0001	0.12	4.1	0.12	0.2	0.2	5
Potassium	0.0005	15	3	15	11.6	11.6	0.09
Silicon	2.2	13,500	660	2,400	3,000	460	51,400
Sodium	7	206,000	70,200	271,000	2,390	2,390	339
Strontium	6	3,040	3,040	11,000	71,000	2,400	500,000
Cobalt	0.07	312	540	60.5	1030	1030	105

Table 1. Elemental composition of drilling fluid constituents (ppm) (Bleier et al., 1993).

A potential source of heavy metals in drilling fluid is from crude itself. Crude oil naturally contains widely varying concentrations of various heavy metals. In the selected well a combination of water based muds (WBMs) and OBMs had used. As shown in Table 2, the major components of WBMs in the investigated site were barite, salt, starch, bentonite, and lime. The metals of greatest concern, because of their potential toxicity and/or abundance in drilling fluids, include chromium, cadmium, and nickel (Neff, 2002). Some of these metals are added intentionally to drilling muds as metal salts or organometallic compounds. Others are present as trace impurities in major mud ingredients, particularly barite and bentonite. One of the major drilling mud additives used in both WBMs and OBMs in the investigated well is barite. The amount of barite used in the investigated well as shown in Table 2 is 702 tonnes. Barite contains variable amounts of heavy metals and it is the main source of heavy metals in the investigated site. Metals concentrations in mud pit of selected well during and after drilling operation are presented in Figure 5. Chromium concentration was detected in the samples at 0–0.08 ppm. Other heavy metals were also at high levels and showed significantly higher values specially by using OBMs: cadmium 0–0.006 ppm, nickel 0–0.024 ppm, and aluminum 0–341 ppm. However, these heavy metal levels are generally above toxic levels. As shown in Figures 5, the concentrations of cadmium, chromium, and nickel increased progressively in the fourth sampling periods because of the contamination of the mud pit with OBMs that was initiated in the fourth sampling period. Concentration of aluminum increased from the first to the third sampling periods, whereas in the fourth period it shows decreased values from 0.05 ppm to 0.006 ppm. Aluminum was not observed in the fifth and sixth sampling periods but maintained an increased value from the seventh to the end of the sampling periods. In the entire study area, chromium levels ranged from 0 to 0.08 ppm but no concentration was observed after the seventh period of the sampling. This can be explained by the storm runoff water at the investigated well site that washes away all these wastes, especially in the mud pits to other locations or seepage from the discharge pits into the surrounding soils. The statistics of the investigated heavy metals are shown in Table 3.

Properties		**24″ hole ≅ 60 m**	**17½″ hole ≅ 1510 m**	**12¼″ hole ≅ 2158 m**	**8½″ hole ≅ 2330 m**
Mud Properties	Mud system	WBM	WBM	WBM	OBM
	pH	10-10.5	10.5-9.8	8-10	9-9.5
	Average salt concentration (mg/l)	2000	185600	297600	380100
	Average calcium concentration (mg/l)	464	2404	3320	231
	YP	11	4-7	6-78	19-27
	PV	35	5-10	8-58	8-12
	Initial Gel	22	3-6	1-13	2
	10 Min. Gel	30	4-8	2-6	3
	Mud lost @ unit (bbl)	0	2588	1252	802
	Density (pcf)	70-62	68-79	79-146	69.5
	Barite (t)	0	27	674.4	0
Mud Material	Salt (t)	2	166	168	15
	Starch (sx)	0	30	727	0
	Bentonite (t)	160	750	0	0
	Lime (sx)	123	69	222	130
	CMS H.V (sx)	0	0	0	17
	IRSATROL(sx)	0	0	0	140
	Diesel (bbl)	0	0	0	615

Note: YP=yield point; PV=plastic viscosity; bbl=barrel; pcf=pound per cubic feet; t=ton; sx=sacks

Table 2. Drilling fluid used in the selected well (Shadizadeh and Zoveidavianpoor, 2010).

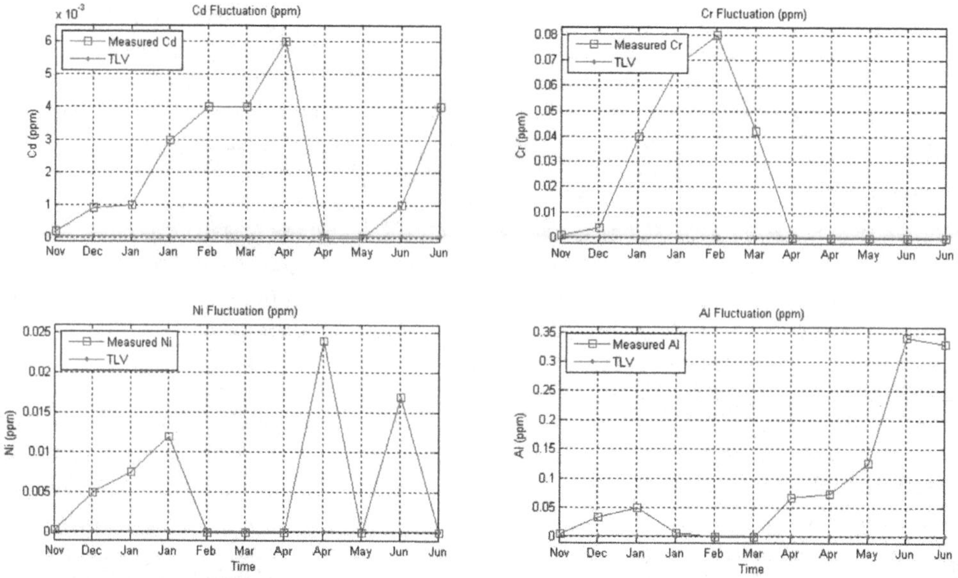

Figure 5. Heavy metals fluctuation during and after drilling operation.

Statistics	Heavy Metals (ppm)			
	Cd	Cr	Ni	Al
Max	0.0060	0.0800	0.024	0.341
Mean	0.0022	0.0214	0.005991	0.09396
Median	1.0000e-003	1.0000e-003	0.0003	0.05
Mode	0.0040	0	0	0
Std	0.0021	0.0306	0.008349	0.1255
Range	0.0060	0.0800	0.024	0.341

Table 3. Heavy metals statistics in the case study

4. Potential effects on natural resources, and minimization strategies

Drilling wastes can harm ecosystems, plants, and animals and cause health problems in humans. Many materials that are released into reserve mud pits also release drilling wastes into the environment, which calls for public awareness as well. When released heavy metals are discharged into unlined pits the toxic substances in the pits can leach directly into the soil and may contaminate groundwater. Additionally, there is no evidence of zero discharge in lined pits. In contrast to most organic pollutants, trace metals are not usually eliminated from aquatic ecosystems by natural processes due to their non-biodegradability. Both toxic and nontoxic heavy metals tend to accumulate in bottom sediments, from which they may be released by various processes of remobilization. Frequently, these metals can move up the biological chain, eventually reaching humans, where they can cause chronic and acute ailments (Ankley et al., 1993). As presented in the previous sections, routine drilling wastes such as drilling muds and cuttings contain a variety of toxic chemicals; they are known to be hazardous to wildlife, livestock, and human health. If pollutants from oil well drilling build up in the food chain, people who consume those natural resources from the contaminated drilled well area could be at risk of health problems such as genetic defects and cancer. For environmental protection, different strategies are considered; (1) restoring the well site to its natural state after drilling, (2) let the liquid to be evaporated, (3) Bioremediation, (4) multi-pit system, and (5) DCRI, which is the focus of this chapter. Because DCRI deal with the initiation and propagation of a fracture in a rock matrix by means of hydraulic pressure, HF will briefly be discussed in the next section.

5. Hydraulic fracturing

Initially, fracturing was a low technology operation consisting of the injection, at low temperature, of a few thousand gallons of napalm into low-pressure reservoirs. Substantially, HF has evolved into a highly engineering and complex procedure. As a technology has improved, so has the number of wells, formations, and fields that can be successfully fractured, increased. The development of high pressure pump units, high strength proppant, and sophisticated fracturing fluids, has meant that deep, low permeability, high temperature, reservoirs can now be fractured (Veatch et al., 1989). This technology is a well-known process, which was originally applied to overcome near wellbore skin damage (Smith, 2006). Since then, it has been expanded to such applications as (1) reservoir stimulation for increase hydrocarbon deliverability, (2) increase drainage area, and decrease pressure drop around the well to minimize problems with asphaltene and/or paraffin deposition, (3) geothermal reservoir recovery, (4) waste disposal, (5) control of sand production, (6) to measure the in-situ stress field and (7) heat extraction (geothermal energy) from deep formations. Obviously, there could be other uses of HF, but the majority of the treatments are performed for the mentioned reasons. HF has made significant contributions to the petroleum industry since its inception (Veatch et al., 1989). By 2009 HF activity has increased 5-fold compared to the investment of a decade earlier and has become the second largest outlay of petroleum companies after drilling (Economides, 2010).

HF is the pumping of fluids at high rates and pressures in order to break the rock. A typical chart of fracturing which shows the common treatment stages is shown in Figure 6. The operation begins with injection of a mixed acid and water named Pre-pad. A mixture of water and a polymer, named Pad, will follows. The fracture will initiated in this stage but contains no proppant. To make the fracture open for fluid flow, a mixture of proppant and the fracturing fluid, which called Slurry will have injected. For more details please consult Daneshy, 2010.

As it clear from section 2, the need has been arises to treat/manage the drill cuttings toward zero discharge by utilization of HF.

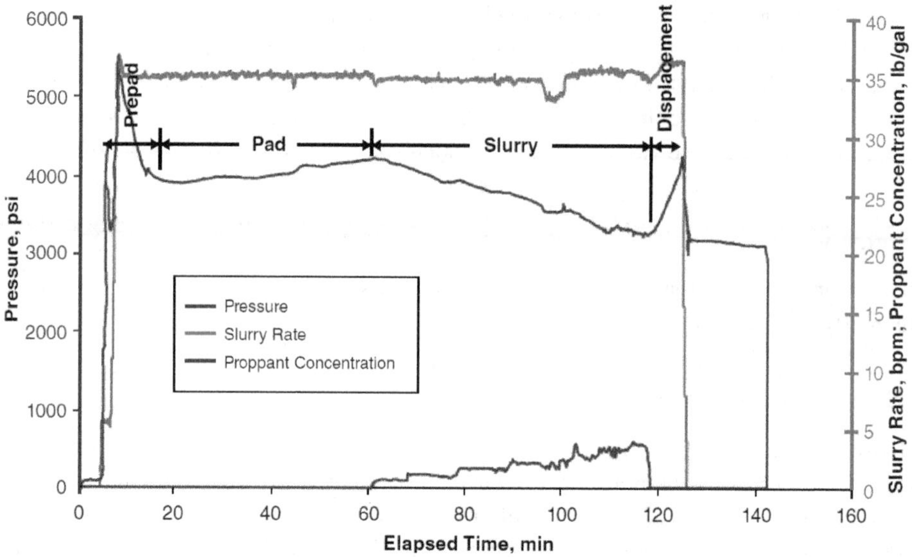

Figure 6. A typical fracturing chart illustrates the steps to HF a well (Daneshy, 2010).

There are both similarities and distinct differences between HF and DCRI which shown in Table 4. More details could be found from Arthur (2010).

Issue	Drill Cutting Re-injection	Hydraulic Fracturing
Target interval	Non Reservoir	Reservoir
Pumping period	Long-term	Short-term
Pumping pressure	Fracture	Fracture
Slurry mixture	Cuttings and fracturing fluids	Proppant and fracturing fluids
Fracture containment study	Essential	Essential

Table 4. Comparison between DCRI and HF.

6. Waste management by DCRI

6.1. An overview

Even though the generation of drill cuttings is a certain result of drilling, those wastes can be treated and/or managed in a number of ways. A summary chart on different drilling wastes management options are presented in Figure 7. As mentioned earlier, the focus of this chapter will be on DCRI.

Valuable literature available regarding the disposal options including: lessons learned concerning biotreating exploration and production wastes (McMillen et al. 2004), successful cases of fixation (Zimmerman and Robert, 1991), converting cuttings into a valuable sources by using vermicomposting (Paulse, 2004), and thermal treatment (Bansal and Sugiarto, 1999).

As summarized in Figure 7, environmental management of drilling wastes may be categorized in three options; waste minimization, recycle/reuse, and disposal. The first and second options are not addressed here. Table 5 shows a comparison among disposal methods which may classified into fixation, thermal treatment, DCRI, and bioremediation/composting. Among the four methods for disposal option that may be considered when deciding on waste management options, the focus of this chapter is on DCRI.

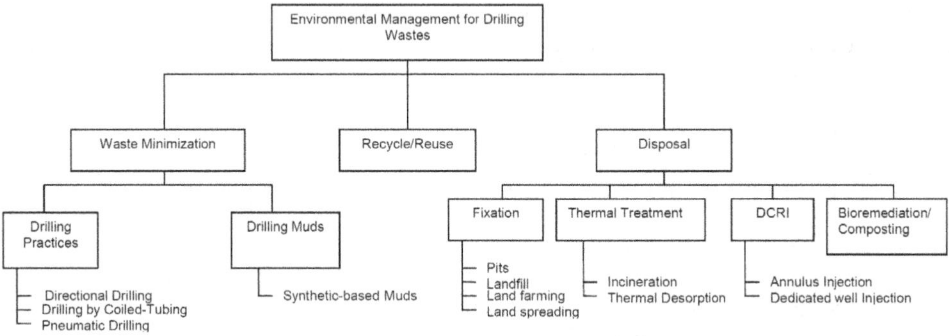

Figure 7. Different approaches in environmental management for drilling wastes

Comparative assessments on alternative disposal options are outlined in Table 5. As clearly shown, environmental impacts and safety risks, which are the most important factors among others, have low level degree and therefore its vulnerability as the best option increases to be adopted as the environmentally friendly drilling waste disposal process. in addition to zero discharge, other advantages of DCRI include; no transportation concerns, no future cleanup responsibilities by the operator, full control over the waste management process, world wide applicability, and its favorable economics. According to Reddoch, (2008): "DCRI is simply the lowest cost, easiest course of action for most drilling operations."

Comparison Factors	Fixation	Thermal Treatment	DCRI	Bioremediation/ Composting
Environmental Impact	Low	High	Low	Medium
Cost	\$9-10/bbl [a]	\$90/metric ton [a]	\$5/bbl [b]	\$500/cubic meter [c]
Safety Risks	High	High	Low	Medium
Technical	Low	Medium	High	Medium

$1m^3$=6.29 bbl (US bbl); 1 metric ton=7.1 bbl (for an oil with 0.88 specific gravity)
[a]Bansal and Sugiarto (1999); [b]Reddoch (2008); [c]McMillen and Gray (1994)

Table 5. Qualitative and quantitative comparison in disposal approaches

The question is raised that what is the relationship between environmental management and DCRI? It's clear that DCRI process will maintain waste containment in a target interval with zero discharge and consequently low HSE risks. Other goals such as cost management and asset management are not covered in this chapter. For more details please consult Bruno et al. (2000).

We can visualize DCRI to loss of circulation of drilling fluids in conventional under balanced drilling operation. Also, it's quite similar to HF operation, because we need to propagate the fractures in the selected horizon and this goal will be achieved by utilization of fracture propagation models which conventionally employed in HF treatment.

Cuttings may be re-injected into the annulus of a well being drilled or into a dedicated well. In annulus injection, cutting would be stored until the desired formation is reached. Whereas in dedicated disposal well, one or more dedicated disposal wells would be drilled and drill waste systems put in place in those wells. A schematic of both types of DCRI is shown in Figure 8.

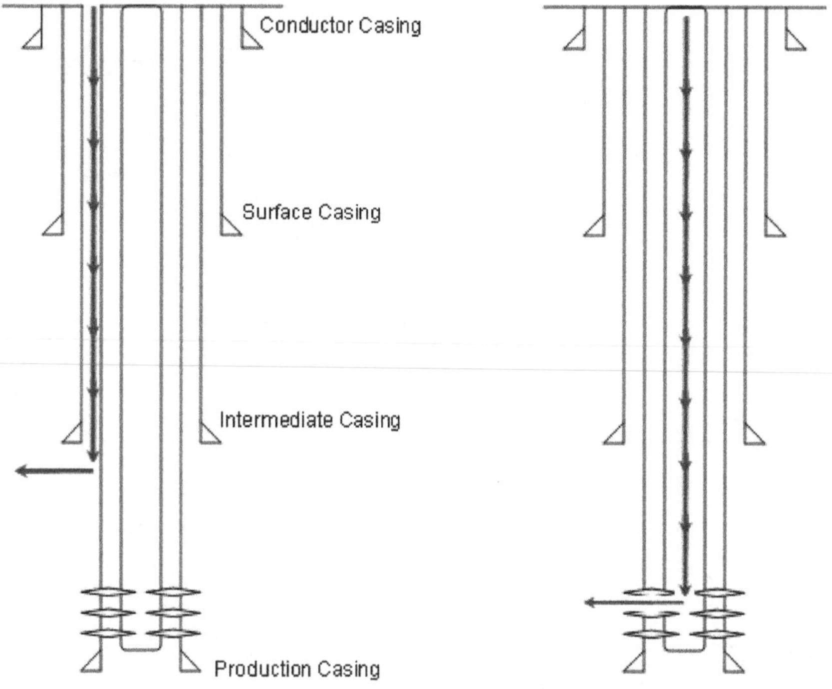

Figure 8. Two major types of DCRI; annulus injection (left) and dedicated well (right).

Drill cuttings may be injected into subsurface geological formations at the drilling site, offshore or onshore and would provide a complete disposal solution. Its worth to note that onshore operations have a wider range of options than offshore operations.

Readers may be asks why this process is called drill cutting re-injection? That's because drill cuttings will be returned back to their origin, deep beneath the Earth's surface.

A sketch of basic setup and flow of DCRI process is shown in Figure 9. Drill cuttings and other oilfield wastes are slurried by being milled and sheared in the presence of water. The resulting slurry is then disposed of by pumping it into a dedicated disposal well, or through the open annulus of a previous well into a fracture created at the casing shoe set in a suitable formation.

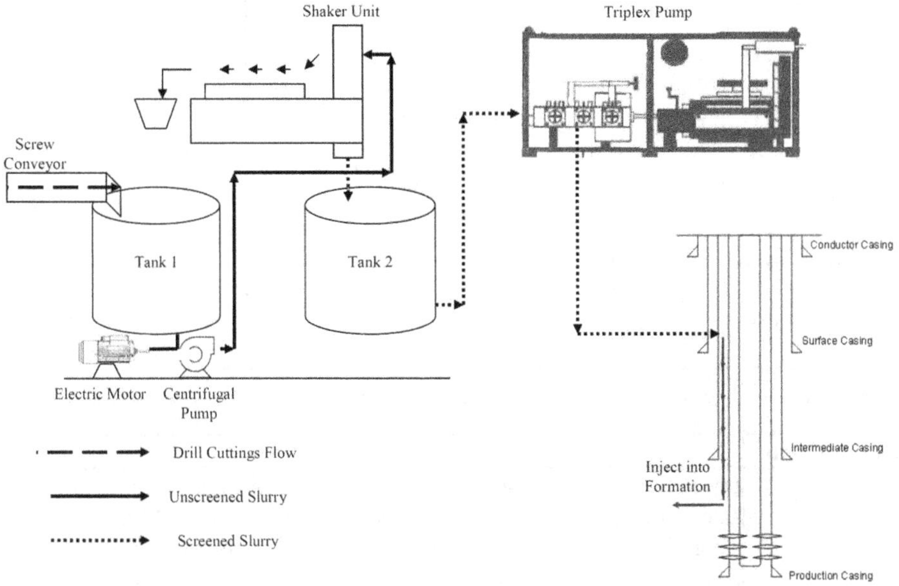

Figure 9. A sketch of basic setup and flow of DCRI

6.2. A case study

In addition to the drill cuttings and drilling fluids, various waste streams need to be handled and disposed of properly include: produced water, contaminated rainwater, scales, and produced sand. DCRI provides a secure operation by injecting cuttings and associated fluids up to several thousand meters below the surface into hydraulically created fractures. In order to guarantee containment within the selected underground formation and perform sufficient design of surface facilities, simulations are performed for the anticipated downhole waste domain.

In this regard, a feasibility study was performed to show the possibility of DCRI in Ahwaz oilfield located in southern Iranian oilfields. The possibility of annular injection and dedicated injection wells was investigated in this study. The objectives were to (1) estimate the volume of drilling waste produced from drilling of each wellbore of the field, (2) select the most appropriate disposal formation in the field, and (3) determine whether the drill wastes can be safely injected into a dedicated well or annular space. Numerous scenarios were considered in the feasibility studies to ensure safe containment of any injected drilling waste. More details could be found by Shadizadeh and Zoveidavianpoor, (2011).

The volumes of drill cuttings and muds, type of utilized mud, and geological information are shown in Table 6. The required data to conduct this study is depicted in Table 7.

Depth (m)	Formation Name	Column	Setting depth (inch≅m)	Lithology	Hole Size (inch)	Cutting volume (bbl)	Mud volume (bbl)	Mud type
1550	Aghajari	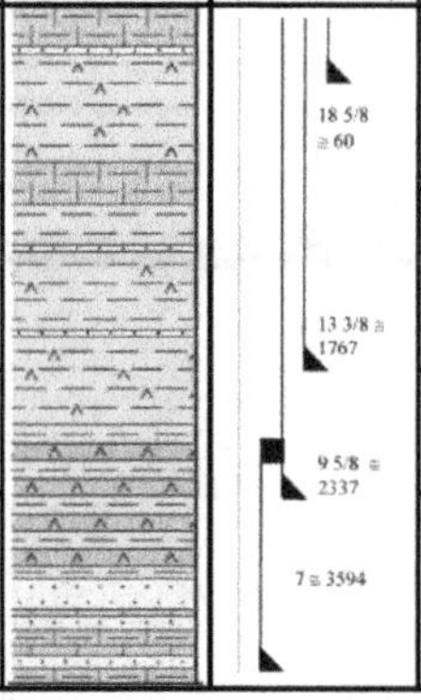		Marl with Sandston bonds	26	132	4400	WBM
1660	Mishan			Marl with Limestone basement	17 1/2	2040	2800	WBM
2332	Gachsaran			Marl, Salt, Anhydrate.	9 5/8	219	3500	WBM
3590	Asmari			Limestone with Sandstone	8 1/2	73	800+400	WBM+ OBM

Table 6. Generalized geologic data along with drill cuttings and mud volumes.

Required data	Description
Injection batch volumes and injection rates	Injection of the slurry is often conducted intermittently in batches into the selected disposal formation, followed by a period of shut-in. depending upon the batch volume and the injection rate, each batch injection may last from less than an hour to several days or even longer.
Minimum in situ stress	Most important in fracture simulation that controls fracture-height growth, fracture azimuth and vertical and horizontal orientation, fracture width, treatment pressures, fracture conductivity, and wastes containment in disposal horizon.
Pore pressure	Very critical parameter to planning and carrying out successful DCRI, because the stress state of the poroelastic medium is directly influenced by pore pressure or reservoir pressure.
Young's modulus	Is the ration of longitudinal stress to longitudinal strain, which has significant effect on fracture geometry, especially on fracture width
Poisson's ratio	Is a measure of the compressibility of material perpendicular to applied stress that has significant effect on fracture geometry
Casing setting depths and injection point	The target which the slurry has to be injected via annulus or dedicated well.
Fluid leak-off data	Means the leaking of fluid from the surface of a fracture into the surrounding rock formation. It's an important parameter controlling the size and geometry of the hydraulically induced fracture.
Slurry rheology	The study of the deformation and flow of matter, that crucial for maintaining zonal isolation.
Fracture toughness	Is an important parameter in fracture modeling and is a measure of a material's resistance to fracture propagation

Table 7. Explanation of required data for DCRI simulation.

In particular, the expense of DCRI requires that the operator knows how the formation will respond to treatment, and whether the treatment design such as selection of pump rates, fluid rheology, accurate rock mechanic properties, pumping schedule and fracture propagation model, will create the intended fracture.

Most 2D models are based on three common models entitles Perkins-Kern-Nordgren (PKN), Khristianovic-Geertsma de Klerk (KGD), and Radial models. The first and second models which assume constant height, are appropriate when the stress contrasts are high between the pay layer and neighboring formations and these contrasts follow lithologic boundaries. For Radial model, its better works in a setting where the fracture grows in a formation of homogeneous stress and mechanical properties so that fracture height is small compared to formation layer thickness. A brief comparison among 2D models is listed in Table 8.

The main advantage of a more advanced method such as pseudo 3D (P3D) over 2D models is that it does not require estimating fracture height, but it does require input of the magnitude of minimum horizontal stress in the zone to be fractured and in the zones immediately above and below.

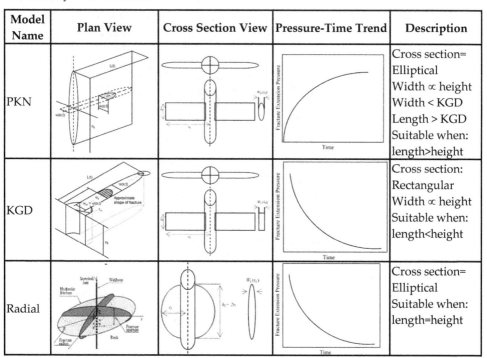

Model Name	Plan View	Cross Section View	Pressure-Time Trend	Description
PKN	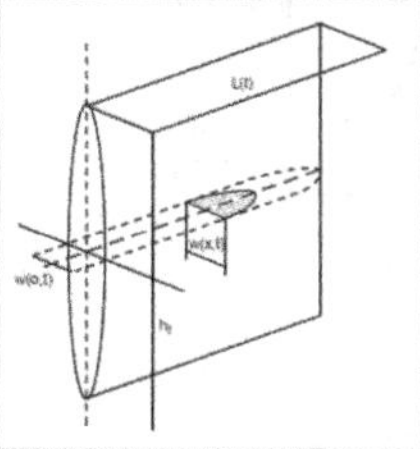	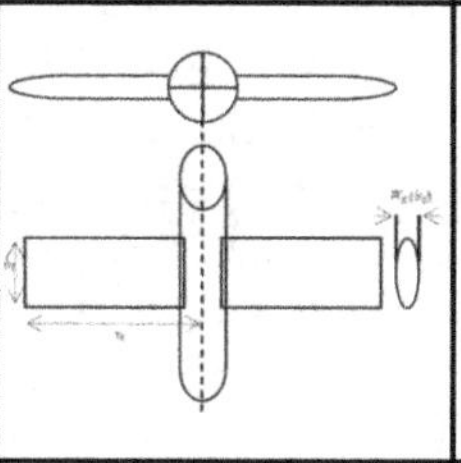	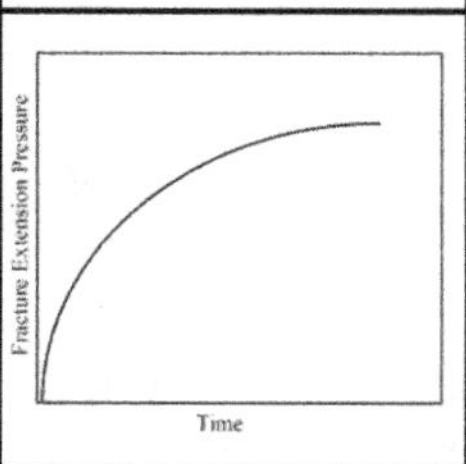	Cross section= Elliptical Width ∝ height Width < KGD Length > KGD Suitable when: length>height
KGD	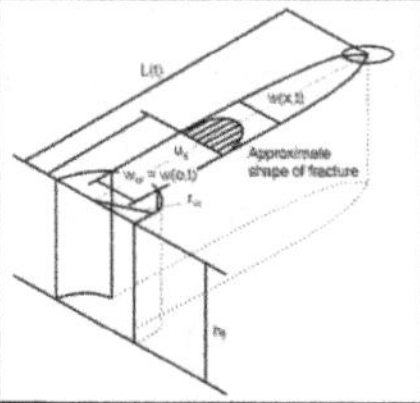	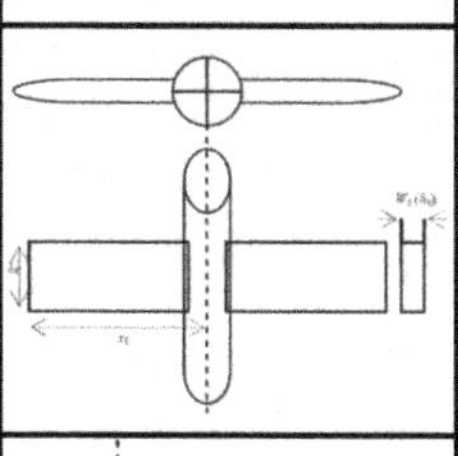	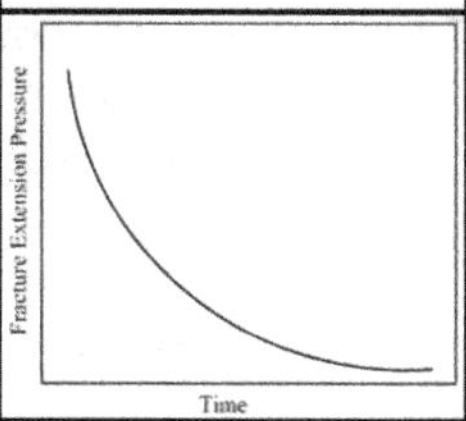	Cross section: Rectangular Width ∝ height Suitable when: length<height
Radial	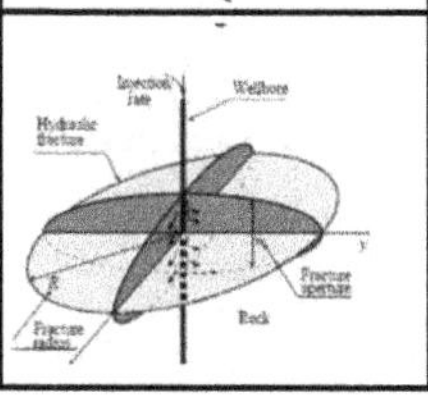	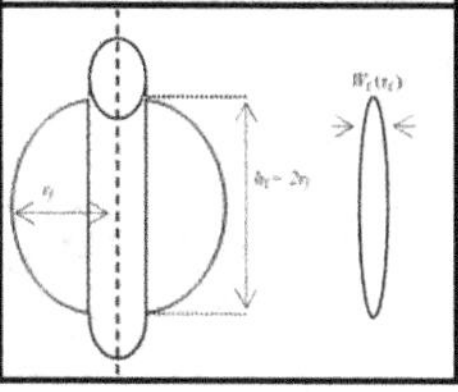	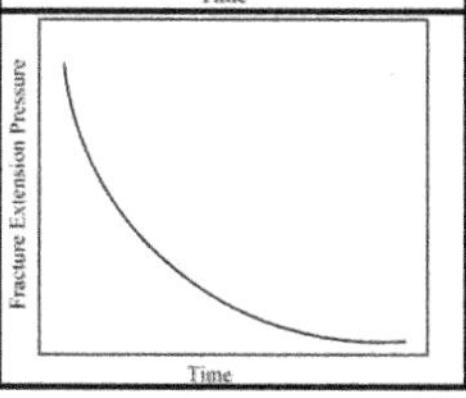	Cross section= Elliptical Suitable when: length=height

Table 8. Comparison of 2D fracture models

6.2.1. Simulation study

Based on the petrophysical logs, from lithological point of view, the relevant formations are fairly marl, sandstone and limestone with an average rock density 2.33gr/cm3. The vertical stress was calculated by integrating the available bulk density with respect to depth. Vertical stress gradient is calculated as Eq. (1):

$$\sigma_v = 0.433\rho_{OB} = 0.433 \times 2.33 = 1psi / ft \tag{1}$$

The values of minimum horizontal stress of Aghajari, Mishan, and Gachsaran formations were 1693, 3847, and 4489, respectively which calculated from Eq. (2) is:

$$\sigma = \frac{\upsilon(\sigma_v D - 2p) + p}{1 - F\upsilon} \tag{2}$$

Elasticity of the formations is determined with the sonic log. Table 9 lists the values of the static elastic Young's modulus, Poisson's ration, leak-off coefficient for the different formation zones shown in Table 5. These values are based on the dynamic elastic Young's module obtained from sonic and density logs. Static elastic Young's module values are often two times smaller than dynamic values derived from sonic logs. The elastic Young's module values that are listed in Table 9 are arbitrarily one-half of their dynamic equivalents. The larger than usual values were used in the analysis for these shallower formations.

Zone Name	Zone Height (ft)	Poission's Ratio*	Pore Pressure* (psi)	Fracture Gradient* (psi/ft)	In-situ Stress (psi)	Young's Modulus (MM psi)	Leak-off Coefficient (ft.min $^{-0.5}$)	Toughness (psi.min $^{-0.5}$)
Aghajari	5250	0.29	1050	0.650	1693	2	0.00081	1000
Mishan	330	0.31	2567	0.714	3847	2	0.00087	1000
Gachsaran	330	0.36	2878	0.780	4489	2	0.00089	1000

Table 9. Formation properties used in fracture simulations.

Slurry rheology design did not performed in this paper and is beyond the scope of this article; however by considering the cuttings brought out of the wellbore and the drilling muds used in Ahwaz oil field, a reasonable result was earned of rheology characteristics of the injection slurry. It was assumed that the cuttings slurry with final rheological condition would behave in a manner similar to the drilling muds used in Ahwaz oil field. Slurry and solid properties are selected from past DCRI operation in literature (Abou-Sayed et al., 2002), which is also near the nature of selected drilling fluids and cuttings lithology of the Ahwaz oilfield and are presented in Table 10.

Density	1.26 SG
Particle Loading	80/100 mesh proppant at a consternation of 2 PPG
Apparent Viscosity	161 cp ≅ 170 1/S
Non-Newtonian power law indices	N=0.26; k=0.15

Table 10. Physical properties of injected cuttings slurry.

For the scenario of casing injection into a dedicated injection well, the intermediate casing can be set on top of Gachsaran formation. The casing is assumed to perforate at a depth about 50 m under the Aghajari formation and the center of the Mishan formation. The initial fracture is assumed to be at the center of the perforated interval.

6.2.2: Simulation results

After determining all required data, a fracture geometry model was selected for use in the simulation. As described previously, the dedicated injection mechanism is more suitable for the Mishan formation because it is deep enough and consists of limestone lithology in a base that is appropriate for reinjection. In each case, the geometry reported indicates the maximum fracture achieved when slurry is pumped continuously. The simulation study is represented for both dedicated wells that consist of two cases and annulus injection well mechanisms.

Dedicated Well Injection Mechanism

Two cases will be presented in this section, which differs in the magnitude of two parameters; Young's modulus and leak-off coefficient.

Case 1: For a case like Ahwaz oilfield in which the vertical distribution of the minimum in situ stress is uniform, a circular fracture is expected. The formations had Young's modulus and leak-off coefficients as shown in Table 9. For this simulation, the fracturing would initiate from the Mishan formation and broke through the Aghajari formation but was still 4,700 ft. below the surface when 50,000 bbl of slurry had been injected continuously. Table 11 summarizes the results of this simulation. Figures 10 and 11 show predicted the fracture shape plot after injection of 50,000 bbl continuously at 5 bbl/min.

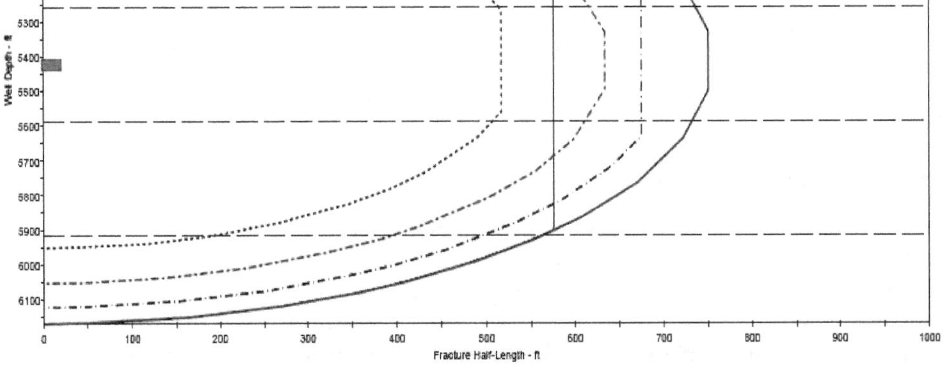

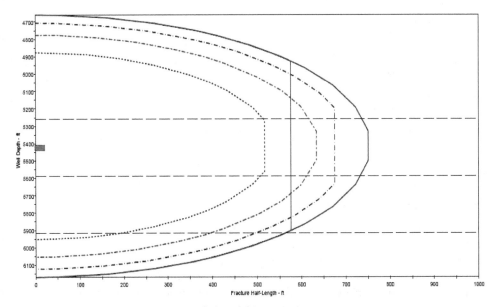

Figure 10. Fracture geometry history- Radial model (case 1).

Case 2: The formations were assumed to have Young's modulus that was twice those listed in Table 9. Also, the leak-off coefficient for formations used was specified as one half of the value listed in Table 9. This extremely large modulus and small leak-off resulted in a much larger fracture. Consequently, this is a very conservative analysis. Even for this very conservative case, the fracture that broke through the Aghajari formation was still 4,550 ft. below the surface when almost 50,000 bbl of slurry had been injected continuously at 5 bbl/min. Table 11 summarizes the results of the fractures created.

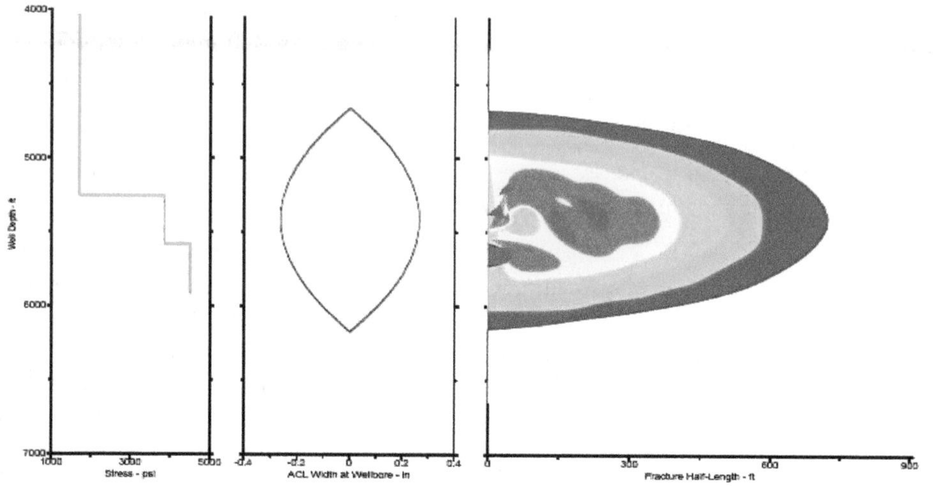

Figure 11. Fracture profile and cuttings concentration- Radial model (Case 2).

Parameters	Case 1	Case 2
Slurry volume (bbls)	50000	50000
Fracture half-length (ft)	576	795
Fracture width at well (in)	0.276	0.237
Net pressure (psi)	71	89
Max surface pressure (psi)	1755	1807
Shut-in time (hrs)	13	26

Table 11. Simulation's results of dedicated well injection.

Annulus Well Injection Mechanism

Annulus injection is only possible if the annulus of an intermediate casing string in an existing well is open to a suitable subsurface formation and this well satisfies a range of screening criteria. The allowable injection pressures for annulus injectors are often lower than the allowable pressures for dedicated wells because of casing burst and collapse limitations for annulus injectors. By considering the lithology and casing design of Ahwaz oilfield, it is concluded that the planned slurry injection would occur in an 18 5/8-in./13 3/8-in. annulus. Other annuli are not possible for injection because they are open to unsuitable subsurface formations. To prevent the upward migration of injected wastes to the surface, the 18 5/8-in. casing string should set at about 1,000 ft. and cement back to the surface, and the 13 3/8-in. string should cement back to 1,500 ft. below the previous casing shoe. This provides a window across the Upper Miocene marl and sandstone of Aghajari formation. For this simulation, the fracturing initiated from the Aghajari formation and grew toward the surface but was still 500 ft. below the surface when 15,000 bbl of slurry had been injected continuously. Table 12 presents the different parameters of the fracture created. Figure 12 shows the predicted fracture shape plot after injection of 15,000 bbl continuously at 5 bbl/min.

Parameters	Radial Model
Slurry volume (bbls)	15000
Fracture half-length (ft)	230
Fracture width at well (in)	643
Net pressure (psi)	2.39
Max surface pressure (psi)	968

Table 12. Simulation results of annular well injection.

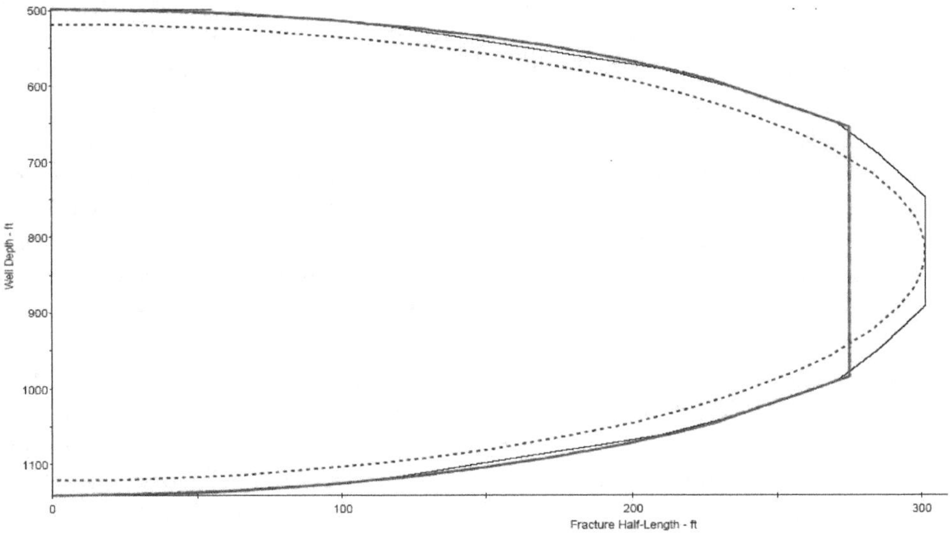

Figure 12. Fracture geometry history- annular injection well (Radial model)

7. Discussion and conclusions

Assessment of environmental impacts of drilling operations and searching for the methodologies to protect nature and resources against negative impacts has become an interesting topic during the last thirty years in upstream petroleum industry. The necessity of environmental management in drilling operation, lessons learned, and a brief list of mitigation options from wastes generated by drilling operations in a southern Iranian oilfield were documented previously (Shadizadeh and Zoveidavianpoor, 2008, 2010). Most of the drilling wastes sources in the oilfields are OMBs and oily cuttings associated with them. Unfortunately, lack of demanding regulations regarding drilling waste discharge leaves room for drilling companies to leave the waste in the nature without treating them (Shadizadeh and Zoveidavianpoor, 2008, 2010). This chapter tried to study the possibilities of waste prevention and zero discharge by utilization of serviceable methods in drilling well sites. So, the feseability study of DCRI at Ahwaz oilfield was initiated and conducted to fulfill the needs of growing upstream petroleum industry in Iran. This article focuses on the design aspect of the technology. Design guidelines are given to include data required for project planning, injection scheme (annulus versus dedicated well) selection, injection well and disposal formation identification, subsurface fracturing simulation, and waste containment. Operational procedures such as slurry rheology were the area of investigation in this study; however, it was determined as input data for simulation that has conformity with the nature of selected drilling fluids and cuttings lithology of the Ahwaz oilfield. Well design requirements and estimation of disposal capacity in each of the injection schemes was performed. This study shows that the DCRI study at Ahwaz oilfield is practical by considering some potential risks involved in any DCRI job. It was determined that by using HF technology, drilling wastes could be reinjected to the Mishan formation or even a shallow formation such as the Aghajari formation without propagation of the fractures to the surface or near wellbores. The thickness of the Aghajari formation provides an appropriate barrier to upward growth of DCRI at the Mishan formation through a dedicated injection well. A dedicated injection well is more typical of longer-term, permanent injection operations and is more common onshore (Keck, 2002). It is simulated that a large amount of drilling waste can be safely injected to Mishan formation. The maximum surface pressure required to inject the slurry is in a range of 1,500 to 2,000 psi, which is completely reasonable with the current surface facilities. The propagation of the fracture to the surface showed to be efficient and safe in the two cases performed in the dedicated well injection scheme. The simulation results confirm that the drilling wastes produced from each wellbore could be injected through annulus of the same wellbore while drilling. The selected annulus for annular reinjection in Ahwaz oilfield is not very favorable because the injection point is close to the surface. As described before, other annuli are not suitable due to abnormal pressure or hydrocarbon bearing. The annular reinjection at Ahwaz oilfield has many serious risks that need a careful job planning. However, the amount of drilling wastes from a typical wellbore is not high and the simulations confirm that 15,000 bbl wastes from a typical wellbore can be injected without serious danger. Advantages and disadvantages of annular and dedicated well injectors are presented in Abou-Sayed and Guo (2001).

It should be noted that the simulations represent upper-bound predictions of the fracture geometry because low leak-off and high Young's modulus is assumed in different formations. In reality, even a very limited change in the amount of fluid leak-off, coupled with intermittent batch injection of slurry, would result in a significantly reduced fracture area. The analyses confirm the integrity and suitability of the injection operations and ensure safe application of this technology at Ahwaz oilfield.

Author details

Mansoor Zoveidavianpoor and Ariffin Samsuri
Universiti Teknologi, Faculty of Petroleum & Renewable Energy Engineering, Malaysia

Seyed Reza Shadizadeh
Petroleum University of Technology, Abadan Faculty of Petroleum Engineering, Iran

Acknowledgement

The authors of this chapter would like to express thier gratitude to University Teknologi Malaysia due to the valuable supports during this study.

8. References

Abou-Sayed, A. S., Quanxin G., McLennan, J. D. and Hagan, J. T. (2000). Case studies of waste disposal through hydraulic fracturing. Presented at the Workshop on Three Dimensional and Advanced Hydraulic Fracture Modeling, held in conjunction with the Fourth North American Rock Mechanics Symposium, July 29, 2000, Seattle.

Abou-Sayed, A. S., and Guo, Q. (2001). Design considerations in drill cuttings re-injection through downhole fracturing. Paper IADC/SPE 72308 presented at the IADC/SPE Middle East Drilling Technology, Bahrain. October 22–24.

Ankley, G. T., Mattson, V. R., Leonard, E. N., West, C. W., and Bennett, J. L. (1993). Predicting the acute toxicity of copper in freshwater sediments: Evaluating the role of acid-volatile sulfide. Environ. Toxic Chem. 12:315–320.

Arthur, J.D., "A Comparative Analysis of Hydraulic Fracturing and Underground Injection", Presented at the GWPC Water/Energy Symposium, Pittsburgh, Pennsylvania. September 25-29, 2010.

Bansal, K. M. and Sugiarto, (1999). Exploration and Production Operations-Waste Management A Comparative Overview: U.S. and Indonesia Cases. Paper SPE 54345 presented at the SPE Asia Pacific Oil and Gas Conference, Jakarta, Indonesia, April 20-22.

Bleier, R., Leuterman, A. J. J., and Stark, C. (1993). Drilling fluids making peace with the environment. J. Petrol. Tech. 45:6–10.

Bruno, M., Reed, A. and Olmstead, S. (2000). Environmental Management, Cost Management, and Asset Management for High-Volume Oil Field Waste Injection Projects. Paper SPE 51119 presented at the IADC/SPE Drilling Conference held in New Orleans, Louisiana, 23–25 February 2000.

Daneshy, A. (2010). Hydraulic Fracturing to Improve Production. The Way Ahead: Tech 101. 6(3), 14-17.

Economides, M. J. (2010). Design Flaws in Hydraulic Fracturing. Paper SPE 127870 presented at the SPE International Symposium and Exhibition on Formation Damage Control. 10-12 February. Lafayette, Louisiana, USA.

Keck, R. G. (2002). Drill cuttings injection: A review of major operations and technical issues. Paper SPE 77553 presented at the SPE Annual Technical Conference and Exhibition, San Antonio, TX, September 29–October 2.

McMillen, S. J. and Gray, N. R. (1994). Biotreatment of Exploration and Production Wastes. Paper SPE 27135 presented at the Second International Conference on Health, Safety & Environment in Oil & Gas Exploration and Production, Jakarta, Indonesia, January 25-27.

McMillen, S. J., Smart, R. and Bernier, R. (2004). Biotreating E&P Wastes: Lessons Learned from 1992-2003. Paper SPE 86794 presented at the 7th SPE International Conference on Health, Safety and Environment in Oil and Gas Exploration and Production, Calgary, Alberta, Canada, 29-31 March 2004.

Neff, J. M. (2002). Bioaccumulation in Marine Organisms. Effects of Contaminants from Oil Well Produced Water. Amsterdam: Elsevier Science Publishers. ISBN: 978-0-08-043716-3. 452 pages

Paulse, J. E., Getliff, J. and Sørheim, R. (2004). Vermicomposting and Best Available Technique for Oily Drilling Waste Management in Environmentally Sensitive Areas. Paper SPE 86730 presented at the 7th SPE International Conference on Health, Safety and Environment in Oil and Gas Exploration and Production, Calgary, Alberta, Canada, 29-31 March 2004.

Reddoch, J. (2008). Why cuttings reinjection doesn't work everywhere—or does it? World Oil, January 2008, 69-70.

Shadizadeh, S. R. and Zoveidavianpoor, M. (2010) 'A Drilling Reserve Mud Pit Assessment in Iran: Environmental Impacts and Awareness', Petroleum Science and Technology, 28: 14, 1513-1526. ISSN: 1091-6466 print/1532-2459 online

Shadizadeh, S. R., and Zoveidavianpoor, M. (2008). Environmental Impact Assessment of Onshore Drilling Operation in Iran, Abadan, Iran: Petroleum University of Technology.

Shadizadeh, S. R., Majidaie, S. and Zoveidavianpoor, M. (2011) 'Investigation of Drill Cuttings Reinjection: Environmental Management in Iranian Ahwaz Oilfield', Petroleum Science and Technology, 29: 11, 1093-1103. ISSN: 1091-6466 print/1532-2459 online.

Veatch, Jr. R. W. Moschovidis, Z. A. (1989). An Overview of Recent Advances in Hydraulic Fracturing Technology. Paper SPE 14085 presented at the International Meeting on Petroleum Engineering. 17-20 March. Beijing, China.

Zimmerman, P. K. and Robert, J. D. (1991). Oil-Based Drill Cuttings Treated by Landfarming. Oil & Gas Journal, 89(32): 81-84.

Chapter 5

Evaluation of Soil Quality Parameters Development in Terms of Sustainable Land Use

Danica Fazekašová

Additional information is available at the end of the chapter

1. Introduction

Soil is a vital natural source and, at the same time, has an economic and eco-social potential. It allows the production of food and raw materials, recycles waste, creates forest-agricultural land, filters and retains water, allows the usage and valorisation of sun energy, ensures the cycle and balance of substances in nature, maintains diversity of plant and animal species. It primarily shapes the quality of the environment; it is the resource and cultural heritage of the Earth; it ensures the life and social being of the population. Agricultural activities realised in landscape affect natural resources. A rational usage of renewable and non-renewable resources which are not retrieved in real time is an essential precondition.

The farming system is the most widespread environmental technology with its positive and negative consequences. It utilises essential natural resources and, at the same time, influences other natural environments. Therefore, ecologisation of farming is a priority of farmers as well as environmentalists. Respecting the principles of soil sustainability and other components of environment is the basic precondition for life sustainability.

A United Nations [UN] conference on environment and development (Rio de Janeiro, 1992) prioritised sustainable development, which is presented in a global development programme for the end of the 20th century and especially for the 21st century (Agenda 21). The concept of sustainable development of agriculture includes such practices in farming which respect ecological aspects in growing plants and ethology of livestock in rearing, do not enhance damage in ecological land stability, respect environmental protection, including surface and underground water and monitor the quality of agricultural produce.

Sustainable agriculture is based on the principle of agriculture being a biological process which, in practice, should imitate key characteristics of the natural ecosystem. It strives to bring diversity into agro-ecosystem, recycle nutrients efficiently and maintain the priority of sunlight as a source of energy for agro-ecosystems.

Specific manifestations of soil require different approaches. In soil protection, these must be ecological (biological) approaches, as this is the only way to achieve sustainable development of ground cover and the resulting economical and social development and environmental balance in society.

Sustainable use of soil takes soil-ecological conditions into consideration and is realised in such a way and in such intensity, which gives rise to neither negative changes in soil, nor establishes trends for the development of negative characteristics in soil. The essential principle of sustainable farming system is its protection from any degradation by natural or man-induced influences. Sustainable development of soil use also encompasses the protection of the soil acreage to such an extent which ensures that all soil functions are employed.

In a number of European countries, sustainable use of soil is realised according to the principles of International Federation of Organic Agriculture Movements [IFOAM] and is referred to as ecological soil management. When introducing ecological systems of soil management, the main criterion is the application of knowledge in the functioning of natural ecosystems, which are typical of plant and animal variety and sunlight is the exclusive source of energy. In cultural (artificial) agro-ecosystems, the structure is disrupted by man drawing the production past the limit of the agro-ecosystem. The ecological system focuses on theoretical elaboration of farming arrangement in sensitive areas (the protection of underground and surface water zones, polluted zones, national parks, protected natural areas and soils heavily endangered by erosion). Continued protection of nature and natural resources is at the forefront; therefore, significant intensifying constituents of conventional agriculture (high dosage of fertilisers, full usage of pesticides, annual subsoil ploughing, major hunts, high ratio of grain crops, intensive breeding, heavy automation) are replaced by technologies with strong economical and ecological components (tillage minimisation, anti-erosive crop rotation, monitoring of plant nutrition, integrated a biological protection of plants, minimal automation, free-range breeding).

According to Organisation for Economic Co-operation and Development [OECD], an indicator is a parameter or a value derived from several parameters. It provides information about a particular observed phenomenon from the viewpoint of its quantitative or qualitative characteristics, present in a give time and area, in the environment as a whole, or its individual components by the qualitative parameters of these components influencing the health condition of the population, as well as the structure and function of the ecosystem in the area in question. From the above stated, it results that there are a number of horizontal and vertical causal links between individual environmental indicators. The "sustainability indicator" can, thus, be defined as a measurable factor, whose imbalance negatively influences the long-term performance of the whole production system. Stable agriculture has a time and space dimension. The time scale depends on the adaptability of the system (usually 5 to 10 years, or more); space can be given by the borderlines of soil-climatic units or areas. Stability indicators should be applicable to the evaluation of the main components of sustainable agriculture. Attention is mainly paid to the level of farming and its productivity regarding the ecological soil potential, maintaining diversity of plant species as well as the protection of natural resources, social-economic viability related to the regional and world economy.

From the viewpoint of agricultural practice, the stability indicators regarding productivity of agricultural production and ecological aspects of farming systems have been explored in most detail. The guaranteed yield on the level of the ecological potential of location (without further input increase), the ability of the system to return to the initial performance in a short period of time after a natural disaster, achieving a relatively high efficiency of water and plant nutrition utilisation, maintaining the soil quality environment (organic mass, soil organisms, nutrients), reliability of the methods used in integrated plant protection, ensuring the quality of water resources, maintaining the level of underground water without major fluctuations, and protecting natural resources are considered quantifiable biophysical indicators of sustainable productivity (Klír, 1997). With regard to the evaluation of ecological sustainability, the most significant indicators are maintenance and improvement of biodiversity in managed, as well as adjacent natural ecosystems, maintaining the environmental quality and avoiding pollution limit excess (Virmani & Singh, 1997 as cited in Fazekašová, 2003).

It is impossible to select universal soil parameters with regard to their suitability for sustainable soil and is subject to specialised discussions. A significant role in the selection of parameters is played by their variability in time, related to parameter stability. The following soil parameters can be distinguished: stable (such as soil depth or granularity), relatively stable (the salt content, the content of organic mass in soil, heavy metal contamination), relatively dynamic (pH, the content of nutrients), and dynamic (soil humidity and temperature, microbial activity, etc.). Stable and relatively stable parameters dominantly influence soil quality, while relatively dynamic and dynamic characteristics are more connected to its short-term changes.

Soil parameters indicate the state of soil ecosystem characteristics, which especially reflect production, buffering, filter and other soil functions. From this view, the structure of soil profile (the soil class), soil type, soil depth, skeletal nature, the content and quality of humus substances, accessible nutrient supply, soil reaction, the content of foreign substances in soil, and soil edaphon seem to be of highest importance.

Soil quality cannot be judged directly; it must be determined from the changes of its parameters. It is more accurate to evaluate the range of appropriate indicators rather than to use a single one. Soil quality is significantly affected by physical, chemical, biological and biochemical properties sensitive to changes in the environment and land management. With regard to physical properties, there are bulk density, porosity, water retention capacity, soil temperature, etc. In the group of chemical characteristics, total carbon and nitrogen content, soil reaction and content of available nutrients are observed. Evaluation of biological parameters focuses on microbial biomass and its activity, soil respiration, potentially mineralised nitrogen, the activity of soil enzymes, etc. Soil enzymatic activity can be used as a microbial indicator of soil quality, since the activity of soil enzymes is closely related to essential soil characteristics. It indicates changes sooner than other soil characteristics and can be an integrating soil-biological index reflecting soil use (Javoreková et al., 2008; Šarapatka, 2002). Wick et al. (2002) considers selected enzymatic activities as suitable indicators for long-term soil monitoring and quality assessment (Miralles, 2007; Geisseler, 2009). A decrease in soil quality is obvious from the values of critical load of risk substances. When evaluating the content of heavy metals in soil, attention must also be paid to their bio-accessibility (Bujnovský & Juráni, 1999).

The chapter deals with a synthetic and comparative analysis of scientific findings regarding the development of soil quality parameters in the conditions of a sustainable farming system. Based on the research carried out between 1997 and 2010 on a model area situated in a marginal region of north-eastern Slovakia (48° 57' N; 20° 05' E), the development of soil indicators are evaluated, focusing on physical (bulk density and soil porosity), chemical (soil pH, inorganic nitrogen, available phosphorus, potassium, magnesium and organic carbon content) and biological parameters (activity of acid and alkaline phosphatase and urease), as well as the presence of risk substances in the soil ecosystem (heavy metal content - Cd, Ni and Pb).

2. Evolution of soil parameters

At present, there is little knowledge with regard to soil development in the conditions of sustainable farming systems whose principles lie in soil maintenance. There is a major effort to increase its natural productivity by as closed a cycle of nutrients as possible with the highest possible reduction of external, mainly energetic and chemical, inputs (Lacko-Bartošová et al., 2005; Fazekašová, 2003). The present findings can hardly be compared to other research due to the different soil-ecological conditions in which they were obtained. The issue of universal methods for all soil types remains a universal problem within the research of soil development. Unless this area is unified, objective comparison will remain on a regional level. Soil parameters are usually determined only in relation to specific topsoil. Certain physical and chemical parameters in subsoil cannot be neglected, since they guarantee soil functions (Fazekašová, 2003).

2.1. Methods

The research project was carried out between 1997 and 2000 and 2008 and 2010 under production conditions in the investigated area situated in a marginal region of north-eastern Slovakia (48° 57' N; 20° 05' E). Here, the ecological farming system has been applied since 1996. The area is situated in the Low Tatras National Park at an altitude ranging from 846 to 1492 m above sea level. In terms of geomorphological division, it is a part of sub-assemblies of the Kráľovohoľské Mountains (Michaeli & Ivanová, 2005). The whole area is situated in a mild zone with a sum of average daily temperatures above 10 ºC ranging from 1600 to 2000 and average precipitation of 700-1200 mm (Fig. 1).

The soil conditions are relatively homogeneous, the largest area being represented by Cambisols, mostly moderate and strongly skeletal, mainly in the subsoil, medium-weight and heavy in granularity (loamy sand, loam, clayey loam). Cambisols are the most common soil type occurring in Slovakia. From an ecological viewpoint, Cambisols are valuable for their irreplaceable ability to retain and accumulate atmospheric fallout and also for their filtration attributes. From the relief viewpoint, the majority of the land is situated on slopes, soil is often eroded and, thus, surface water resources are threatened. With regard to pollution, there is an assumption that heavy metals are transported to crops (due to the acidity of these soils). In the current crop structure, cereal acreage represents 33.3 %, potatoes 16 % to 18 % and fodder crops 49.8%. Crops are rotated as follows: perennial fodder (clover mixture) → perennial fodder (clover mixture) → winter crops (winter wheat, winter rye, triticale and winter barley) → root crops (potatoes) → spring crops (spring barley, oats) → annual mixture (oats pea, peas, ryegrass). Arable land is fertilised with manure dosage of approximately 30 t ha^{-1} once in two years. The permitted phosphorous and potassium mineral fertilisers have not been added in the past five years. The permanent grassland and arable land were fertilised with liquid organic fertiliser in the spring season, 3 000 l ha^{-1} (minimum nutrients content: total nitrogen expressed as N in dry mass at least 15 %, total phosphorus as P_2O_5 in dry mass less than 0.2 %, total potassium as K_2O in dry mass less than 0.4 %, total sulphur as S in dry mass at least 16.5 %).

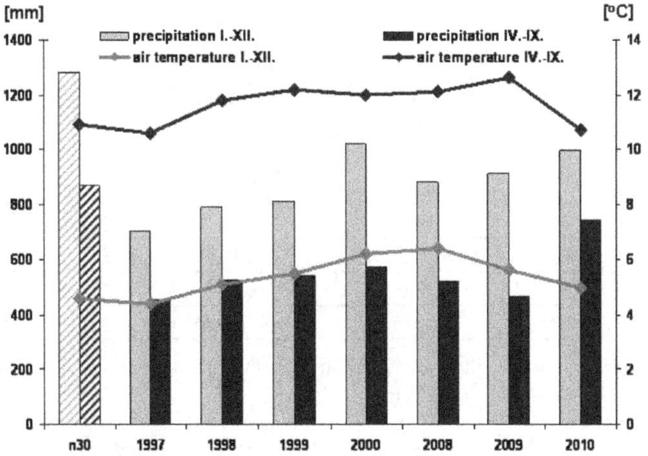

Figure 1. The course of average air temperatures (°C) and sum of precipitation (mm) during the observed period in the observed area situated in a marginal region of north-eastern Slovakia

Soil samples for physical, chemical and biological soil properties and heavy metal content determination were obtained in spring time in a connected stand on five permanent research sites, from the depth of 0.05 m to 0.15 m. Part of the soil samples were air-dried, sieved (sieve with 2 mm size opening), homogenised prior to the analysis and used for measurements of chemical and biological soil characteristics and heavy metal content. From the physical soil properties, soil bulk density and soil porosity were studied and evaluated in a Kopecky physical cylinder with a capacity of 100 cm^3 (Fiala et al., 1999). From the chemical soil characteristics, soil pH in 1M $CaCl_2$ solution was monitored and evaluated, as well as inorganic nitrogen, available phosphorus, potassium, and magnesium with Mehlich III and organic carbon content (Fiala et al., 1999). The available heavy metal content (Cd, Ni and Pb) of the samples was determined in 2M HNO_3 solution using atomic absorption spectrophotometer (Matúšková & Vojtáš, 2005). The following biological soil characteristics were monitored: activity of acid and alkaline phosphatase (Grejtovský, 1991) and urease (Chaziev, 1976). The obtained data were tested by mathematical-statistical methods from which analysis of variance and regression analysis were used (the Statgraphics software package).

2.2. Evolution of physical soil parameters

The changes in physical characteristics of soil not only result from meteorological factors, yearly farming plan, or from the course of vegetation, but also depend on the employed farming system. Larson and Pierce (1991) confirmed that soil quality can be evaluated and the sustainability of a system assessed on the basis of essential physical indicators.

Soil granularity, and especially the ratio of clay particles, primarily influences physical, hydro-physical and chemical characteristics. The soils in the monitored localities according to the content of clay particles based on Novák's classification (Fulajtár, 2006) are of loamy-sandy, loamy and clay-loamy category (Tab. 1.).

Diameter of particles [%]	Studied locality				
	I.	II.	III.	IV.	V.
> 0.25	31.3	14.5	11.5	32.5	16.0
0.25–0.05	21.6	15.5	18.9	13.9	14.9
0.05–0.01	27.8	32.4	24.3	22.2	31.3
0.01–0.001	15.5	29.3	3.3	24.0	2.6
< 0.001	3.8	8.3	11.0	7.4	8.2
I. Category	19.3	37.6	45.3	31.4	37.8
Soil	loamy sand	loam	clayey loam	loam	loam

Table 1. Particle grain-size composition of soil [%] in the monitored area situated in a marginal region of north-eastern Slovakia in depth 0.05–0.15 m.

Bulk density as an integral value of soil granularity, humus content and anthropogenic impacts on soil should not exceed the limits given for individual soil types (Tab. 2.).

Soil texture	Sandy	Loamy sand	Sandy loam	Loam	Clayey loam and clay	Clay
Bulk density	≥ 1.70	≥ 1.60	≥ 1.55	≥ 1.45	≥ 1.40	≥ 1.35
Porosity	≤ 38	≤ 40	≤ 42	≤ 45	≤ 47	≤ 48

Table 2. Critical values of bulk density soil [$t.m^{-3}$] and porosity [%] for different of soil texture (Líška et al., 2008)

Long-term research has shown that ecological soil farming regulates bulk density of soil. The measured values of bulk density were in the range of 0.94 to 1.35 $t.m^{-3}$ (Fig. 2.), in 1997 to 2009, there was a moderate decrease and values comparable to average figures for the given soil type and category according to Líška et al. (2008) were achieved (Tab. 2.), with the exception of 2010, when a mild increase in bulk density was measured. At the same time, this parameter proved to change under the influence of the water content and meteorological exposure (Kotorová, Šoltýsová & Mati, 2010). In 2010, in comparison to the previous years, precipitation reached higher values (Fig. 1.).

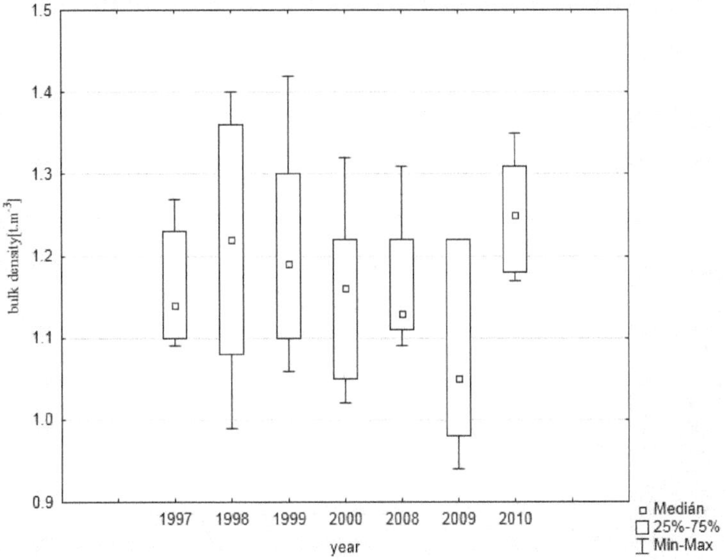

Figure 2. Bulk density of soil in the monitored area situated in a marginal region of north-eastern Slovakia expressed by descriptive statistics

General porosity is closely related to bulk density. From the total pore volume, which should not fall below 38 % for sandy soil and below 48 % for clay-loam soil (Líška et al., 2008), the share of non-capillary pores rapidly releasing gravitational water and allowing good air exchange between soil and climate should be sufficient. The share of non-capillary pores (Pn) in comparison to capillary pores (Pk) should be higher in heavy soils.

As can be seen from Fig. 3., the values show that, in the observed timeframe, porosity levels ranged between 46.43 and 64.49 %. Considering this parameter, optimum conditions were created for the growth of most arable crops, which are given by general porosity between 55 and 65 % and 20 and 25 % soil air content (Rode, 1969).

A statistically significant effect in the monitored year and locality on all observed soil physical parameters was confirmed by an analysis of variance (Tab. 3.).

Parameter	Min.	Max.	Mean	Standard error	Source of variability	d. f.	F-Ratio	P
bulk density [t.m^{-3}]	1.04	1.35	1.18	0.016347	year	6	8.33	++
					locality	4	28.39	++
porosity [%]	49.15	60.56	55.52	0.617864	year	6	8.24	++
					locality	4	28.41	++

Table 3. Analysis of variance of soil physical parameters in the monitored area situated in a marginal region of north-eastern Slovakia
++P< 0,01 +P< 0,05

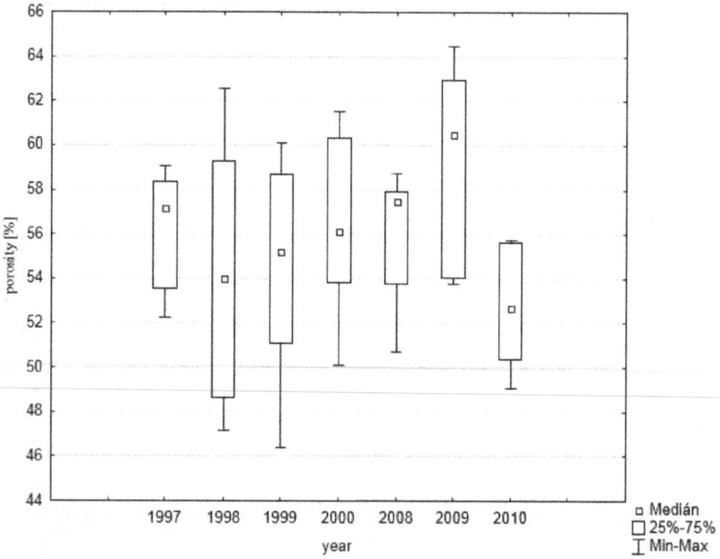

Figure 3. Porosity of soil in the monitored area situated in a marginal region of north-eastern Slovakia expressed by descriptive statistics

2.3. Evolution of chemical soil parameters

Chemical parameters are considered relatively dynamic (pH, nutrient content) and, in terms of plant growth and development, vital. Their deficiency is reflected in crop production. At the same time, they serve as indicators of additional inputs in the form of fertilisers. Sustainable farming systems exclude, or reduce, the use of artificial fertilisers; therefore, it is necessary to pay attention to the dynamics of chemical soil parameters in order to prevent one-way draining of nutrients, particularly phosphorus and potassium.

The soil pH is an important factor for soil fertility despite the fact that its values change dynamically, depending on so-called internal and external factors. It influences the buffering and filtering capacities, the quality of organic substances, nutrient accessibility for plants and the production of biomass in most crops grown. A majority of arable crops suit the range of slightly acidic to slightly alkaline soil pH – 6 to 7.5 (Krnáčová, Račko & Bedrna,

1997). A pH value lower than 5.5 is undesirable and requires ameliorative lime treatment. Similarly, from the viewpoint of productivity, alkaline soils (pH>8.4) are limiting and require appropriate measures.

In the course of monitoring the model area, the values of soil pH ranged between 5.1 and 7.2. The average values of soil pH increased moderately and were in the category of slightly acidic to neutral soil pH (6.3 – 6.9) (Fig. 4.).

This can be assigned to the ecological farming system, as physiologically acidic mineral fertilisers were not applied. On the contrary, organic fertilisers (manure at the dosage 30 t ha^{-1} and liquid organic fertilisers at the dosage 3000 l ha^{-1}) were applied. The organic matter positively influences the buffering capacity of soil, which is why the soil reaction was stabilised. Nevertheless, it is necessary to pay continuous attention to soil reaction, since soil is naturally acidified through acid atmospheric fallout as well as calcium intake by plants.

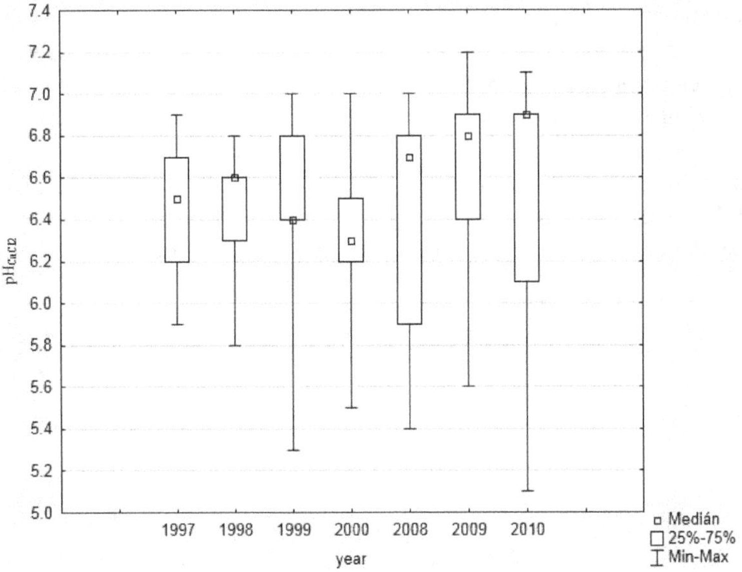

Figure 4. Soil reaction (pH/$CaCl_2$) in the monitored area situated in a marginal region of north-eastern Slovakia expressed by descriptive statistics

Nitrogen, phosphorus and potassium are the most important nutrients. Their supply to soil can be realised in various ways; fertilisation being of most importance. A lack of essential nutrients is rapidly reflected in the level of plant production. Nitrogen in soil is restored as part of its natural cycle. Its additional supply is necessary for intensified harvest, when its natural supply is not sufficient in order to achieve the targeted harvest. The supply of phosphorus and potassium by fertilisation is related to their supply in soil. Their supply in soil is not exhaustless; moreover, when constantly utilised, they are not naturally renewable.

According to Bielek (1998), there is a small probability that an increase in the total nitrogen content has a positive effect on soil fertility. This only applies to productive and highly productive soils. For soils with low production capacity, a reciprocal ratio between the total nitrogen content and soil fertility is typical. From the total nitrogen in soil, 95 % to 98 % is bound in organic forms; fertility functions determine mechanisms of its accessibility to plants. It is mainly organic nitrogen mineralisation, or, more specifically, that part of mineralisation which prevails over carbon immobilisation related to fertility. Inorganic nitrogen only represents a small part of total nitrogen and its content in the season is subject to frequent and fast changes, resulting from natural and anthropic factors. The concentrations of the main forms of mineral nitrogen (ammonia, nitrates) result from pure mineralisation and frequent nitrification of nitrogen in soil. In our research carried out in natural conditions, medium to highly favourable content of inorganic nitrogen in topsoil has been observed (Fig. 5.) in spite of the fact that in the soil-ecological conditions of the investigated area (a mild zone with a sum of average daily temperatures above 10 ºC ranging from 1600 ºC to 2000 ºC and average precipitation of 700-1200 mm), the nitrogen mineralisation is less intensive (the optimum temperature for an intensive process is 28-30 ºC); therefore, even with a high total content of nitrogen, the content of mineral (i.e. immediately available) nitrogen may not be high. The assumption is that by adding high doses of organic fertiliser, the total nitrogen content will increase. However, including legumes in the crop rotation can increase the content of immediately available nitrogen. These crops leave high amounts of nitrogen in soil (more than 100 kg ha^{-1} N), which are later available for the crops grown in the following period (Jurčová & Torma, 1998; Kováčik, 2001).

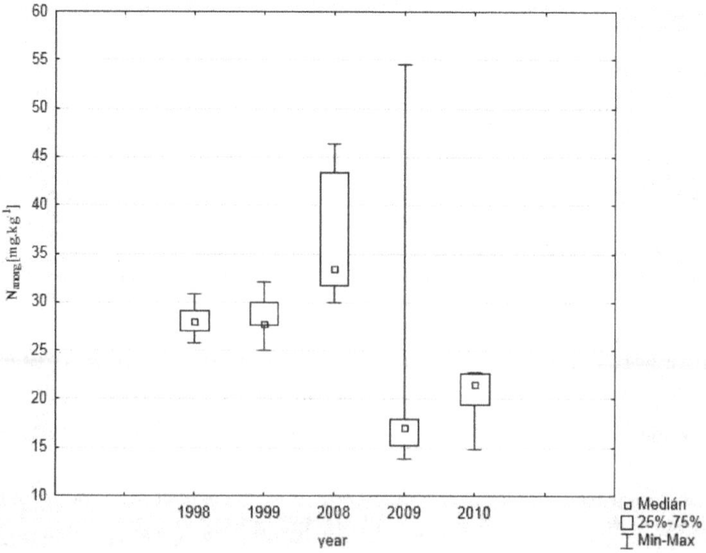

Figure 5. N_{anorg} content of soil in the monitored area situated in a marginal region of north-eastern Slovakia expressed by descriptive statistics

Phosphorus is firmly fixed in soil and its proportion is relatively stable and dependent on soil reaction values. Between 1997 and 2010, the value of soil pH did not significantly change in the investigated area. With regard to the above, the proportion of available phosphorus changed only minimally (Fig. 6.).

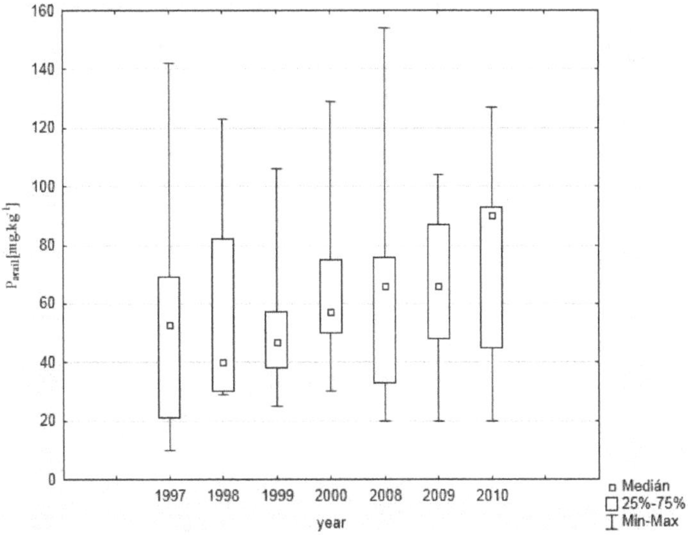

Figure 6. P_{avail} content of soil in the monitored area situated in a marginal region of north-eastern Slovakia expressed by descriptive statistics

In the observed period, the proportion of potassium and magnesium was relatively stable (Fig. 7. and Fig. 8). Due to the grain structure of the soils (medium and heavy soils), these nutrients are bound to soil particles and are not prone to soil washing in spite of high precipitation throughout the year.

Organic mass determines the soils quality, as it binds soil particles, stabilises soil (by which the risk of erosion decreases), increases water retention and cationic exchange capacity and reduces the negative impact of pesticides, heavy metals and other pollutants. A high proportion of organic carbon alone cannot guarantee a high yield; however, the influence of soil carbon on productivity increases when the levels of carbon decrease below 1%. With a content C_{org} 1.0 to 1.5, productivity decreases by 15 % and with content C_{org} under 1.0 %, it decreases by as much as 25 %. The most significant parameter is the ratio of humin acids and fulvene acids. This ratio is considered highly favourable, if it is higher than 2, satisfactory in the range between 1 and 2, and unfavourable if lower than 1. With regard to biological activity of soil, so-called non-specific humus substances play a significant role. These are a source of nutrients for soil microorganisms, participating in important cyclic biochemical processes (Hraško & Bedrna, 1988).

The content of humus in soil is a parameter prone to significant changes in the long-term. The application of high amounts of organic fertilisers and incorporating perennial fodder crops in the crop rotation influenced the preservation of humus content. The measured C_{ox} values ranged from 2.16 to 3.92 (Fig. 9.), which, when conversed to humus (conversion coefficient 1.724), are medium to good humic soils (Vilček et al., 2005).

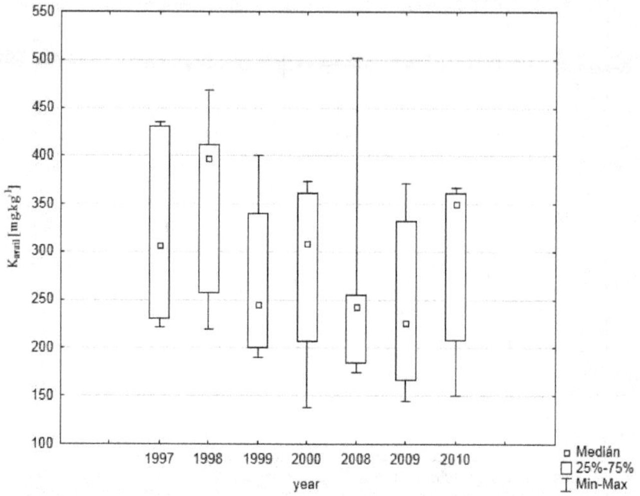

Figure 7. K_{avail} content of soil in the monitored area situated in a marginal region of north-eastern Slovakia expressed by descriptive statistics

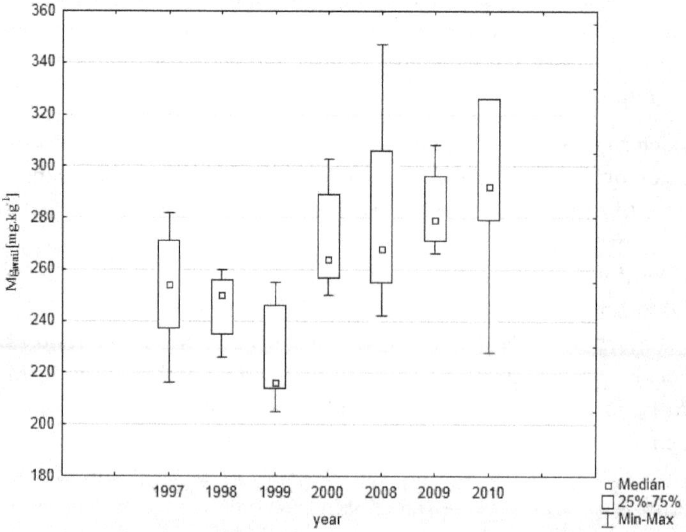

Figure 8. Mg_{avail} content of soil in the monitored area situated in a marginal region of north-eastern Slovakia expressed by descriptive statistics

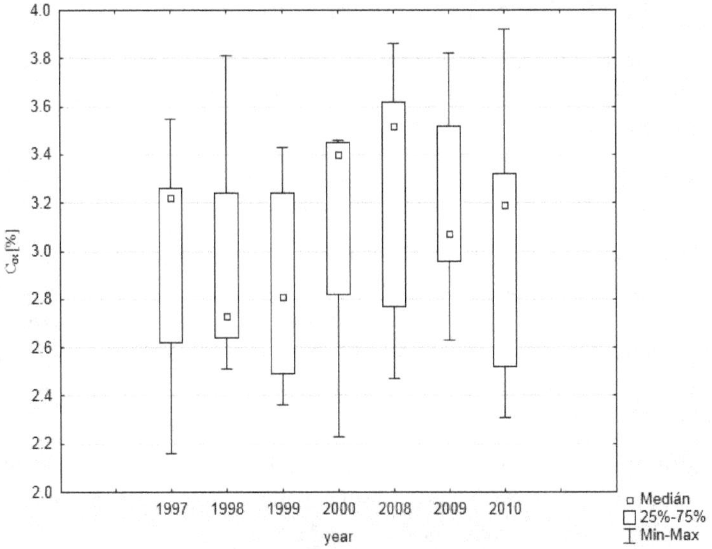

Figure 9. C_{ox} content of soil in the monitored area situated in a marginal region of north-eastern Slovakia expressed by descriptive statistics

A statistically significant effect of the monitored year on all observed soil chemical parameters was confirmed by an analysis of variance (Tab. 4.). The influence of the monitored locality on soil chemical parameters was also statistically significant, with the exception of N_{anorg}.

Parameter	Min.	Max.	Mean	Standard error	Source of variability	d. f.	F-Ratio	P
$pH/CaCl_2$	5.77	7.13	6.41	0.083124	year locality	6 4	0.82 24.51	- ++
C_{ox} [%]	2.25	3.61	3.03	0.084802	year locality	6 4	1.46 23.99	- ++
N_{anorg} [$mg.kg^{-1}$]	16.76	40.50	27.52	1.698623	year locality	6 4	12.77 2.22	++ -
P_{avail} [$mg.kg^{-1}$]	19.97	127.88	64.63	3.494827	year locality	6 4	3.02 121.06	++ ++
K_{avail} [$mg.kg^{-1}$]	168.59	427.98	290.91	12.5772	year locality	6 4	5.87 43.46	++ ++
Mg_{avail} [$mg.kg^{-1}$]	215.98	301.43	265.0	4.918103	year locality	6 4	17.52 3.73	++ ++

Table 4. Analysis of variance of soil chemical parameters in the monitored area situated in a marginal region of north-eastern Slovakia

++P< 0,01 +P< 0,05

2.4. Evolution of biological soil parameters

The information on biological soil parameters is not as plentiful as it is in the case of physical and chemical parameters, despite the fact that the effect of edaphon on biochemical processes in soil, nutrients balance, soil structure, etc. is proven in general.

There are a great number of enzymes in soil, depending on the diversity of soil organisms and the conditions of organic substances transformation.

Soil enzymes regulate the functioning of the ecosystem and play key biochemical functions in the overall process of organic matter decomposition in the soil system (Sinsabaugh et al., 1997). They are important in catalysing several important reactions necessary for the life processes of micro-organisms in soils and the stabilisation of soil structure, the decomposition of organic wastes, organic matter formation and nutrient cycling (Dick et al., 1994 cited in Makoi & Ndakidemi, 2008).

Enzymes are present in the cells of living organisms in soil (bacteria, fungi, algae, and soil fauna) and plant roots. Micro-organisms are the major source of enzymes in soil. The amount and quality of enzymes in soil is dependent on their characteristics, volumes and forms of organic matter and the activity of micro-flora. Enzymatic soil activity is higher in fertile soils with plenitudes of organic matter. The highest proportion of various enzymes can be found in the humus soil horizon (Pejve, 1966). The activity of soil enzymes can be enhanced by using organic fertilisers (Burns, 1978; Iovieno et al., 2009; Chander et al., 1997). The urease enzyme belongs to the hydrolases group of enzymes and is responsible for the hydrolysis of urea fertiliser applied to the soil into NH_3 and CO_2 with the concomitant rise in soil pH. This, in turn, results in a rapid N loss to the atmosphere through NH_3 volatilisation. Due to this role, urease activities in soils have received a lot of attention since it was first reported, a process considered vital in the regulation of N supply to plants after urea fertilisation (Makoi & Ndakidemi, 2008).

Soil urease originates mainly from plants and micro-organisms. It can be found as a free enzyme in soil solution, and yet more often firmly bound to soil organic mass or minerals, as well as inside living cells (Klose & Tabatabai; 2000; Alef & Nannipieri, 1995). Its activity depends on soil humidity (Baligar et al., 2005), pH, humus proportion and quality (Tabatabai & Acosta-Martínez, 2000) and the total nitrogen content (Nourbakhsh & Monreal, 2004). At the same time, an increased sensitivity to excess content of heavy metals (Kromka & Bedrna, 2000) and a negative effect of triazine herbicides on the activity of enzymes (Belińska & Prangal, 2007) was shown.

Phosphatases are a broad group of enzymes that are capable of catalysing hydrolysis of esters and anhydrides of phosphoric acid. In soil ecosystems, these enzymes are believed to play critical roles in P cycles (Speir et al., 2003) as evidence shows that they are correlated to P stress and plant growth. Apart from being good indicators of soil fertility, phosphatase enzymes play key roles in the soil system (Dick et al., 2000 cited in Makoi & Ndakidemi, 2008).

Soil phosphatase has certain typical characteristics. It depends on the substratum and its concentration. Two optimums levels, acidic and alkaline, are often present (Burns, 1978). An optimum pH of soil phosphatase is influenced by a great number of factors.

Soil pH differs from the pH optimal for phosphatase activity. Soil phosphatase can be inactive if the differences between soil pH and optimum enzyme pH are too great (Chaziev, 1976). The activity of soil phosphatase is higher in soils with high humidity in comparison to dry soils or soils with normal humidity. Phosphatase activity declines with an increasing soil depth, which is caused mainly by lower biological activity in lower soil profiles. Inorganic phosphate, copper, mercury and vanadium also have a considerable inhibitory effect on soil phosphatase activity (Burns, 1978; Speir et al., 2003).

There was minimum fluctuation in the measured values of soil enzyme activity in the observed period. The urease values ranged from 0.43 to 0.67 mg $NH4^{+}$ - $N.g^{-1}.24\ hour^{-1}$, and the values of acidic and alkaline phosphatase between 236.8 and 336.5 μg $P.g^{-1}.3\ hour^{-1}$ (Fig. 10., Fig. 11. and Fig. 12.). These are values typical for sparse-vegetation soils (Burns, 1978).

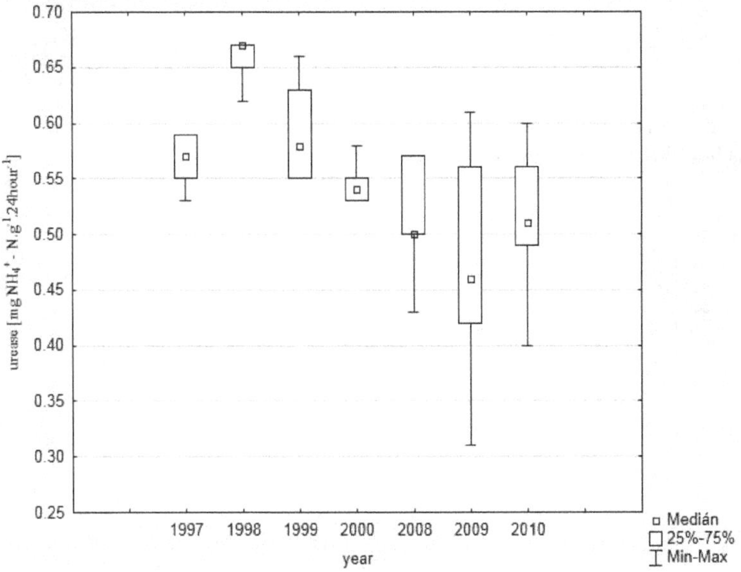

Figure 10. Urease activity of soil in the monitored area situated in a marginal region of north-eastern Slovakia expressed by descriptive statistics

At the same time, a higher activity of soil enzymes in lower temperatures was confirmed (the area is situated in a mild district with a sum of average daily temperatures above 10 °C ranging from 1600 to 2000 and average precipitation between 700 and 1200 mm) and organic fertilisers and soil organic mass stimulate the activity of soil phosphatase and significantly enhance the protection of natural soil urease (Chaziev, 1976; Bremner & Mulvaney, 1978).

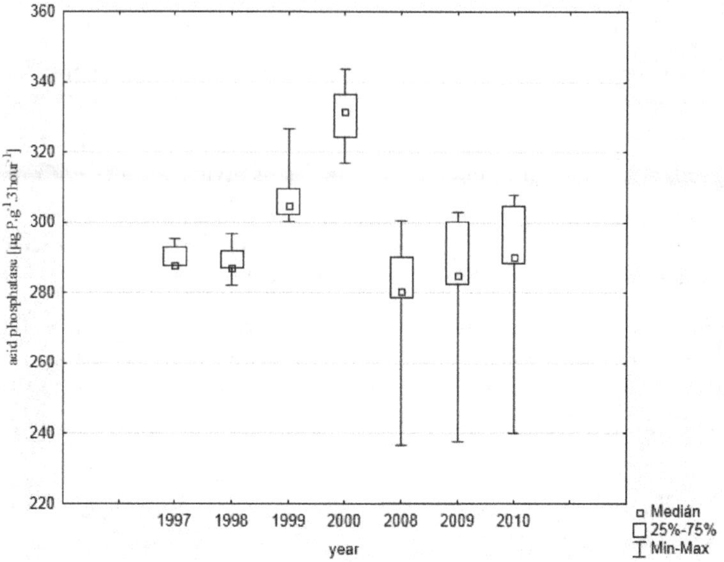

Figure 11. Acid phosphatase activity of soil in the monitored area situated in a marginal region of north-eastern Slovakia expressed by descriptive statistics

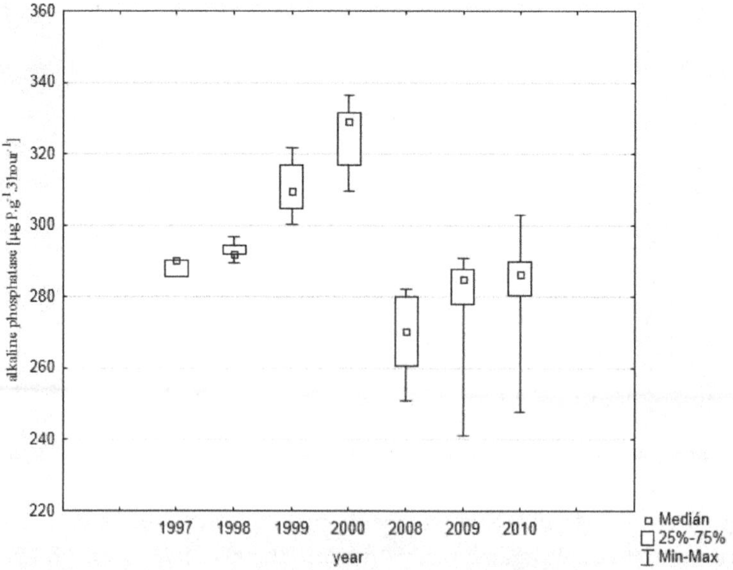

Figure 12. Alkaline phosphatase activity of soil in the monitored area situated in a marginal region of north-eastern Slovakia expressed by descriptive statistics

A statistically significant effect of the monitored year and locality on all observed soil biological parameters was confirmed by an analysis of variance (Tab. 5.).

Parameter	Min.	Max.	Mean	Standard error	Source of variability	d. f.	F-Ratio	P
urease [mg NH_4^+ - N.g^{-1} .24hour^{-1}]	0.454	0.674	0.551	0.007954	year locality	6 4	44.07 33.12	++ ++
acid phosphatase [μg P.g^{-1} .3hour^{-1}]	271.23	306.77	294.91	2.65001	year locality	6 4	36.16 15.85	++ ++
alkaline phosphatase [μg P.g^{-1} .3hour^{-1}]	264.35	329.33	291.97	1.96567	year locality	6 4	75.50 16.39	++ ++

Table 5. Analysis of variance of soil biological parameters in the monitored area situated in a marginal region of north-eastern Slovakia
++P< 0,01 +P< 0,05

2.5. Concentration of heavy metals in soil

An increase in inputs employed in the farming system has gradually brought about the need for studying and evaluating their potential negative influence on soil environment and production quality. Monitoring soil contamination with various degrees of biotoxicity is an important area. Fertilisers, especially industrially produced, are considered (including rock slackening, atmospheric decline and waste stock) a significant source of risk elements in soil (Beneš, Benešová, 1993). The system of farming, including the use of fertilisers, can also indirectly affect the acceptability of risk elements for plants (Beneš, 1993).

Loading agricultural soil with harmful substances is serious, since soil is not only the key to agricultural production but also has filtration and buffering capacities. Soil considerably influences the composition and quality of underground water and provides a living environment for soil micro-organisms (Tischer, 2008; Gulser, 2008). It could be assumed that accumulating higher concentrations of heavy metals in soil is a potentially serious danger to the food chain (Torma et al., 1997). It is especially toxic elements and organically highly-persistent substances that are among harmful substances entering soil.

Heavy metals as a large group of polluters are a serious problem in all components of the environment, including soil. As a great number of these have considerable toxic effects, their highest allowed concentrations are defined for the soil system, similarly to those for air and water. It is extremely difficult to define limit concentrations of heavy metals for soil, since, in contrast to air and water, soil is an extremely heterogeneous system and mobility of inorganic contaminants, closely related to the intake by plants, depends on several soil factors. The approaches towards the determination of metal concentration limits in soil vary significantly in individual countries. In some countries, the definition of limits for heavy metals concentrations is based on soil use (these are defined as so-called trigger and action values), or, possibly, on eco-toxicological data in so-called standard soil and limit values for the total and dissolvable concentration of heavy metals in soil (Barančíková, 1998; Makovníková et al., 2006).

Toxicity of heavy metals varies; it decreases in the following line Hg>Cd>Ni>Pb>Cr and their influence is enhanced by their non-degradability. Soil is only presented as a passive acceptor of heavy metals; it becomes the source of polluting other components of the environment and the food chain. Changes in soil properties are responsible for the mobilisation of metals, especially pH, humus content and quality and the proportion of clay fraction (Barančíková, 1998).

With regard to the above findings, the content of the following risk elements was observed in the conditions of sustainable use of soil: lead, cadmium and nickel (in the leachate 2M HNO_3) (Fig. 13., Fig. 14. and Fig. 15.). The evaluation showed that the content of dangerous elements in soil did not reach maximum permitted values for the Slovak Republic (Act No. 220/2004 Coll.) and the measured values corresponded with natural contents of the observed elements in soil and base rocks (Makovníková et al., 2006). At the same time, in ecological systems, no anthropogenic pollution by applying chemical substances and sediments in soil is present.

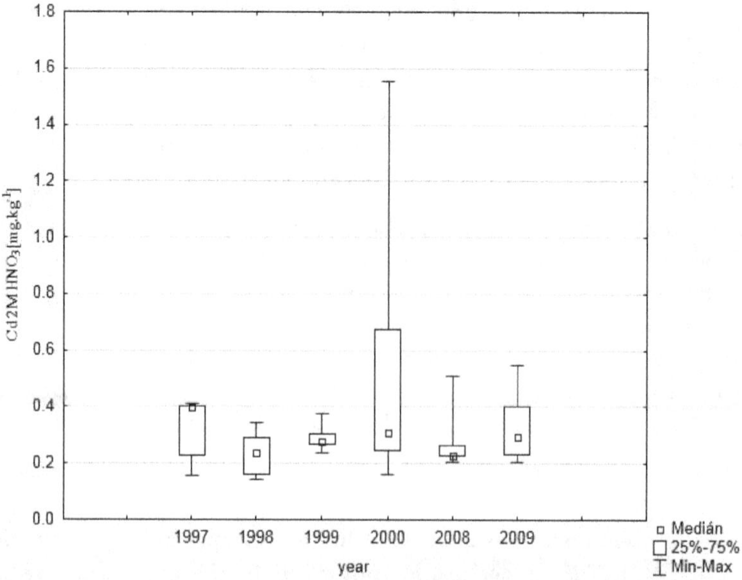

Figure 13. Cd content of soil in the monitored area situated in a marginal region of north-eastern Slovakia expressed by descriptive statistics

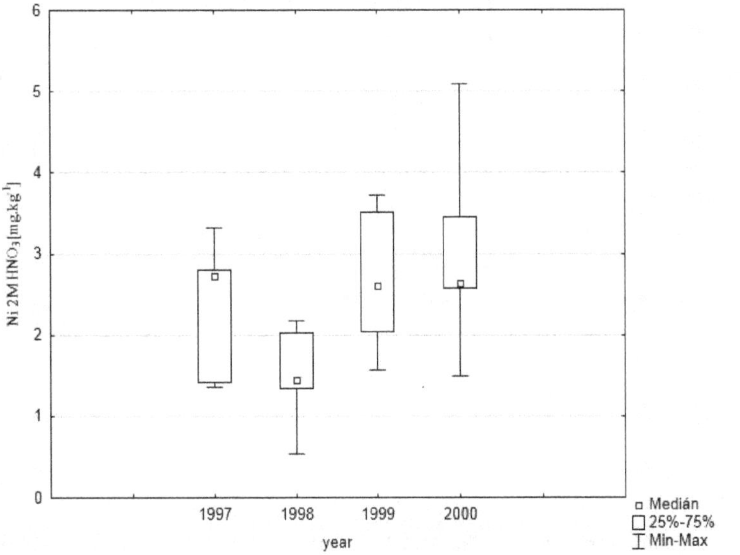

Figure 14. Ni content of soil in the monitored area situated in a marginal region of north-eastern Slovakia expressed by descriptive statistics

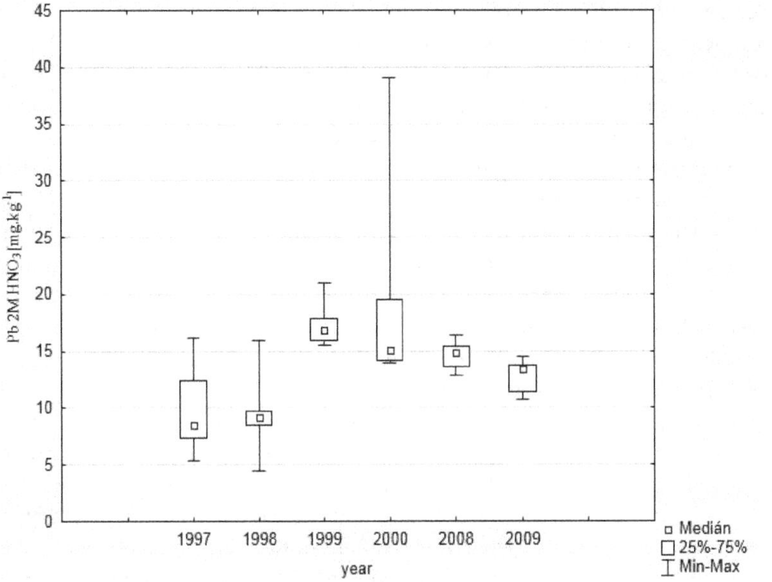

Figure 15. Pb content of soil in the monitored area situated in a marginal region of north-eastern Slovakia expressed by descriptive statistics

A statistically significant effect of the monitored year and locality on observed heavy metal content of the soil was confirmed by an analysis of variance (Tab. 6.).

Parameter	Min.	Max.	Mean	Standard error	Source of variability	d. f.	F-Ratio	P
Pb 2M HNO_3 [$mg.kg^{-1}$]	7.77	22.18	14.11	0.823555	year locality	6 4	12.30 22.01	++ ++
Cd 2M HNO_3 [$mg.kg^{-1}$]	0.129	0.697	0.343	0.037233	year locality	6 4	9.37 19.69	++ ++
Ni 2M HNO_3 [$mg.kg^{-1}$]	0.934	3.436	2.392	0.140375	year locality	6 4	24.08 25.20	++ ++

Table 6. Analysis of variance of the heavy metal content of the soil in the monitored area situated in a marginal region of north-eastern Slovakia
++P< 0,01 +P< 0,05

3. Conclusion

The farming system is the most widespread environmental technology with its positive and negative consequences. It utilises essential natural resources and, at the same time, influences other natural environments. Therefore, ecologisation of farming is a priority of farmers as well as environmentalists. Respecting the principles of soil sustainability and other components of the environment is a basic precondition for life sustainability. Sustainable agriculture is based on the principle of agriculture being a biological process which, in practice, should imitate key characteristics of the natural ecosystem. It strives to bring diversity into agro-ecosystems, recycle nutrients efficiently and maintain the priority of sunlight as a source of energy for agro-ecosystems. Sustainable use of soil takes soil-ecological conditions into consideration and is realised in such a way and in such intensity, which gives rise to neither negative changes in soil, nor establishes trends for the development of negative characteristics in soil. It is impossible to select universal soil parameters for sustainable soil, which is why the area is subject to specialised discussions. A significant role in the selection of parameters is played by their variability in time, related to parameter stability. Stable (such as soil depth or granularity), relatively stable (the salt content, the content of organic mass in soil, heavy metal contamination), relatively dynamic (pH, the content of nutrients) and dynamic (soil humidity and temperature, microbial activity, etc.) parameters are more connected to its short-term changes. Soil quality cannot be judged directly; it must be determined from the changes of its parameters. It is more accurate to evaluate the range of appropriate indicators rather than to use a single one. Soil quality is significantly affected by physical, chemical, biological and biochemical properties sensitive to changes in the environment and land management. At present, there is little knowledge with regard to soil development in the conditions of sustainable farming systems. The present findings can hardly be compared to other research due to the different soil-ecological conditions in which they were obtained. The issue of universal methods for all soil types remains a universal problem within the research of soil development. Unless this area is unified, objective comparison will remain on a regional level. Soil parameters are usually determined only in relation to specific topsoil. Certain physical and chemical parameters in subsoil cannot be neglected, since they guarantee soil functions.

The present results showed development of selected soil parameters during long-term monitoring on a model area situated in a marginal region of north-eastern Slovakia where an ecological farming system was applied. Soil physical properties change not only under the influence of weather conditions, crop year, vegetation pass, but also under the influence of applied management systems. During the year and growing season, bulk density value also varies depending on water availability in the soil, weather and farming methods. The research showed that soil physical properties get adjusted after long-term application of an ecological farming system and the measured values were stabilised, reaching levels comparable with the average values for the soil type. Agrochemical soil characteristics did not change significantly during the research period. High doses of organic fertilisers had a positive effect on soil productivity, and, thus, indirectly on maintaining soil pH, the available nutrient content and retention of humus in soil. In spite of this, it is necessary to continuously pay attention to soil reaction, because soil is naturally acidified through acid atmospheric fallout as well as calcium intake by plants. Values of selected heavy metals in the monitored period did not exceed the limit values published in Act No. 220/2004 Coll. The values of activity of phosphatase and urease changed minimally during the research period and they refer to values typical for soils with sparse vegetation. At the same time, it was proven that increasing the content of soil organic matter promotes natural protection of soil enzymes. Analysis of variance confirmed a statistically significant effect of the monitored year on all observed soil parameters. The effect of the observed locality, with the exception of $pH/CaCl_2$, C_{ox} and N_{anorg}, on other soil parameters was also statistically significant.

Author details

Danica Fazekašová
University of Prešov, Slovakia

Acknowledgement

The study was supported by VEGA 1/0601/08 *Effect of biotic and abiotic factors on ecosystem sustainability*, VEGA 1/0627/12 *Diversity, resiliency and health of ecosystems in different farming system and polluted territories in anthropogenic land* and KEGA 012PU-4/2012 *Preparation and realization of the research focused on creating teaching aids for education of environmental subject.*

4. References

Act. No. 220/2004 on Soil Protection and Agricultural Soil Using

Agenda 21 and Indicators of Sustainable Development, MŽ SR, Bratislava, Slovak Republic, ISBN 8088833035

Alef, K. & Nannipieri, P. (1995). Soil Sampling, Handling, Storage and Analysis, In: *Methods in Applied Soil Microbiology and Biochemistry,* K. Alef, P. Nannipieri, (Eds.), 49–115, Academic Press, ISBN 978-0-12-513840-6 London, Great Britain

Baligar, V. C. ; Wright, R. J. & Hern, J. L. (2005). Enzyme Activities in Soil Influenced by Levels of Applied Sulfur and Phosphorus, *Communications in Soil Science and Plant Analysis,* Vol.36, No.13-14, pp 1725-1735, ISSN 00103624

Barančíková, G. (1998). The Proposal of Special Classification of Agricultural Soils of Slovakia from Viewpoint of their Sensibility to Contamination by Heavy Metals, *Plant Production,* Vol.44, No.3, pp. 117 – 122, ISSN 0370-663X

Belińska, E. J. & Prangal, J. (2007). Enzymatic Activity of Soil Contaminated with Triazine Herbicides, *Polish Journal of Environmental Studies,* Vol.16, No.2, pp 295-300, ISSN 1230-1485

Beneš, Š. & Benešová, J. (1993). Balance of Risk Elements in the spheres of the Environment, *Plant Production,* Vol.39, No.10, pp. 941-958, ISSN 0370-663X

Beneš, Š. (1993). *The Contents and Balance of Elements in the Spheres of the Environment,* 1. part, MZ, Praha, Czech Republic

Bielek, P. (1998). *Nitrogen in Agricultural Soils of Slovakia,* Soil Science and Conservation Research Institute, ISBN 80-85361-44-2, Bratislava, Slovak Republic

Bremner, J. M & Mulvaney, R. L. (1978). Urease Activity in Soils, In: *Soil Enzymes,* R.G.Burns, (Ed.), 149 – 196, Academic Press, New York, USA

Bujnovský, R. & Juráni, B. (1995). *The Subsoil,* Soil Science and Conservation Research Institute, ISBN 80-85361-49-3, Bratislava, Slovak Republic

Burns, R. G. (1978). *Soil Enzymes,* Academic Press, ISBN 0-12-145850-4, New York, USA

Caldwell, B. A. (2005). Enzyme Activities as a Component of Soil Biodiversity, *Pedobiologia,* Vol. 49, pp. 637 – 644, ISSN 0031-4056

Chander, K.; Goyal, S.; Mundra, M. C. & Kapoor, K. K. (1997). Organic Matter, Microbial Biomass and Enzyme Activity of Soils under Different Crop Rotations in the Tropics, *Biol Fertil Soils,* Vol.24, No.3, pp 306-310, ISSN 01782762

Chazijev, F. CH. (1976). *Soil Enzyme Activity,* Nauka, Moskva, Russia

Fazekašová, D. (2003). *Sustainable Use of Soil – Definition and Evaluation of Indicators and Parameters of Soil Development,* FHPV PU, ISBN 80-8068-228-3, Prešov, Slovak Republic

Fazekašová, D.; Kotorová, D.; Balázs, P.; Baranová, B. & Bobuľská, L. (2011). Spatial Varialibility of Physical Soil Properties in Conditons of Ecological Farming in Protected Area, *Ekológia (Bratislava),* International Journal for Ecological problems of the Biosphere,Vol.30, No.1, p. 1-11 ISSN 1335-342X

Fiala, K.; Barančíková, G.; Brečková, V.; Búrik, V.; Houšková, B.; Chomaničová, A.; Kobra, J.; Litavec, T.; Makovníková, L.;Pechová, B. & Váradiová, D. (1999). *Partial Monitoring System – Soil,* Binding methods, 1st ed.; Soil Science and Conservation Research Institute, ISBN 80-85361-55-8, Bratislava, Slovak Republic

Fulajtár, E. (2006). *Physical Parameters of Soil,* Soil Science and Conservation Research Institute, ISBN Bratislava, Slovak Republic

Geisseler, D. & Horwath, W. R. (2009). Short-term Dynamics of Soil Carbon, Microbial Biomass, and Soil Enzyme Activities as Compared to Longer-term Effects of Tillage in Irrigated Row Crops, *Biol Fertil Soils,* No.46, pp. 65-72, ISSN 1432-0789

Grejtovský, A. (1991). Influence of Soil Improvers on Enzymatic Activity of Heavy Alluvial Soil, *Rostl. Výr*, 37, pp 289–295, ISSN 1214-1178

Gulser, F. & Erdogan, E. (2008). The Effects of Heavy Metal Pollution on Enzyme Activities and Basal Soil Respiration of Roadside Soils, *Environ Monit Assess*, Vol.145, No1-3, pp. 127-133, ISSN 1573-2959

Hraško, J. & Bedrna, Z. (1988). *Applied Pedology*, Príroda, Bratislava, Slovak Republic

Iovieno, P. ; Morra, L. ; Leone, A. ; Pagano, L. & Alfani, A. (2009). Effect of Organic and Mineral Fertilizers on Soil Respiration and Enzyme Activities of two Mediterranean Horticultural Soils. *Biol Fertil Soils*, Vol.45, No.5, pp. 555-561, ISSN 01782762

Javoreková, S.; Králiková, A.; Labuda, R.; Labudová, S. & Maková, J. (2008). *Soil Biology in Agroecosystems*, Slovak University of Agriculture in Nitra, ISBN 978-80-552-0007-1, Nitra, Slovak Republic

Jordan, D. ; Kremer, R. J. ; Bergfield, W. A. ; Kim, K. Y. & Cacnio, V. N. (1995). Evaluation of Microbial Methods as Potential Indicators of Soil Quality in Historical Agricultural Fields. *Biol Fertil Soils*, Vol.19, No.4, pp 297-302, ISSN 01782762

Jurčová, O. & Torma, S. (1998). *Methodology for Quantifying Potential Nutrient Plant Residues*, Soil Science and Conservation Research Institute, ISBN 8085361892, Bratislava, Slovak Republic

Klír, J. (1997). *Sustainable Agriculture*, Department of Agricultural and Food Information, ISBN 80-86153-18-5, Praha, Czech Republic

Klose, S. & Tabatabai, M. A. (2000). Urease Activity of Microbial Biomass in Soils as Affected by Cropping Systems. *Biol Fertil Soils*, Vol.31, No.3-4, pp 191-199, ISSN 01782762

Kotorová, D. ; Šoltýsová, B. & Mati, R. (2010). *Properties of Fluvisols on the East Slovak Lowland at their Different Tillage*, 1. ed. CVRV, ISBN 978-80-89417-25-4, Piešťany, Slovak Republic

Kováčik, P. (2001). *Methodology of Nutrient Balance in Ecological Farming System*, Slovak University of Agriculture in Nitra, ISBN 80-7137-957-3, Nitra, Slovak Republic

Krnáčová, Z.; Račko, J. & Bedrna, Z. (1997). Indicators of Sustainable Development Use of Agricultural Soils in Slovakia. *Acta Environmentalica Universitatis Comenianae*, Supplement, pp. 165-172, ISSN 1335-0285, Bratislava, Slovak Republic

Kromka, M. & Bedrna, Z. (2000). *Soil Hygiene*, Comenius University in Bratislava, ISBN 80-223-1602-4, Bratislava, Slovak Republic

Lacko-Bartošová, M. ; Cagáň, Ľ.; Čuboň, J. ; Kováč, K. ; Kováčik, P. ; Macák, M. ; Moudrý, M. & Sabo P. (2005). *Sustainable and Ecological Agriculture*, 1st ed. Slovak University of Agriculture in Nitra, ISBN 80-8069-556-3, Nitra, Slovak Republic

Larson, W.E. & Pierce, F.J. (1991). Conservation and Enhancement of Soil Quality. *Evaluation for Sustainable Land Management in the Developing World*, Vol.2, No.12, pp. 175-203, Technical Papers. IBSRAM Proc.

Líška, E.; Bajla, J.; Candráková, E.; Frančák, J.; Hrubý, D.; Illeš, L.; Korenko, M.; Nozdrovický, L.; Pospišil, R.; Špánik, F. & Žember, J.; (2008). *General Crop Production*, Slovak University of Agriculture in Nitra, ISBN 978-80-552-0016-3, Nitra, Slovak Republic

Makoi, J. H. J. R. & Ndakidemi, P. A. (2008). Selected Soil Enzymes: Examples of their Potential Roles in the Ecosystem, *African Journal of Biotechnology*, Vol. 7, No.3, pp. 181-191, ISSN 1684–5315

Makovníková, J.; Barančíková, G.; Dlapa, P. & Dercová, K. (2006). Inorganic Contaminants in Soil Ecosystems, *Chem. Listy,* Vol.100, No.6, pp. 424 – 432, ISSN 0009-2770

Matúšková, L. & Vojtáš, J. (2005). *Guideline for Detecting Soil Contamination,* Soil Science and Conservation Research Institute, Bratislava, Slovak Republic

Michaeli, E. & Ivanová, M. (2005). Regional Geoecological Structure of Landscape and Primary Development Potential of the Prešov Self-governing Region. *Folia Geographica.* Vol. 18, No.8, pp. 116-142, ISSN 1336-6157

Miralles, I.; Ortega, M.; Sánches-Marañón, M.; Leirós, M. C.; Trasar-Cepeda, C. & Gil-Sotres, F. (2007). Biochemical Properties of Range and Forest Soils in Mediterranean Mountain Environments, *Biol Fertil Soils,* No.43, pp. 721-729, ISSN 1432-0789

Nourbakhsh, F. & Monreal, C. M. (2004). Effects of Soil Properties and Trace Metals on Urease Activities of Calcareous Soils. *Biol Fertil Soils,* Vol.40, No.5, pp. 359-362, ISSN 01782762

Pejve, J. (1966). *Soil Biochemistry,* SVLP, Bratislava, Slovak Republic

Rode, A. A. (1969). *Basic Science of Soil Moisture,* Tom II, Gidrometeorologič. izd., Leningrad, Russia

Šarapatka, B. (2002). Possibilities of Using Enzyme Activities as Indicators of Productivity and Systems Quality, *Biological indicators of soil quality,* pp. 26-31, ISBN 80-7157-64-25, MZLU, Brno, Czech Republic

Sinsabaugh, R. L.; Findlay, S.; Franchini, P. & Fischer, D. (1997). Enzymatic Analysis of Riverine Bacterioplankton Production, *Limnol. Ocecmogr.,* Vol. 48, No.1, (1997), pp. 29 – 38, ISSN 0024-3590

Speir, T. W. ; Van Schaik, A. P. & Lloyd-Jones, A. R. (2003). Temporal Response of Soil Biochemical Properties in a Pastoral Soil after Cultivation Following High Application Rates of Undigested Sewage Sludge. *Biol Fertil Soils,* Vol.38, No.6, pp. 377-385, ISSN 01782762

Tabatabaj, M. A. & Acosta-Martínez, V. (2000). Enzyme Activities in Limed Agricultural Soil. *Biol Fertil Soils,* Vol.31, No.1, pp. 85-91, ISSN 01782762

Tischer, S.; Tanneberg, H. & Guggenberger, G. (2008). Microbial Parameters of Soils Contaminated with Heavy Assessment for Ecotoxicological Monitoring, *Polish Journal of Ecology,* Vol.56, No.3, (2008), pp. 471 – 479, ISSN 15052249

Torma, S. (1999). *Potassium - an Important Nutrient in the Soil and Plant, Pedo disertationes,* Soil Science and Conservation Research Institute, ISBN 80-85361-51-5, Bratislava, Slovak Republic

Trasar-Cepeda, C. ; Leirós, C. ; Gil-Sotres, & F. Seoane S. (1998). Towards a Biochemical Quality Index for Soils: An Expression Relating Several Biological and Biochemical Properties. *Biol Fertil Soils,* Vol.26, No.2, pp. 100-106, ISSN 01782762

Vilček, J. ; Hronec, O. & Bedrna, Z. (2005). *Environmental Pedology,* Soil Science and Conservation Research Institute, ISBN 80-8069-501-6, Bratislava, Slovak Republic

Wick, B.; Kűhne, R. F. & Vielhauer, K. (2002). Temporal Variability of Selected Soil Microbiological and Biochemical Indicators under Different Soil Quality Conditions in South-western Nigeria, *Biol Fertil Soils,* No.35, pp. 115-121, ISSN 1432-0789

A Geologic and Geomorphologic Analysis of the Zacatecas and Guadalupe Quadrangles in Order to Define Hazardous Zones Associated with the Erosion Processes

Felipe de Jesús Escalona-Alcázar, Bianney Escobedo-Arellano, Brenda Castillo-Félix, Perla García-Sandoval, Luz Leticia Gurrola-Menchaca, Carlos Carrillo-Castillo, Ernesto-Patricio Núñez-Peña, Jorge Bluhm-Gutiérrez and Alicia Esparza-Martínez

Additional information is available at the end of the chapter

1. Introduction

In the State of Zacatecas, Mexico (**Figure 1**), the environment is not usually taken into account as a critical variable for the urban growth and development planning. The expansion of the cities of Zacatecas and Guadalupe in the study area is merely based on the land use change according to the Urban Development Code[1] and the Urban Development Plan 2004-2030 [2]. The Code states the urban growth policies which apply to the whole state; whereas the Plan is a compilation of documents related to the urban growth tendency, population distribution, and basic population services (i.e. water supply). Regardless of the scale, the criteria for the land use change policies are unclear. The Urban Development Plan [2] suggests avoiding urban growth towards areas geologically and topographically unstable and those with flood potential. In every case, the slope should be less than 30°. There are no available maps that indicate where these areas are located, so the criteria for land use will remain unclear.

Although the geology and geomorphology are mentioned [2], their value as critical variables is not taken into account in practice for any purpose. The geologic and geomorphologic variables defined [3] are only indicated, but not located on a map for planning development. Moreover, it is suggested [2] that there is a necessity for a detailed mapping of the geomorphic agents defined [3].

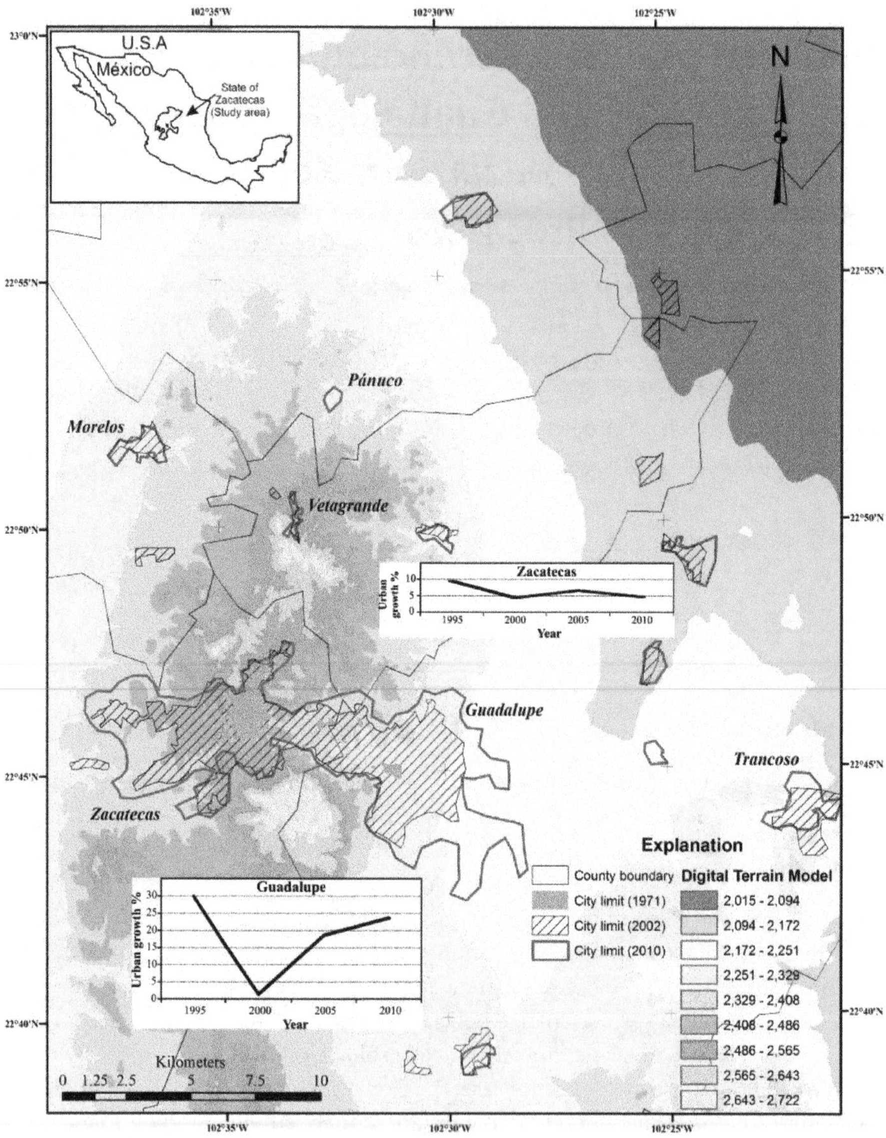

Figure 1. Digital Terrain Model of the study area. The area includes six municipalities whose names are in bold-italic. The largest cities are Zacatecas and Guadalupe. The graphs show the population growth during the last 20 years [8-10], Guadalupe is among the fastest growing municipalities in Mexico (~25% in 2010). The 1971 city limit was obtained from INEGI [11-12]; the growth of these cities until 2002 is a compilation made by the first author; while for 2010 it was a combination between a Google Earth image (August 5th, 2009) and field work. Inset is the location of the State of Zacatecas and the study area.

Due to the land use change and the subsequent landscape modification, the erosion processes are currently becoming active in places usually considered to be stable. The effects are: fractured streets, roads, and houses, voids under the streets and buildings, slope instability, and rock blocks falling next to the roads themselves (**Figure 2**). The primary erosion agent is rainwater. The geomorphic processes are slow but steady contributors, due to the semiarid climatic conditions [3]. The average annual precipitation is 500 mm/yr; while the average annual temperature is 22°C. Therefore, the erosion processes develop slowly, but successfully. Until now, the hazardous zones have only been reported in specific places; the effects of which are the mass removal under houses or streets with active fracturing and, in worst case situations, their slow collapse (**Figure 2**). The exact location of these areas has not been taken into account as a social, economic or environmental problem. The natural hazards recognized by the authorities are mostly related to the mining industry and their products (i.e. mine tailings, open pits); on the other hand, the erosion and its effects [2] are considered of minor importance. The landscape, geology, land use, soil cover, and their modifications are barely considered to be serious enough to promote the development of dangerous areas once the original conditions are changed.

Figure 2. Images showing the erosion hazards associated with different lithology: a) and b) are in the granitic facies of the Zacatecas Conglomerate; c) moderately consolidated facies of the Conglomerate; and d) deformed lava flows from Las Pilas Complex. Yellow arrows point toward the first attempt of damage repair. Purple arrows are the sites where a second repair phase has been attempted; while the green arrow shows a third repair phase. The red arrows point to the voids created due to the soil removal by the erosion processes.

For several years now, attempts have been made in order to identify and describe the geomorphic processes acting in the Zacatecas and Guadalupe cities and their relationship with geology and geomorphology [3-7]. In this paper we define the hazardous zones by means of a GIS analysis that integrates the geologic and geomorphologic mapping combined with the digital slope modeling, land use, and soil type.

1.1. Local geology overview

The stratigraphic sequence of the study area is composed by units ranging from the Early Cretaceous Period to Present Age (**Figure 3**). The oldest unit is the Early Cretaceous Zacatecas Formation[13]. The Zacatecas Formation is made up by Greenschists Facies metamorphic rocks whose protoliths are wacke, mudstone, limestone, as well as, interstratified conglomerates with lava flows and tuffs. The whole sequence is cut by dikes, sills, and dioritic laccoliths. Its upper contact is transitional with the Early Cretaceous Las Pilas Complex. This unit is composed of mafic lavas with pillowed and massive structure, commonly foliated and deformed (**Figure 4a and b**). Lava flows contain interbedded wacke, greywacke, mudstone, and minor tuff and limestone. It is considered that laccoliths and dikes of the Las Pilas Complex are part of the same volcanic sequence. Moderate to intense hydrothermal alteration affects the entire Mesozoic record. Both, the Zacatecas Formation and the Las Pilas Complex are known as the Zacatecas Group. This group shows the results of a Late Cretaceous compressive deformation stage and at least five extensional deformation phases that have occurred from the Oligocene to Recent [16]. The association of deformed structures with faulting and steep slopes promotes rock falling.

The Las Pilas Complex is unconformably covered by the Eocene Zacatecas Conglomerate that crops out in the Zacatecas and Guadalupe cities. This conglomerate is composed of five facies named according to their clast-rich abundance and their physical characteristics (**Figure 4c and d**). The sandstone-rich layers are more easily erodible than the conglomeratic ones. This characteristic promotes differential erosion processes that, in combination with moderate to steep slopes, favors the generation of hazardous areas. The conglomerate was deposited in a WNW-ESE fault-bounded basin whose deformation is less intense than the one showed by the Zacatecas Group.

At the top of the stratigraphic column the Zacatecas Conglomerate is in transitional contact with the Oligocene-Miocene Volcanic Sequence that is composed of interbedded ash-flow and air-fall tuffs, breccias, and rhyolitic flows and domes [3, 17]. They commonly develop cliffs that, when associated with steep slopes, encourage the falling of the rocks. The cliffs next to the main faults and steep slopes are also suitable areas for the rock falling process to happen.

1.2. Geomorphic features and unconsolidated deposits

The study area belongs to the Basin and Range extensional province [16] that is formed by NNW-SSE normal faults forming horsts and grabens. The main geomorphologic feature is the Sierra de Zacatecas that is oriented NNW-SSE (**Figure 5**) bounded to the west by the

Calera Valley and to the east by the El Palmar Valley. A minor range, Sierra de Tolosa, is located to the east; both ranges are separated by the El Palmar Valley.

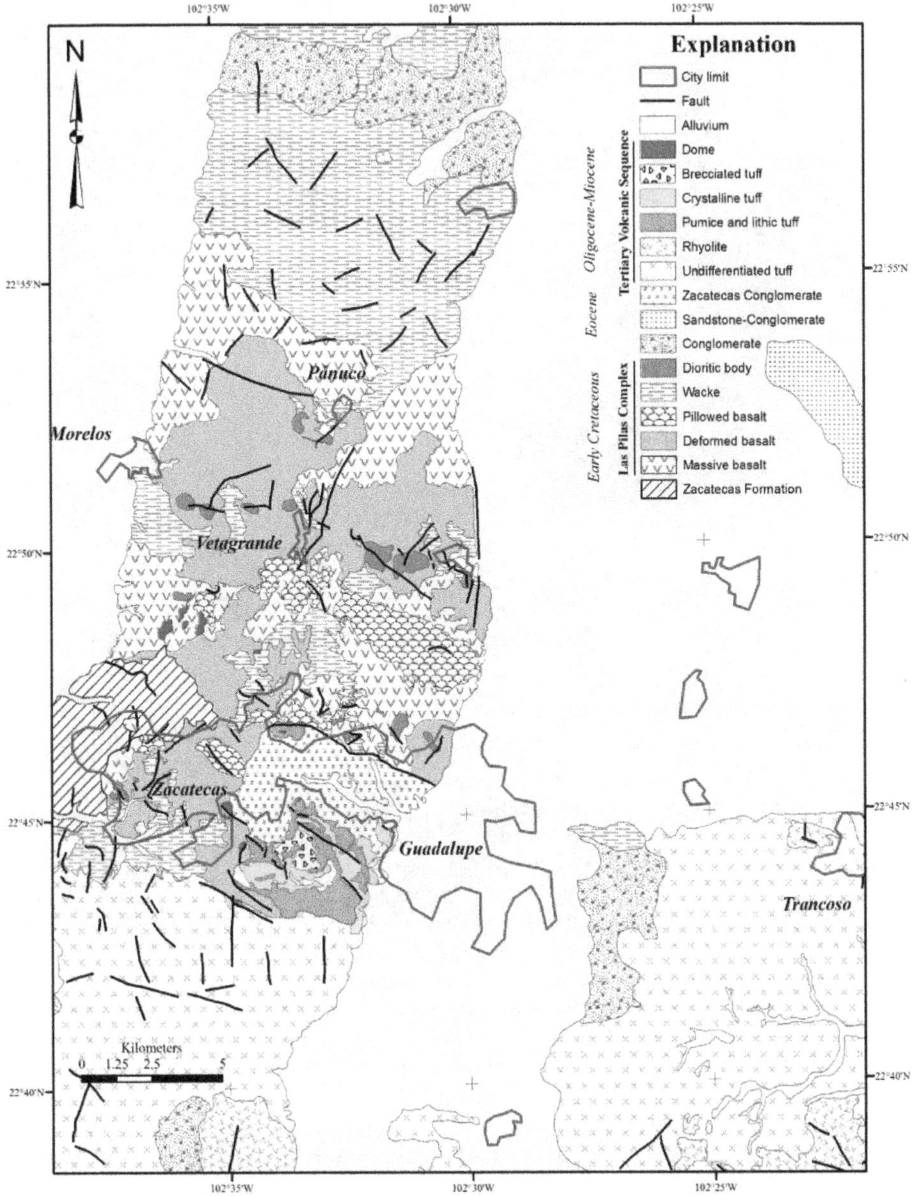

Figure 3. Geologic map of the Zacatecas and Guadalupe quadrangles. Detailed mapping from the Mesozoic sequence is taken from [13-14], while the Tertiary Volcanic Sequence is from [3, 15]. The names are from the municipalities of the study area.

Figure 4. Images showing the main physical characteristics of lithological units that crop out in the Zacatecas and Guadalupe quadrangles. The Las Pilas Complex can be deformed (a) or massive (b), while the Zacatecas Conglomerate can be moderately consolidated (c) with differential erosion (d).

The main erosional agent in the area is rainwater; it affects unconsolidated to moderately consolidated sediments either from slope or fluvial deposits, as well as alluvial fans. These geomorphic features occur in the neighborhood of the present city limit (**Figure 5**). The slope deposits are made of interbedded sandstone and conglomerate. The sandstone thickness is less than 50 cm, whereas the conglomerate thickness varies from a few centimeters up to 30 meters. Commonly, both show normal gradation and, sometimes there are interbedded pumice and/or lithic tuffs. Since the sandstone more easily eroded than the conglomerate, it generates a differential erosion process (**Figure 4d**).

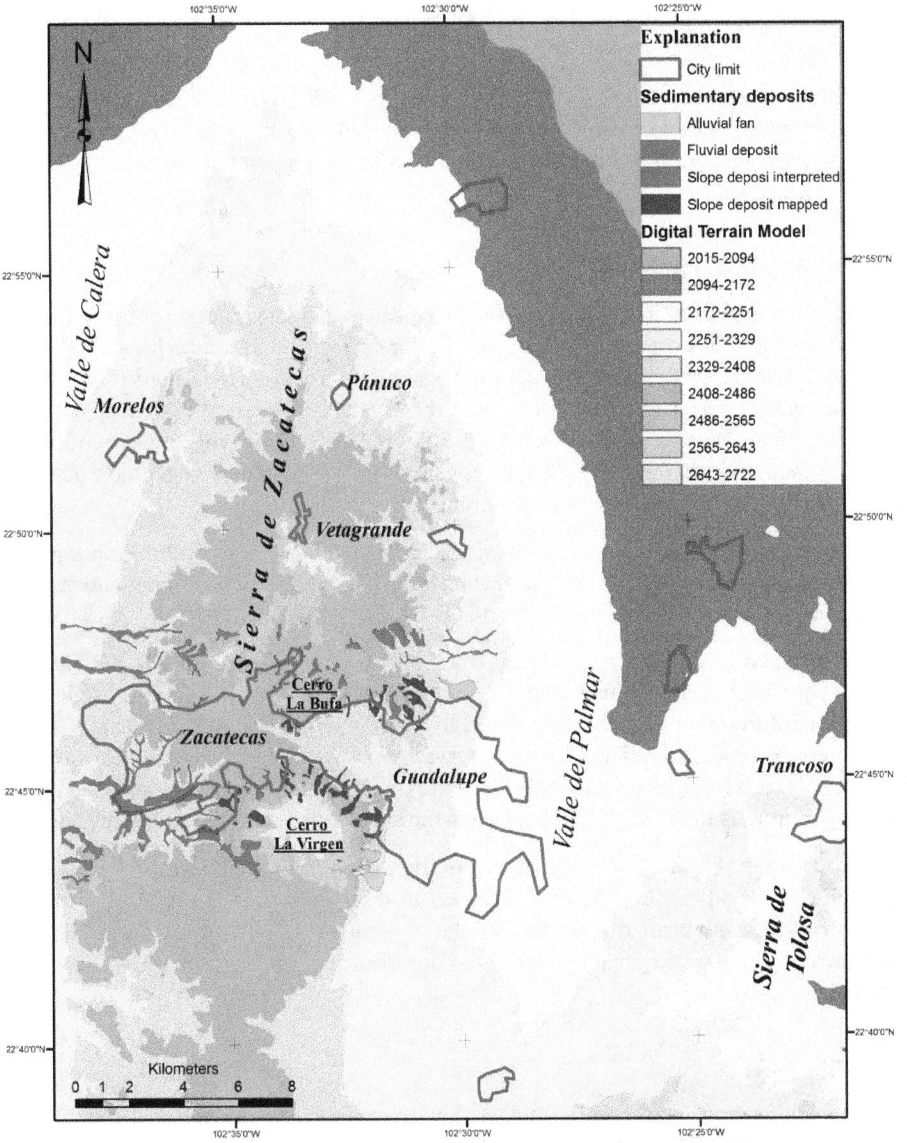

Figure 5. Digital Terrain Model showing sedimentary deposits that are in the neighborhood of the city limit of Zacatecas and Guadalupe. Horizontal bold-italics names are the municipalities; while in italics are the ranges and valleys. The main topographic features are underlined.

The slope deposits are of unknown age. Their morphologic expression is masked by the relief. The cities of Zacatecas and Guadalupe are currently growing directly on these deposits (**Figure 5**). Regarding the alluvial fans, there are a few of them in the eastern side of the Sierra de Zacatecas because most of the valleys are used for agricultural activities. The fluvial deposits are common in the western side of this sierra; however, the arroyos are currently modified due to city growth.

The buildings constructed on moderate to loosely consolidated sediments and lithological units are more vulnerable than those with well consolidated materials. In the sedimentary deposits shown in **Figure 5** the erosion of loose materials generate hazardous zones if the terrain slope is over 20°.

1.3. Land use and vegetation

The study area is located in a semi-desertic area with an average annual precipitation of less than ~500 mm/yr so most of the water supply for any all human activities is taken from underground wells. In these conditions the vegetation varieties are limited (**Figure 6**). The valleys are mostly used for agriculture. The thorn scrub and nopal (*Opuntia*) occupy the gentle hills of the sierras de Zacatecas and Tolosa. This type of vegetation requires little water and their roots extend laterally. The bushes are either natural vegetation outside the city limit or reforested areas inside those limits.

Since the original data was generated during the decade of 1970 [18-19], the situation now is different. The area shown as "Natural bushes" (**Figure 6**) is currently being substituted by thorn scrub and nopal. However, there is no up-to-date cartography.

The land use shown in **Figure 7** is taken from [18-19]; as well as in **Figure 6**, the valleys are used for agriculture. Cattle use is now greater, while the forestry area has decreased. Though the information on the criteria used to define "forestry use" is unclear, informally, these areas were considered for land conservation. However, since there are no written rules, the land use changes and modifications are based on unknown criteria. What has been seen is a continuous modification of landscape for urban purposes next to the city limit.

The land use map (**Figure 7**) was created from the INEGI (Instituto Nacional de Estadística, Geografía e Informática), institution depository of most of the cartographic information in the country. These maps are the basis for all the projects that require cartography. This being said, the conditions of the maps vary. The older ones were made different than the ones made today.

1.4. Edaphology

The soil classification used here is that of the United Nations Educational, Scientific and Cultural Organization [20]. In the Sierra de Zacatecas, where most municipalities are located, the Lithosol Eutric is dominant (**Figure 8**). It is composed by local, scarcely transported, rock clasts whose size varies from coarse sand to gravel; usually it is poor in organic matter. Its thickness is less than 15 cm.

The fluvisols are found along the arroyos; they show well-developed bedding, with normal gradation and variable amounts of organic matter. Their thickness is unknown since they are filling the valleys. The change in their texture and composition allows us to define the subclass.

The xerosols are dominant in the valleys. They have variable amounts of organic matter. A whitish layer at the top is characteristic of this soil, and it is usually due to the carbonate or sulfate accumulation.

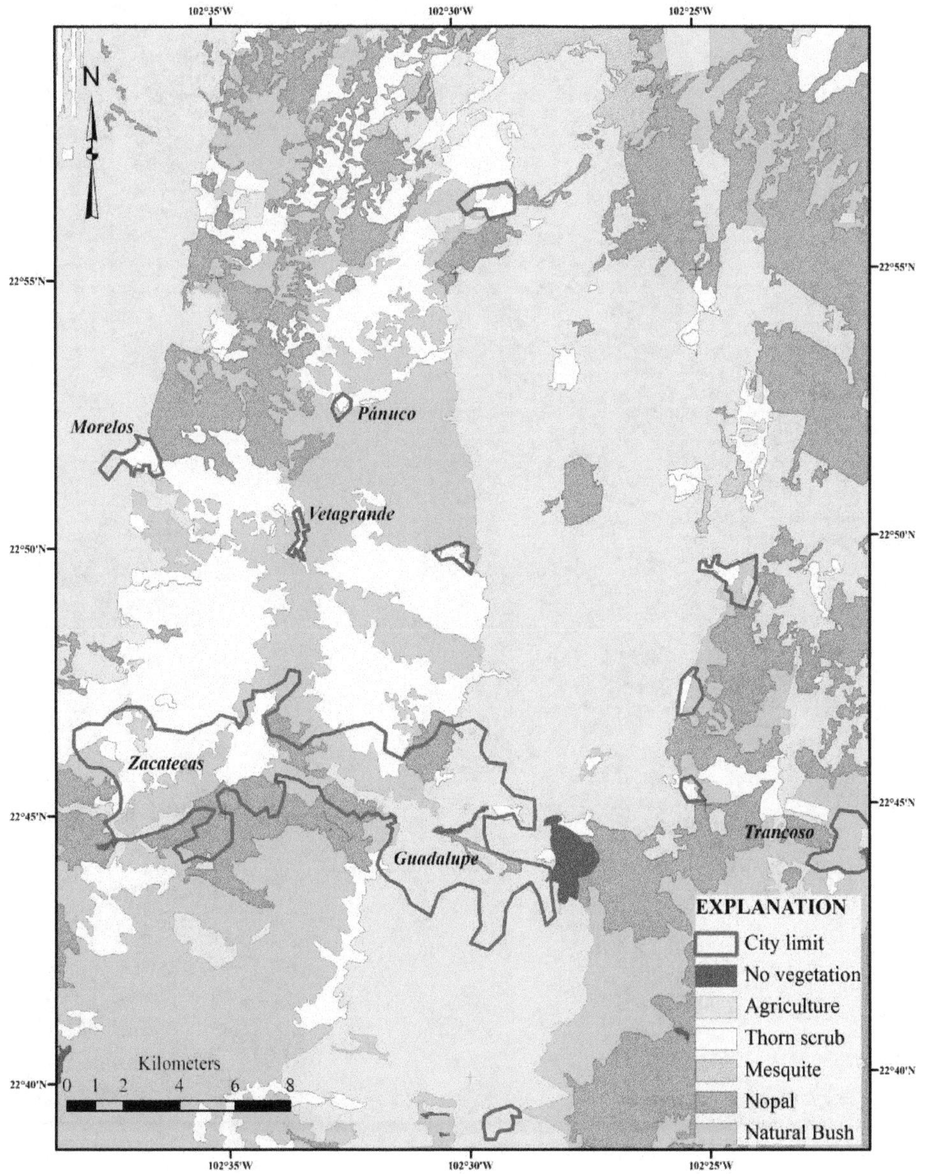

Figure 6. Vegetation of the study area, modified from [18-19]. The city limit corresponds to 2010. The names in bold, italics are the municipalities. The places with no vegetation are artificial dams. The original data are from the decade of 1970 [18-19].

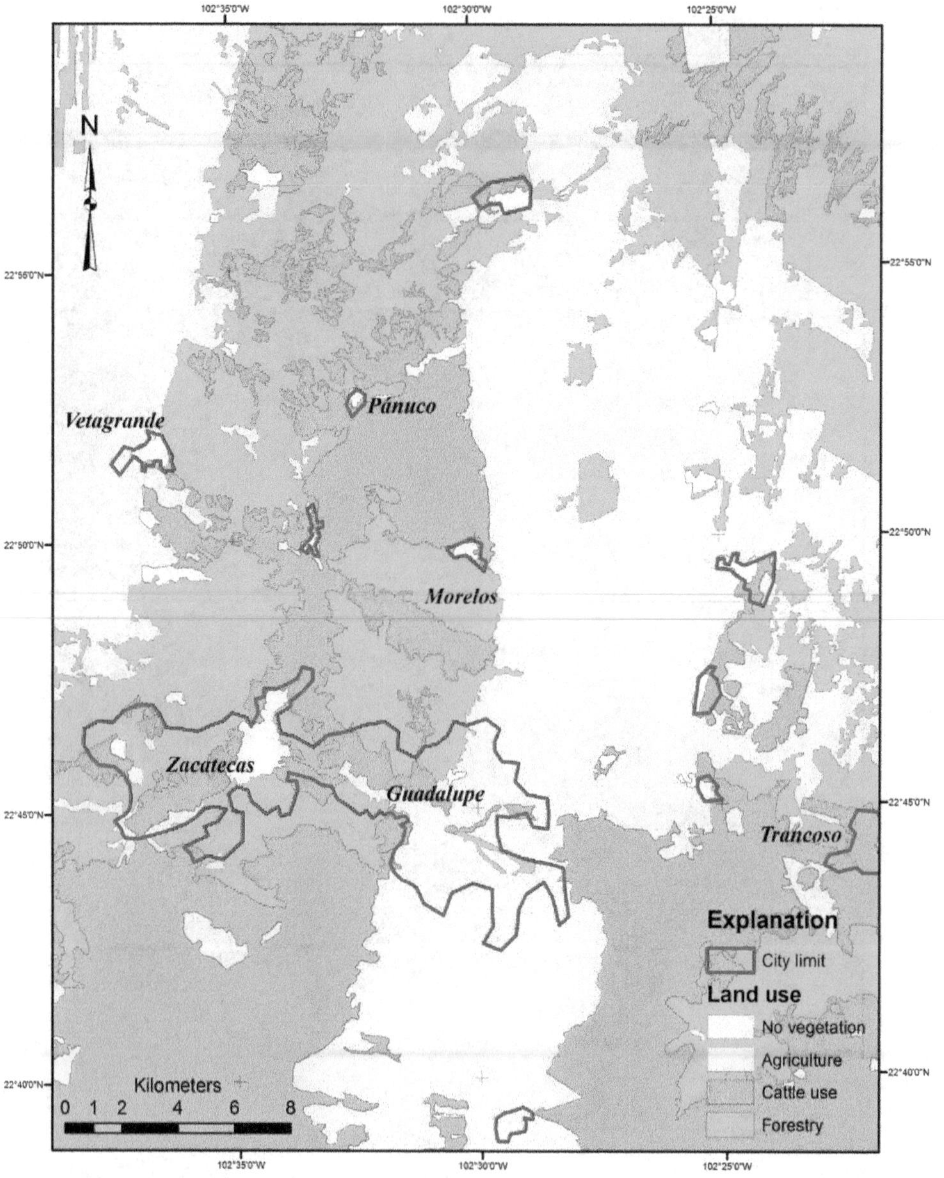

Figure 7. Land use proposed to the study area from [18-19]. The city limit is from 2010. The names in bold-italics are the municipalities. The places with no vegetation are the city limits in 1971. The original land use data are from the decade of 1970 [18-19].

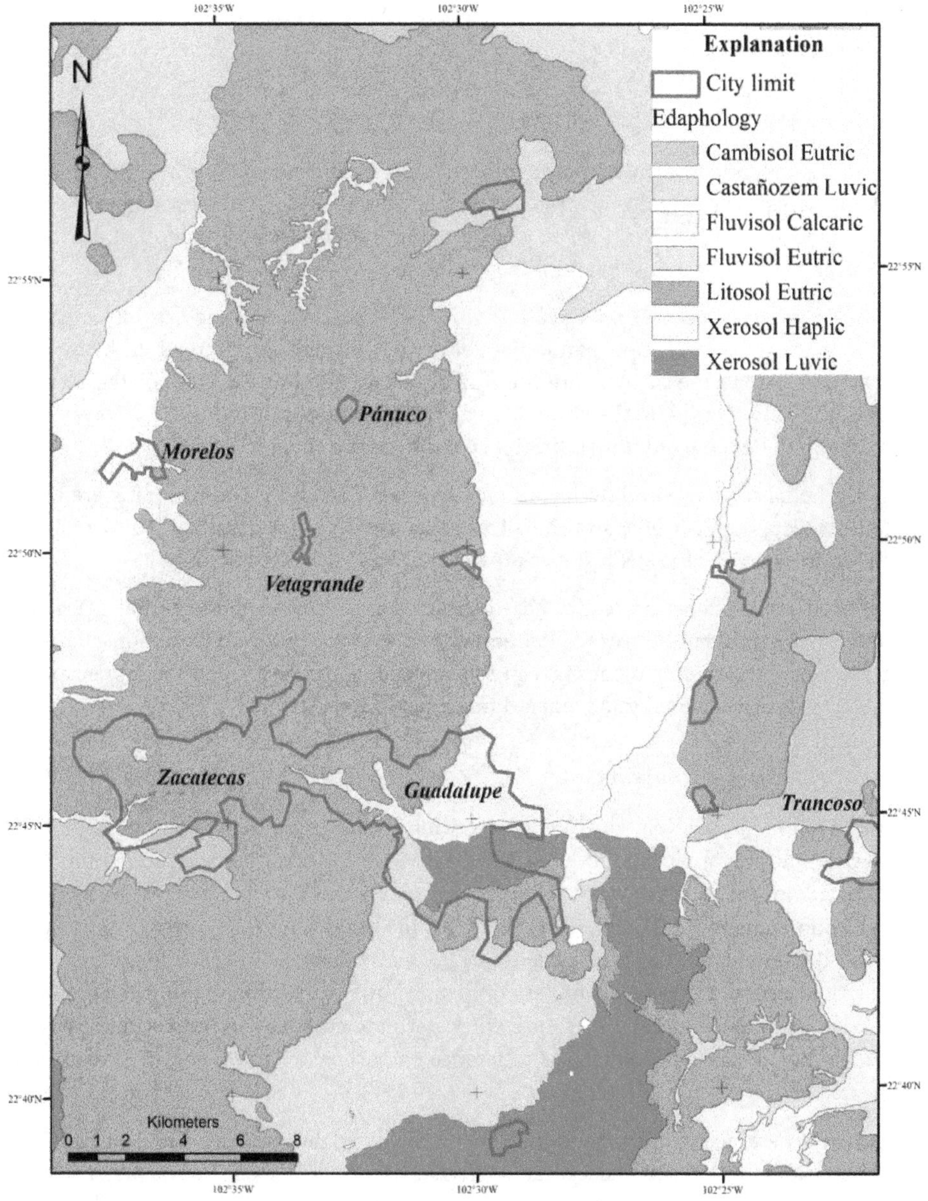

Figure 8. A simplified edaphologic map of the study area. Modified from [21-22].

2. Methodology

This chapter describes the elements used in the geologic and geomorphologic analysis. It starts with the field mapping of the elements where the hazards are occurring. These parameters were the basis for a geomorphologic study and finally all the information was integrated and analyzed in a GIS.

2.1. Field work

The risk areas are associated with slope deposits. The field work was directed to recognize the risk elements in the field and map them at scale 1:10,000. These elements are ancient landslide deposits and fallen rocks. They occur along arroyos, close to the top of the highest hills, and can be traced downslope until they reach the city limit.

The hazardous areas are not only related with the slope deposits; the erosion effects are accelerated where the landscape is modified, either by cutting through the hills or by filling in the arroyos. Along the arroyos are the fluvial terraces and at the edge of the Sierra de Zacatecas, the alluvial fans. In the study area all unconsolidated or moderately consolidated materials tend to be removed if their original conditions are changed.

The stratigraphic sequence and faults are mainly taken from [3, 13]. The rock falls are related with faults, steep slopes (> 20°) and rain. The rocks usually move less than 15 meters away from the source because the steep slopes are normally less than 10 m high.

In places where the cities are expected to grow, the landscape modification starts with vegetation removal, then the surface flattening, and eventually the construction phase. The cartography of the elements mentioned in this subchapter, unless when the landscape was modified artificially, where digitalized and managed in ArcGIS v. 9.3.1.

2.2. Geomorphologic analysis

The geomorphologic analysis was made according to the procedure described by [23]. The method uses a topographic chart scale 1:50,000 which is divided into squares; for this study the length side of each square was 1 km. In each square, four parameters were measured: 1) the dissection density (DD) that is defined as the total length of arroyos per square kilometer; 2) the general dissection density (GDD) which is the sum of the lengths of all the topographic curves per square kilometer; 3) the maximum dissection depth (MDD) which is the elevation difference measured from a creek perpendicular to the nearest highest point; and 4) the relief energy (RE) that is the difference between the highest and the lowest point in each square.

The measured parameters in each square were stored in a database in ArcGIS software. Each value is considered to be in the center of the square [23]. For each parameter, a Kriging interpolation procedure was used to define a raster image showing the spatial distribution of the variable. Since each parameter is in GIS Image format, the values can be managed for classification.

Based on field mapping, the slope deposits originated where DD> 10 km/km^2, GDD> 25 km/km^2, MDD>130 m and RE > 160. The areas defined this way were called "high erosion zones". The function used is a mathematical logical union of variables. If the variables change from 8.5 to 10 km/km^2, 20 to 25 km/km^2, 100 to 130 m and 130 to 160, respectively; they belong with the main body of the slope deposit, so they are "medium erosion zones". Whereas values ranging from 7 to 8.5 km/km^2, 15 to 20 km/km^2, 70 to 100 m, and 100 to 130, respectively, are located at the tip of the slope deposit and they are called "low erosion zones". Lower values are considered as "very low erosion zones".

The erosion areas defined this way do not take the slope into account. Therefore, due to the interpolation method used, "high erosion zones" can be located in a flat area, as well as an area with slope.

2.3. GIS analysis

The digital slope model (DSM) was obtained from the digital terrain model. The slopes were divided, according with field observations, in: 1) 0° to 5° semi-plain, 2) 5.01° to 10° gentle slope, 3) 10.01° to 20° hillside, 4) 20.01° to 30° ramps, and 5) >30.01° scarp or cliff. Since the DSM and the geomorphologic data are in the same coordinates system (NAD27; UTM-13N) the maps can overlay for spatial analysis. We used an intersection function of the selected DSM data with the geomorphologic analysis. This way, the DSM has redefined the hazardous areas by combining the slope ranges with the erosion zones. The intense erosion zones are located only in the scarps; while the medium erosion zones occur in the ramps and scarps. The low erosion zones are located in hillsides and ramps.

Once a result was obtained from the combination mentioned in the above paragraph, the next step was to combine the edaphology and land use information. All these areas are in the same kind of soil: Lithosol Eutric. This soil is less than 15 cm thick, rich in gravel and sand with variable amounts of organic matter. The vegetal coverage consists of bushes, nopals and grazing vegetation. When using the edaphology and land use data, the obtained results did not modify the previous outcome.

The analysis for rock falling was making a buffer of 15 meters length of the "Geology map". The structures were the result of the combination of the buffer on the fault and the slope range and direction.

All the information layers were spatially analyzed in ArcGis software, Ver. 9.3.1, and the results were verified in the field. For proof, we selected places where there were two or more erosion zones. Thus, the changes by erosion promoted by the lithology, sedimentary deposits, geologic structures, vegetation, or soil types could be observed.

3. Results

In this chapter all the mapping and digital analysis is integrated to get the definition of erosion zones and their relationship with hazardous zones. The obtained results were checked in the field to verify that our model is a reliable tool for urban development planning.

The basis for the analysis was the Digital Slope Model (DSM) (**Figure 9**). The cell size used was 20 m since, after testing larger and smaller sizes, that was the best dimension that defines the landscape.

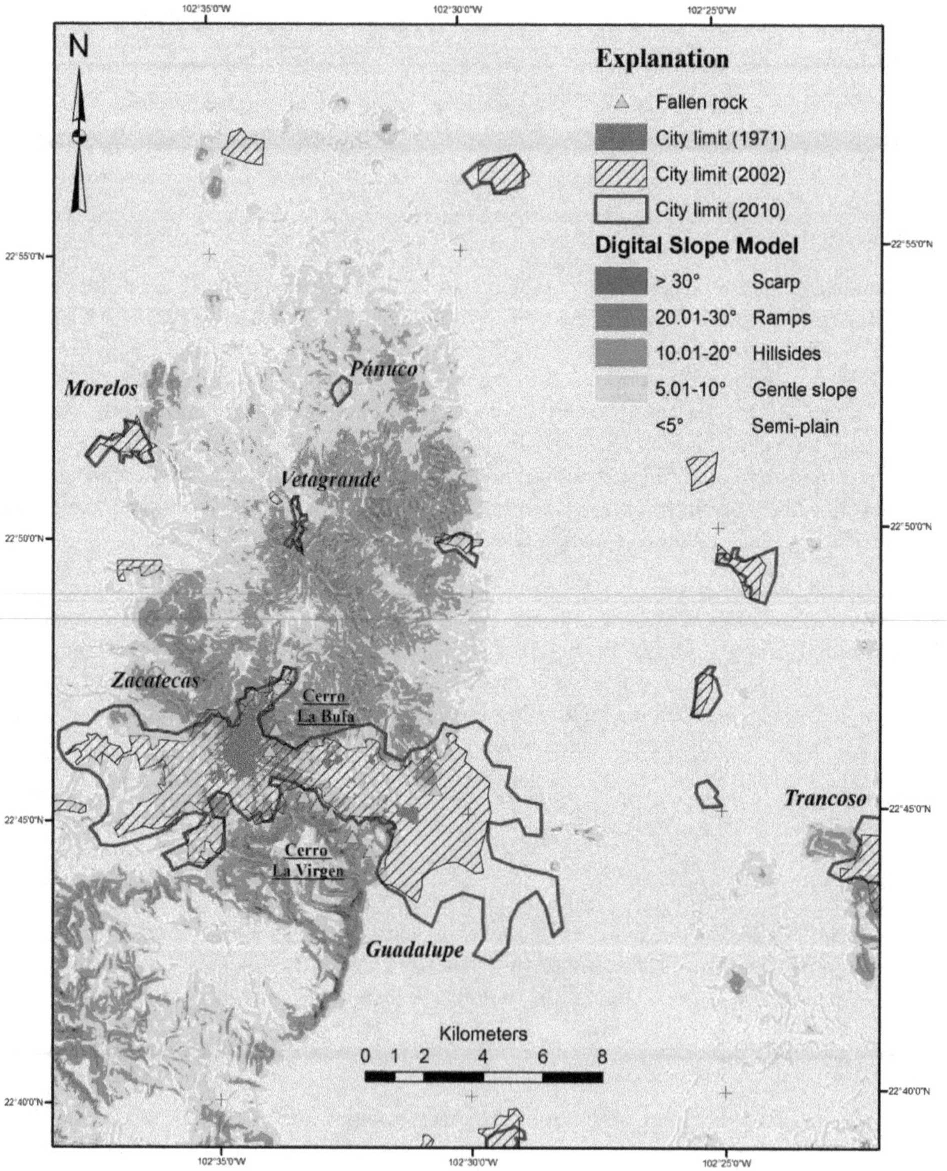

Figure 9. A digital slope model of the study area. The model was built using the contour lines each 10 m. The fallen rocks are those whose diameter is larger than 3.5 m. They are mostly related with ramps and scarps on Cerro La Virgen and Cerro La Bufa.

The **Figure 9** depicts the Zacatecas and Guadalupe cities growing in the gentle slope and semi-plain areas up to 2002; the use of hillsides and ramps was restricted to a few areas in Guadalupe. Due to population growth and continuous pressure on the territory, after 2002, the cities advanced toward the hillside areas. Moreover, due to the pressure on the territory, the arroyos became urbanized. The modification of the original conditions of the territory on hillsides and ramps favors the erosion on unconsolidated to moderately consolidated materials, both natural and artificial.

After 2002 the city grew in the direction of higher slopes, thus, aiding the territory to be affected by erosion. Furthermore, the city limit now is closer to scarps or cliffs where the fallen rocks become a common feature. The **Figure 9** shows the location of rocks whose diameter is over 3.5 m; however, the common ones are those of ~ 1 m that can be found upslope.

3.1. Geologic analysis

Rock falling is a common phenomenon in the study area. It is related with rock type and its fabric. Those rocks that are massive and deformed, if cut by faults, are most likely to be suitable for the development of this process. Additionally, the slope plays an important role in the location of hazardous areas associated with rock falling.

The analysis was made according to the observations made in modern and ancient fall rock. At present time the fallen rocks (~ 50 cm in diameter or less) observed are less than 15 m away from the fault, if there are scarps and ramps. If there is a combination of flowing water after rain with the slope, the displacement could be as far as ~ 100 m away from the source. Whereas the largest distances seen are ~ 1,000 m away from the faults (**Figure 10**).

The spatial analysis was made considering the following parameters:

1. The location of the mapped faults.
2. The lithologic units whose fabric is massive or deformed; they are: Zacatecas Formation, Deformed andesite and Undifferentiated tuff.
3. The slopes: scarp, ramp and hillslope.
4. The travel distance recorded by the fallen rocks.

The analysis is based on the location of the faults and making buffers at 15 and 100 m. The areas defined by the buffers, if they intersect ramps and scarps, are classified as hazardous zones. Due to the steep slopes no rock type was included.

The historical record indicates that a rock can travel as far as 1,000 m. For the faults mapped, this distance is in the hillslope area. During analysis, the first step was to define the buffers at 1,000 m. The next step was to perform a logical operation. If the buffer intersects a hillslope, and the lithology is considered, then the intersection identifies the area which has the possibility to have fallen rock. In **Figure 10** the areas for each distance are irregular polygons with voids inside of them. The voids are areas where the above mentioned conditions are not satisfied.

The results shown in **Figure 10** indicate the areas susceptible to have, or have the potential to be affected by fallen rocks. The field verification of these results indicates that our model defines vulnerable areas for rock falls. However, the results do not indicate the recurrence time or periodicity of the phenomena, It is merely starting where it could happen.

The hazardous areas are mostly located outside of the city limits. Additionally, **Figure 10** can be used for planning urban growth and, if necessary, make the proper preventative arrangements to avoid possible damages to the population and/or infrastructure prior to the landscape modification.

3.2. Geomorphologic analysis

In the Zacatecas and Guadalupe area the geomorphic agents, in relevance order, are: rainwater, gravity, wind and ice[3]. Since water is the main geomorphic agent, considering the semiarid climatic conditions, the effects of the erosion are most evident during the rainy season in the summer. In the winter, the wind and ice can increase their erosive effects in the loose materials.

Figure 11 shows the distribution of each geomorphologic parameters. It can be noticed that during the last 40 years the cities grew on medium to low erosion zones. However, this tendency has recently changed. Nowadays, the growth is close to the limits of Sierra de Zacatecas, getting closer to the "high erosion zones". The Dissection Density is the only parameter that barely has "High erosion" values. This is because the slope deposits start where the creeks do too; so the length of the "high erosion" is short.

The geomorphologic parameters are joined using a logical expression: If two or more "high erosion" areas intersect and the slopes and ramps are scarp, then they define an "Intense erosion zone" (**Figure 12**). The "Medium erosion zones" are defined if two or more "medium erosion" intersects with hillslope. The "Low erosion" areas are defined if two or more "low erosion" parameters intersect the gentle slope and semi-plain.

The erosion zones are defined on the basis of geomorphologic analysis and the DSM (**Figure 12**). The effects on the different lithological units, as well as the land coverage, soil type, and sedimentary deposits are verified in the field. If the natural vegetation and soil are preserved, the erosion processes do not play an important role independently of where they are located. If this is case, the surface creeping is the only geomorphic process. The vegetation and soil removal, along with the modification of arroyos and basin modifications and/or filling with unconsolidated materials, are the starting point for the erosion processes that affects the landscape.

When the original conditions are changed, the zonation that is defined here then applies. The effects are in the unconsolidated to moderately consolidated materials. There is a slow, but continuous, removal of sediments; mainly sand. The erosion occurs mainly during the rainy season in the summer. However, due to the low precipitation (~ 500 mm/yr) the monthly amount of rain could be so low that its effect as erosive agent could be minor. During the rainy season, the clay content in the sediments promotes the hydration and dehydration that, together with scarps and ramps, favors the formation of gullies.

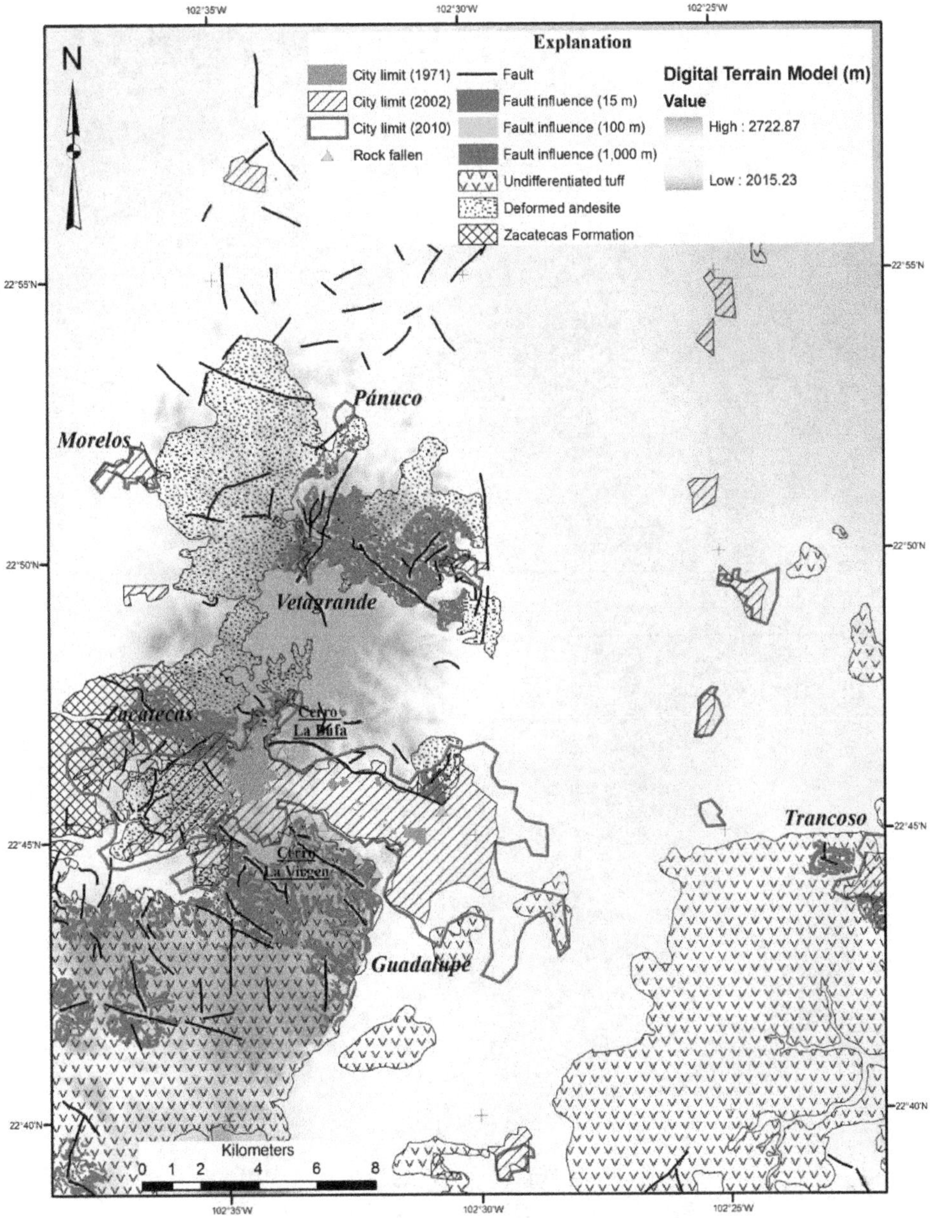

Figure 10. A digital terrain model showing the hazards associated with rock falling in scarps and ramps (red), hillslope close to the faults (green) and hillslope away from faults (magenta). It can be noticed that there are some faults not associated with erosion areas. This is because they are in gentle slopes or semi-plain areas.

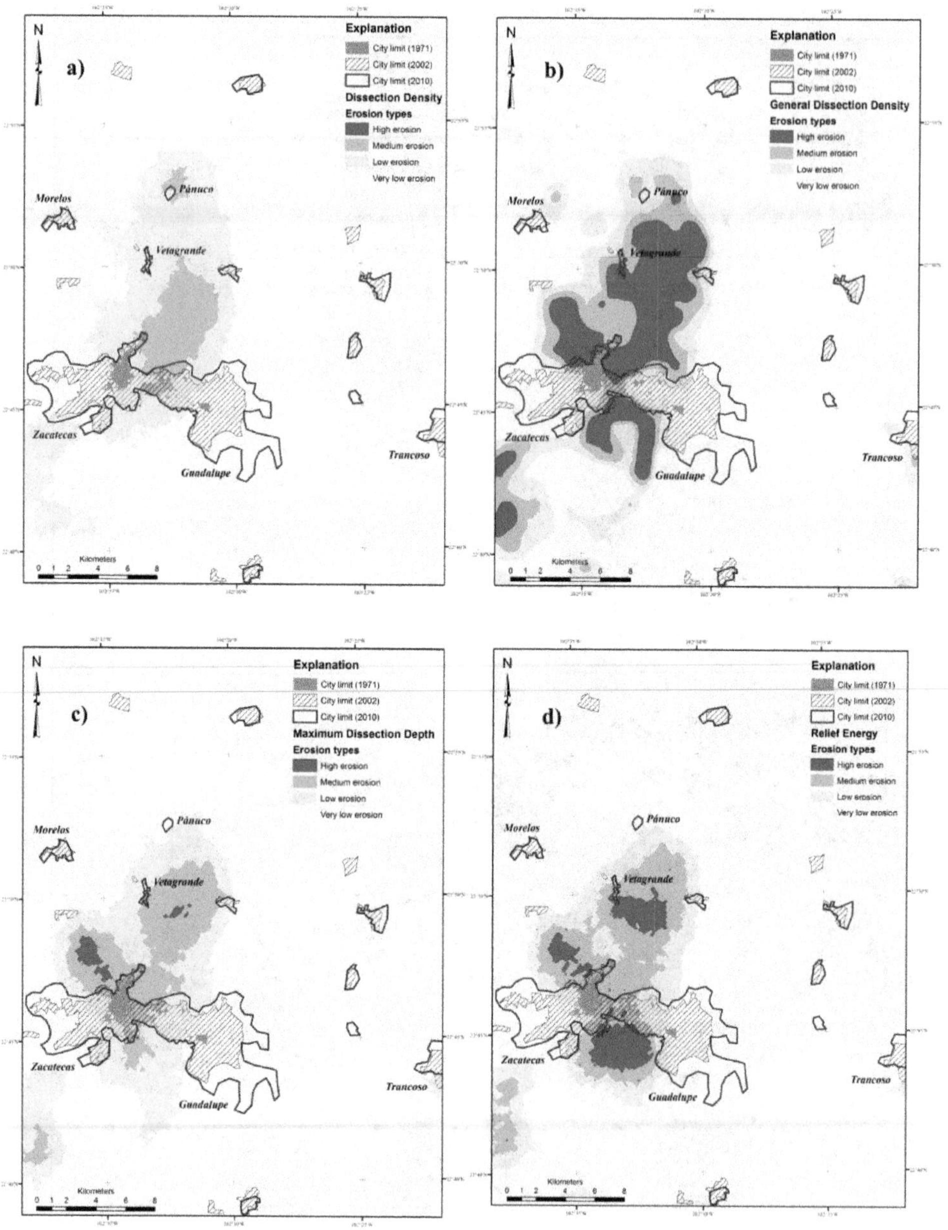

Figure 11. Maps showing the distribution of the geomorphologic parameters: a) Dissection density, b) General Dissection Density; c) Maximum Dissection Depth; d) Relief Energy. The erosion type values were defined according to section 2.2. Up to now, the city has grown in moderate to low erosion zones, although the "High erosion zone" is getting closer.

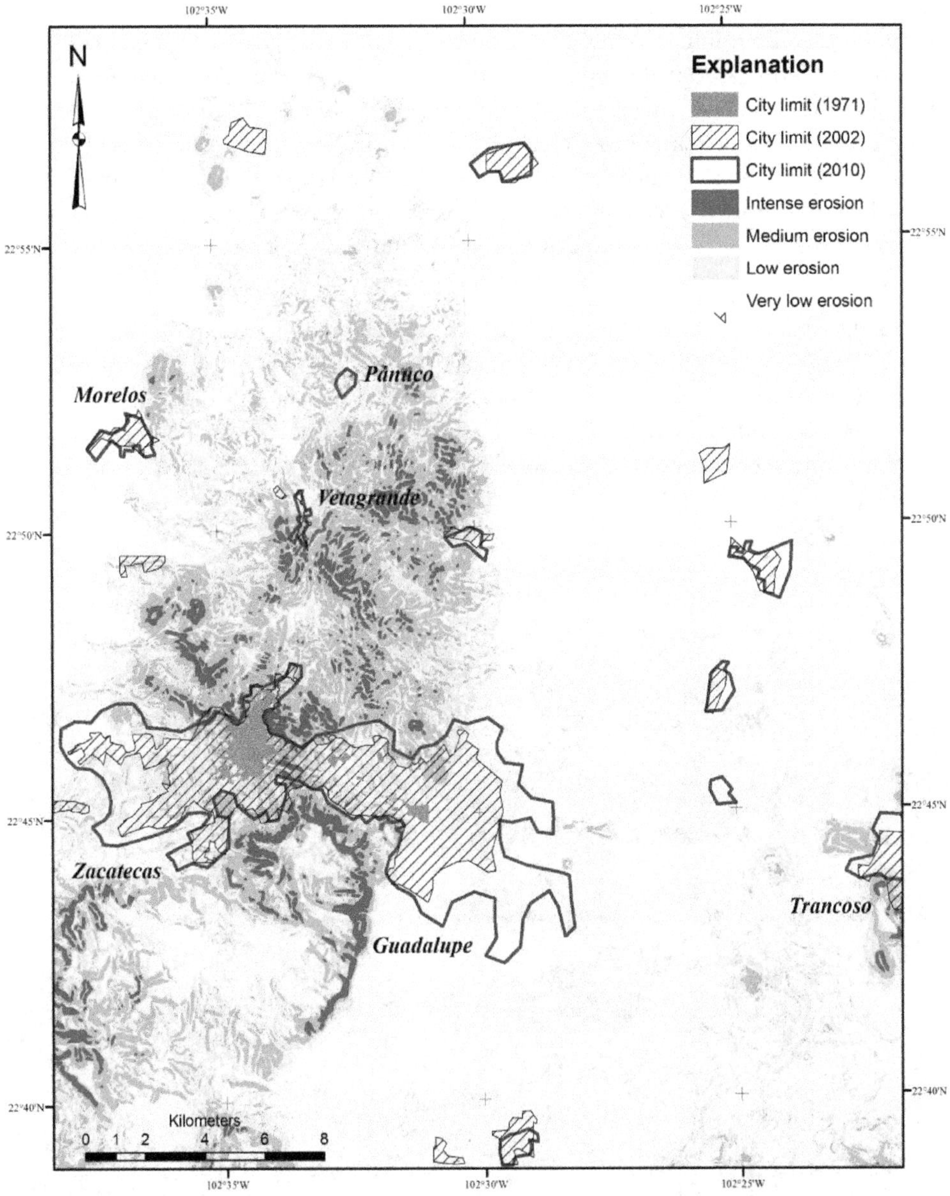

Figure 12. Erosion zones defined for the Zacatecas and Guadalupe quadrangles based on the geomorphologic and the digital slope model analyses. The "Intense erosion" areas indicate that if the natural conditions are changed, either by removing or adding new materials, they can be easily eroded according to the rain intensity. In the "Medium erosion" zones occurs the same event as in the previous one, but in a slower way. The "Low erosion" zones are located in the semi plain and gentle slope areas; their effects depend on the amount of running water after a rain.

The consequences of erosion in the rocks from the Mesozoic sequence and the Tertiary Volcanic rocks (both lithologic units well consolidated) are related with the faulting and fracturing patterns associated with scarps and ramps in intense and medium erosion zones. In the low erosion zones the effects are negligible. Most of the Zacatecas and Guadalupe cities are built on the Zacatecas Conglomerate whose facies composition and structure define differential erosion zones. Clearly, the sand and clay-rich facies can be more easily affected by the erosion processes than those well consolidated. The differential erosion removes the sandy rich strata, leaving unstable the conglomeratic ones or large rock fragments. With time, the unbalanced materials fall down.

The results shown in **Figure 12** were verified in the field and they are presented in **Figure 13**. The erosion removes unconsolidated materials if the original conditions are changed; otherwise the process is very slow. The velocity at which erosion acts depends on the zone they are in. The model here defined should be taken into account for the urban development planning. This model locates areas potentially affected by erosion if the landscape is modified.

Figure 13. Field verification of the proposed model for hazardous zones associated with erosion. a) Moderate erosion zone; in the filling deposit can be seen gullies formed after the rainy season; whereas the conglomerate is affected by differential erosion. b) Intense erosion zone. An excavation was made

for unknown reasons. The effects after the rainy season were that the wall and sidewalk fall. c) A Low erosion zone in a semi plain area; more than one year after the road was built, large gullies developed. d) An intense to moderate erosion zone. The undisturbed area is in the intense erosion where no visible effects are seen; while the moderate one is where the streets are traced.

4. Conclusions

If the landscape is not modified, the only erosion process acting is the surface creeping. This is independent of the location of the erosion zone. In the intense to medium erosion zones the mobility of loose materials is mainly achieved by rainwater when: 1) the landscape is modified, 2) the road cuts have a high angle or high angle slopes, 3) in the moderately consolidated facies of the Zacatecas Conglomerate, and 4) in the sedimentary deposits. The vegetation, land use and edaphology seem not to have any significant outcome in the definition of the erosion zones.

In the well consolidated rocks, the effects of the high erosion zones are associated with faults and fractures. In the historical record, rocks of less than 1 m in diameter can travel as far as 1 km away from the source rock in scarps to hillslopes. The model here presented detect the areas with strong possibilities of having fall rocks fallen.

The low erosion areas only have an effect in loose materials; while in the very low erosion zones the effects are along the arroyos.

5. Further research

After this study several questions were answered, but new ones arise for further research such as:

1. Until now no geophysics method has been used to define the extent and depth of the slope deposits, fluvial terraces, alluvial fans and artificially filled places. A method that could be used is surface waves, in this way the elastic parameters of the unconsolidated sediments could be obtained and may be used in construction regulations. The surface waves could also be used to locate buried slope deposits.
2. The slope deposits are of unknown age; they could be dated by looking for fossils, using U-Th-He, cosmogenic isotopes and/or paleomagnetism. The knowledge of their age and recurrence could be useful in defining the hazards′ recurrence.
3. The geomorphologic parameters used allowed us to define the erosion zones if the topography is abrupt; however they are not designed to evaluate the almost flat surfaces of the valleys. A further work is to look for the parameters that could be used to evaluate erosion in the valleys.
4. In the places damaged by the erosion processes, it is necessary to develop a mitigation plan. This should be made by an interdisciplinary professional team.
5. It is necessary to define the unconsolidated material loss; either from soil, fillings, sedimentary deposits or unconsolidated materials. In this way it could be possible to

make a precise evaluation of the sediment removal and the location of the more likely places where it will occur.

Author details

Felipe de Jesús Escalona-Alcázar*, Bianney Escobedo-Arellano, Brenda Castillo-Félix, Perla García-Sandoval, Luz Leticia Gurrola-Menchaca, Carlos Carrillo-Castillo, Ernesto-Patricio Núñez-Peña, Jorge Bluhm-Gutiérrez and Alicia Esparza-Martínez
Unidad Académica de Ciencias de la Tierra,
Universidad Autónoma de Zacatecas Calzada de la Universidad, Fracc. Progreso, Zacatecas, México

Acknowledgement

This research was partly financed by the project PROMEP No. UAZ-PTC-133 granted to Felipe Escalona. Partial support was also given by the Unidad Académica de Ciencias de la Tierra (UACT) of the Universidad Autónoma de Zacatecas, México. The authors acknowledge José de Jesús Fernández-Ávalos, Director of the UACT, for their support. Comments from Juan Carlos García y Barragán helped to improve this manuscript. The authors acknowledge the Copyediting/Proofreading by Charlotte Presley.

6. References

[1] Código Urbano para el Estado de Zacatecas (1996) Official Journal of the Zacatecas State Government. Ordenance No. 81, Published on September 11th, 1996, 122 pp.
[2] Acuerdo de la Declaratoria de Reservas de Suelos Derivada del Programa de Desarrollo Urbano de la Conurbación Zacatecas-Guadalupe (2004) Official Journal of the Zacatecas State Government. Agreement signed on August 6th, 2004, 273 pp.
[3] Escalona-Alcázar FJ, Suárez-Plascencia C, Pérez-Román AM, Ortiz-Acevedo O., Bañuelos-Álvarez C (2003) La secuencia volcánica Terciaria del Cerro La Virgen y los procesos geomorfológicos que generan riesgo en la zona conurbada Zacatecas-Guadalupe. GEOS 23(1): 2-16
[4] Atlas de Riesgos de la Ciudad de Zacatecas (2007) Ministry of Social Development. 119 pp. (Unpublished)
[5] Enciso-de la Vega S (1994) Crecimiento urbano de la Ciudad de Zacatecas y sus asentamientos humanos en zonas mineralizadas polimetálicas. Revista Mexicana de Ciencias Geológicas 11(1): 106-112.
[6] Escalona-Alcázar FJ (2010) Evaluación preliminar de los riesgos debidos a la geomorfología de la zona urbana Zacatecas-Guadalupe y sus alrededores. GEOS 29(2): 252-256.

* Corresponding Author

[7] Escalona-Alcázar FJ, Delgado-Argote LA, Rivera-Salinas AF (2010) Assessment of land subsidence associated with intense erosion zones in the Zacatecas and Guadalupe quadrangles, Mexico. In: Carreón-Freyre D, Cerca M, Galloway D, editors. Land Subsidence, Associated Hazards and the Role of Natural Resources Development; October 17-22, Queretaro, Mexico. IAHS Publication 339: 210-212

[8] XI General Census of Population (1990) National Institute of Statistics, Geography and Informatics.

[9] XII General Census of Population (2000) National Institute of Statistics, Geography and Informatics.

[10] XIII General Census of Population (2010) National Institute of Statistics, Geography and Informatics.

[11] Zacatecas Topographic Map scale 1:50,000 (F13B58) (1973) General Direction of Studies of the National Territory.

[12] Guadalupe Topographic Map scale 1:50,000 (F13B68) (1973) General Direction of Studies of the National Territory.

[13] Escalona-Alcázar FJ, Delgado-Argote LA, Weber B, Núñez-Peña EP, Valencia VA, Ortiz-Acevedo O (2009) Kinematics and U-Pb dating of detrital zircons from the Sierra de Zacatecas, Mexico. Revista Mexicana de Ciencias Geológicas 26(1): 48-64

[14] Zacatecas Geologic Map scale 1:50,000 (F13B58) (1998) National Institute of Statistics, Geography and Informatics.

[15] Guadalupe Geologic Map scale 1:50,000 (F13B68) (1998) National Institute of Statistics, Geography and Informatics.

[16] Aranda-Gómez JJ, Henry HD, Luhr JF (2000) Evolución tectonomagmática post-paleocénica de la Sierra Madre Occidental y la parte meridional de la provincia tectónica de Cuencas y Sierras. Boletín de la Sociedad Geológica Mexicana LIII: 59-71

[17] Ponce BS, Clark KF (1988) The Zacatecas Mining District: A Tertiary caldera complex associated with precious and base metal mineralization. Economic Geology 83: 1668-1682

[18] Zacatecas Land Use and Vegetation Map scale 1:50,000 (1977) National Commission on National Territory Studies.

[19] Guadalupe Land Use and Vegetation Map scale 1:50,000 (1977) National Commission on National Territory Studies.

[20] FAO/UNESCO Soil map of the World 1:5,000,000 Paris (1974) United Nations Educational, Scientific and Cultural Organization

[21] Zacatecas Edaphologic Map scale 1:50,000 (1971) National Commission on National Territory Studies.

[22] Guadalupe Edaphologic Map scale 1:50,000 (1971) National Commission on National Territory Studies.

[23] Lugo-Hubp J (1988) Elementos de geomorfología aplicada. 1st Edition. Mexico: UNAM. 128 pp.

Chapter 7

The Compost of Olive Mill Pomace: From a Waste to a Resource – Environmental Benefits of Its Application in Olive Oil Groves

Beatriz Gómez-Muñoz, David J. Hatch, Roland Bol and Roberto García-Ruiz

Additional information is available at the end of the chapter

1. Introduction

Olive oil farming is a significant feature of land use in Mediterranean regions, covering over five million hectares in the EU Member States. The main areas of olive oil production are in Spain (2.4 million ha), followed by Italy (1.4 million ha), Greece (1 million ha) and Portugal (0.5 million ha) [1]. Whilst olive plantations are found over most of the Mediterranean region, around 65% of the Spanish olive oil area is located in Andalusia (southern Spain), representing 30% of the total EU olive oil production and about 35% of the utilized agricultural area. Therefore, olive oil farming in Andalusia is of great social, economic and environmental significance and any change in the olive oil industry in terms of management practices and post-processing options could be of wide importance, at least at the regional scale.

The olive oil industry generates large quantities of by-products. Almost all of the olive mills in Spain use the two–phase centrifugation system for oil extraction to reduce wastewater generation and lower the contaminant load, compared with the three–phase centrifugation system [2] which is currently used by other Mediterranean countries. The main by–product of the two–phase extraction system is olive mill pomace (OMP, hereafter), which in Mediterranean areas is produced during a short period over the winter, from November to February [3], the amount generated varying between 7 and 30 million m^3 per year [4-6]. Typically, OMP is a semi–solid to semi–liquid by-product resulting from the mix of "alpechin", the main by–products resulting from the older three–phase extraction procedure, and "orujo". This by-product is made mainly with water, seed and pulp and is a potentially harmful by–product for the environment, because of the phytotoxic and antimicrobial properties, low pH, relatively high salinity and organic load, and the phenolic and lipid constituents [7-11]. Direct application to rivers or soil is not allowed under most of the national regulations of the producer countries. The main physico–chemical characterization of OMP can be found in other reports [10,12,13]. According to these studies, OMP is acidic, with a very high content of organic matter and carbon, rich in potassium (K), poor in phosphorus (P), with intermediate levels of nitrogen (N) and may also contain phenolic and lipid compounds.

Some economic (due to costs associated with disposal) and environmental problems arise from the disposal of OMP. The various options for the fate of the large amounts of OMP which are produced annually in Andalusia can be very diverse: a general description of these can be found in [14]. Briefly, one alternative for the disposal of this large amount of OMP could be in evaporative ponds, but large areas would be required for this option which might also pose several potential environmental problems such as bad odour, leaching and insect proliferation. Another major preferred option would include the generation of renewable energy taking advantage of the relatively high calorific value of OMP. Other important uses include the transformation of OMP into an organic fertiliser and soil conditioner through composting. The process of composting OMP consists of mixing it with a blend of natural organic residues (e.g. olive leaves and twigs collected after cleaning the olive fruit in the mill, and/or straw, or manures), which is then allowed to decompose in aerated piles for 7 to 9 months. This means of re-utilization can help to improve soil fertility in olive oil farms which are characterized by low organic matter, reduce the cost of inorganic fertilisers or, for a commercial enterprise, can provide an additional source of revenue for the olive oil mill economy. The main reasons for composting are that OMP has a semi–solid consistency which makes it difficult to manage, and to eliminate any phytotoxical effects by composting for at least 18 weeks [15]. Composting OMP enables it to be sanitized; the mass and volume of the product are reduced and stabilized prior to land spreading. As already mentioned, before composting, OMP is mixed with bulking agents such as olive tree leaves, which are gathered along with the olive fruit, twigs and small branches, straw [16], cotton waste [17] and manure to increase the nutrient content, or any other materials of animal or plant origin which are available locally.

In Andalusia, composted olive mill pomace (COMP, hereafter) production has increased exponentially during the last seven years from 1000 tonnes in 2003 to 70000 in 2011 [18] and there are about 14 olive mills which are producing COMP in this region. Despite this rapid increase, there are no published studies on the main agrochemical properties and the effects of OMP application to soil. Indeed, there are few studies on the main physico-chemical changes in OMP during composting [19-21], or on the chemical characterisation of the final OMP composted product [22]. Moreover, these studies have been undertaken using only a limited number of OMP composts, which were produced in only low experimental quantities and at a small experimental scale. The main aims of this chapter are to review the information from other studies and our own on the agrochemical characterization of COMP currently produced in Andalusia and on the short- and long-term effects of its application on the soil physico-chemical and biological properties.

2. Agrochemical characterization of composted olive mill pomace

The relatively high diversity of the bulking agents used (such as olive mill pomace, olive leafy material (OLM), manure or straw) as well as the variable proportion in which they are mixed is responsible for the highly heterogeneous nature and variability of the quality of the COMP.

2.1. Composition of commercially produced composted olive mill pomace

As far as we know, there are no studies on the characterization of currently commercially produced COMP. In one of our own study, seven ready to apply OMP composts were collected from different olive mills located in several provinces of Andalusia. COMP samples differed in composition and in the proportions of primary materials such as OMP, OLM, manure and straw, as shown in Table 1 and Picture 1.

Composted olive mill pomace	OMP	OLM	Manure	Straw
COMP1	80%	7%	Sheep M 13%	
COMP2	75%	12%	Poultry M 13%	
COMP3	60%		Sheep M 40%	
COMP4	70%	8%	Sheep M 12%	10%
COMP5	85%	5%		10%
COMP6	80%	20%		
COMP7	80%	13%	Poultry M 7%	

Table 1. Composition of different composted olive mill pomaces described in this chapter. OMP and OLM refer to olive mill pomace and olive leafy materials, respectively.

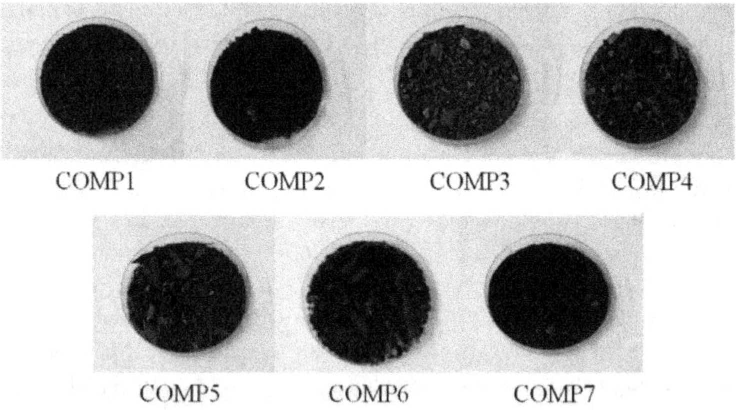

Picture 1. Visual appearance of the composted olive mill pomace used in this study.

COMP pH ranged 7.45 to 8.34 and were adequate for most agricultural purposes. Those COMPs made of manure tended to have a higher pH. Mean COMP pH was 8.03, a value similar to those reported by [20,23] for other experimentally–produced composts made of OMP, and is within the pH range considered as optimal for the activity of microorganisms and plant growth [24,25]. In all cases, electric conductivity was lower than the 10 dS m^{-1}, threshold established as indicator of possible phytotoxic/phyto–inhibitory effects on plants or in soil [26]. COMP has a high content of organic matter (60.5%, on average) and carbon (30.7%, on average) (Table 2). These values are higher than those reported for cow, sheep and poultry manures and similar to those found for horse, pig and rabbit [27]. The highest values were found in those composts which included OLM and straw with little or no manure (COMP4 through to COMP7) and may be due to incomplete organic matter degradation of the larger particle sizes and the higher lignin content of OLM. A low organic matter degradation rate during composting has been reported [20], mainly because the high lignin content of the OMP and the high moisture of the initial mixture which limits sufficient aeration. For these COMPs, application to soil could be a good strategy to increase the organic matter content of soils of olive oil farms which in the Mediterranean basin are usually depleted in organic matter and are exposed to progressive degradation processes. Furthermore, the content of labile carbon in the COMPs are relatively low, indicating that respired carbon derived from compost after application to soil would be expected to be low. Therefore, COMP application to soil could increase C storage due to the high total carbon content and expected low rate of C mineralisation (see below).

Total nitrogen (TN) averaged 1.5%, and was higher for those COMP which included relatively high amounts of manure during composting (Table 2). This values is within the range of that found for compost made from plant residues [28,29] and similar to that reported for compost produced experimentally with OMP plus rabbit or sheep manure, and rice straw or almond shells [22].

The C:N ratio ranged from 27.2 to 35.8 for COMP4, COMP5 and COMP6, which did not include, or contained only limited amounts of manure, while the ratio was much lower (at 10.5 – 18.7) for COMP1, COMP2, COMP3 and COMP7, which included sheep or poultry manures (Table 2).

The total K (TK) in the composts averaged 1.7% and was highest for COMP2, which included a high proportion of poultry manure and lowest for COMP4 and COMP5, both of which included straw (Table 2). These values were within the range of values reported by other authors [20,23] for experimentally–produced compost and also similar to those of OMP [13]. This indicates that K was not lost through leaching during the composting process. The total P of the composts averaged 0.41% and was highest for those amended with manure (COMP1, COMP2 and COMP3) and lowest for those which included only OLM or straw (Table 2). TP values were lower than those of municipal solid wastes and sewage sludge, although similar to other vegetable wastes and manures [30].

Those COMPs with a high percentage of manure had lower lignin contents, but lignin content tended to increase in those COMPs with a higher proportion of OLM and straw. Polyphenol contents of the COMPs were less than 2%, even though OMP is usually characterized by high levels of polyphenols (Table 2). Polyphenols from olive oil mill waste water have been found to be toxic for some soil microorganisms [31], and this is one of the main reasons why direct application to soil cannot be recommended. The low polyphenol contents in the composts agree with values found by other researchers [32], who showed that the polyphenol content decreases during composting. In all cases, the polyphenol contents were lower than 4%: the limit which has been established where there is a shift between net N mineralization and immobilization during decomposition [33]. Those composts which are currently produced in Andalusia and which included manure showed typical lignin and polyphenol contents lower that 20% and 2%, respectively and therefore are suitable as organic fertilisers. Lignin and polyphenol contents of residues have been shown to be robust indices for the prediction of N mineralisation from residue–N after incorporation in soil [34,35], with typical thresholds for immediate net N mineralisation being < 15% lignin and < 3 – 4% total extractable polyphenol contents [36-38].

COMPs phytotoxicities (Zuconni test) were typically higher than 50%, except for COMP3 and COMP5, suggesting that a relatively high percentage of the currently produced COMPs are mature enough to be applied in the field.

	COMP commercially produced (min –mean- max)	**Others authors[1]**
Organic matter (LOI) (g kg^{-1})	272 - 605 – 879	465 - 621
Total C (g kg^{-1})	184 - 307 - 390	301 - 491
Total N (g kg^{-1})	10.7 - 15.0 - 20.0	14.0 - 27.8
C:N	10.5 - 21.9 - 35.8	14.0 - 22.7
Total P (%)	0.19 – 0.41 -1.19	0.5 - 1.5
Total K (%)	1.06 – 1.73 -2.39	20.6 - 39.5
Lignin (g kg^{-1})	76.0 - 218 - 313	410 - 426
Polyphenols (%)	0.94 – 1.33 - 2.1	-
Labile organic C (g kg^{-1})	4.6 - 16.5 - 25.0	7.3 - 10.2

Table 2. Main physico-chemical properties of seven commercially produced composted olive mill pomaces (Table 1) and results from bibliographic review of different authors for composted olive mill pomace. Values are mean of four replicates.

[1]Other authors [19,20,21,22,23] for experimentally produced COMPs.

2.2. Nutrient distribution in different particle size fractions of composted olive mill pomace

The separation and application of different COMP particle sizes could provide for better optimization of COMP management, because a fairly clear relationship between particle–size distribution of an exogenous source of organic matter and the C and N dynamics in soils for sludge compost [39] (among others) has been demonstrated. Similarly, C mineralisation and turnover was seen to differ according to the particle–size fraction in a cattle slurry compost [40,41], also from a sludge–straw mixture [42] and in an aerobically digested sewage sludge composted along with screened green waste compost, stored yard trimmings and crushed wood pallets [39]. Moreover, the nitrogen availability in compost has also been shown to be related to particle size, increasing as particle size decreased in sludge compost [43]. Generally, N mineralisation is greater in the fine and water–soluble fractions than in coarser fractions, at least for manure and city refuse composts [44] and composted sewage sludge [39]. These results suggest that the size of fractions in compost contribute substantially to the total C and N dynamics of mineralisation after compost is incorporated into soil. This is of particular interest, because depending on the range in particle sizes, composts could therefore provide a means for storage of C in soil (i.e. from the larger sized compost fractions), or a source of available N (i.e. from the finer compost fractions). The diversity and variable amount of raw materials used to obtain COMP make the final products very heterogeneous and therefore, it is expected that the nutrients and the main physico-chemical properties also differ in the particle size fractions of COMP. However, no studies existed to corroborate this, therefore in a separate study we examined whether or not this expectation about COMP was correct.

To estimate the nutrient distribution in particle size fraction of COMP, seven commercially produced COMPs were studied (Table 1) by sieving successively through to obtain 3 different sized fractions: <0.212, 0.212 – 1.0 and 1.0 – 60 mm. About 52% of COMP particles were larger than 1mm for those COMPs with a high content of manure, whereas, in contrast for COMPs without manure and with straw the consisted for 80% of these larger particle sizes. . The percentage of fine particles (<0.212 mm) doubled for COMPs with manure. These COMPs showed similar percentages of particle size fractions reported for duck manure [45], suggesting that during composting, manures generate a relatively high content of finer fractions. In general, the content of different compounds rich in organic C, such as organic matter, total carbon and lignin increased with particle size, being significantly higher for fractions between 1 to 6 mm. In sewage sludge compost made with screened refuse, yard trimmings and pallets [39], and for two types of dairy slurries others found [41], as in this study, an increase in the total C content with larger particles, although these differences were less distinct.

In contrast, nutrients contents including total N, mineral N, total P or total K were significantly higher in the fine fractions (<0.212 mm). More than 40% of the total N, P and K found in the original COMPs were in the <1 mm particle fractions in COMPs which included sheep or poultry manure during composting, whereas for COMP5, 6 and 7, the contribution of <1 mm fractions was lower than 30%. This finding agrees with those for manure and city refuse composts [44], animal slurry [41,45-47] and sludge compost [39], from workers who observed that most of the total N and various forms of N available in the short-term were mostly in the finer and water–soluble fractions.

As a result of lower N, but higher C, the C:N ratio decreased as particle size increased (Figure 1). Similar relationships between C:N ratio and particle size fractions were obtained for cattle slurry [46,47] and for sludge compost [39]. Typically, the C:N ratio for particles finer than 1 mm was lower than 18.2 in all COMPs, except those which included straw (COMP4 and COMP5), suggesting higher N availability of the organic N from these particles. During decomposition, manures with C:N ratios below 15 are likely to result in positive (net) N mineralisation after application to soil [48].

Overall, the germination index was higher than 80% in the <1 mm fractions, whereas it was lower than 50% (e.g. with some degree of phytotoxicity) for fractions 1.0 – 6.0 mm, although this was not true for COMPs made without manure (COMP5 and 6 mixtures) (Figure 1). Thus, the fractionation of the compost produced in olive mills containing <1 mm phytotoxicity–free fractions is highly recommended for commercial purposes.

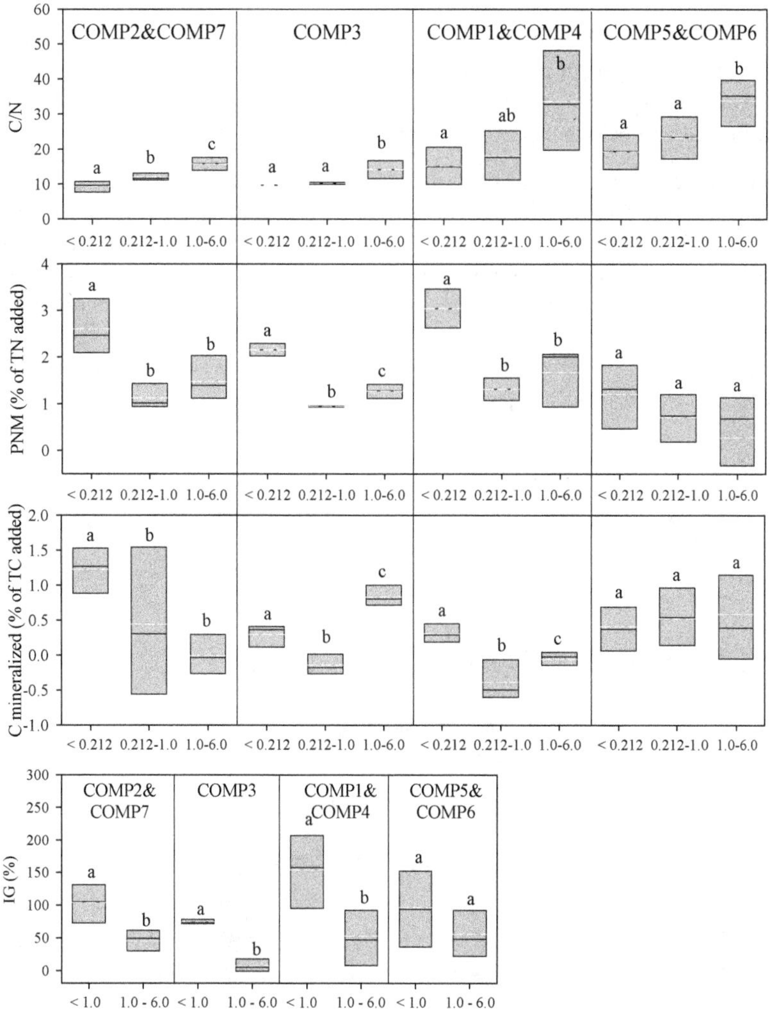

Figure 1. Box–plot of C:N, potential mineralisable N (PNM), mineralisable C and germination index of particle size fractions of combinations of COMPs according to the Principal Component Analysis. Different letters stand for significant differences (P <0.05) among particle size fractions.

The main differences among seven COMP tested according their quality was shown in the Principal Component Analysis (Figure 2). First principal component (PC1), which was negatively correlated with COMP quality indicators (e.g. total and available nutrients, C:N ratio,...), categorized the COMPs according to the pool of nutrients and organic matter and C, showing the high influence of raw materials in the quality of individual COMPs. Those COMPs made with poultry manure resulted in a high quality product, followed by COMPs made with sheep manure, whereas the quality of those COMPs made with a high proportion of OLM and straw but no manure was low. In addition, there was a trend for the quality to increase in relation to the content of finer fractions.

This highlighted the large influence of the raw materials in the final composition and quality of COMP product; a high proportion of manure increased the abundance of finer particles with high nutrients contents, whereas the addition of large amounts of OLM caused an increase in the content of larger particles with a high content of C compounds.

Overall, the separation and application of different COMP particle sizes could be useful for better optimization of COMP management, considering that the smallest particles (<1 mm) have a much higher germination index value, total N, P and K and a higher potential to mineralise N, providing a more homogeneous, high quality compost. On the other hand, larger particles (>1 mm) had higher contents of organic matter and C and tended to have the lowest C mineralisation and thus, on application to the soil in olive groves as a soil conditioner, could improve the structure and increase the organic matter of the poorer soils and increase the storage of soil organic C.

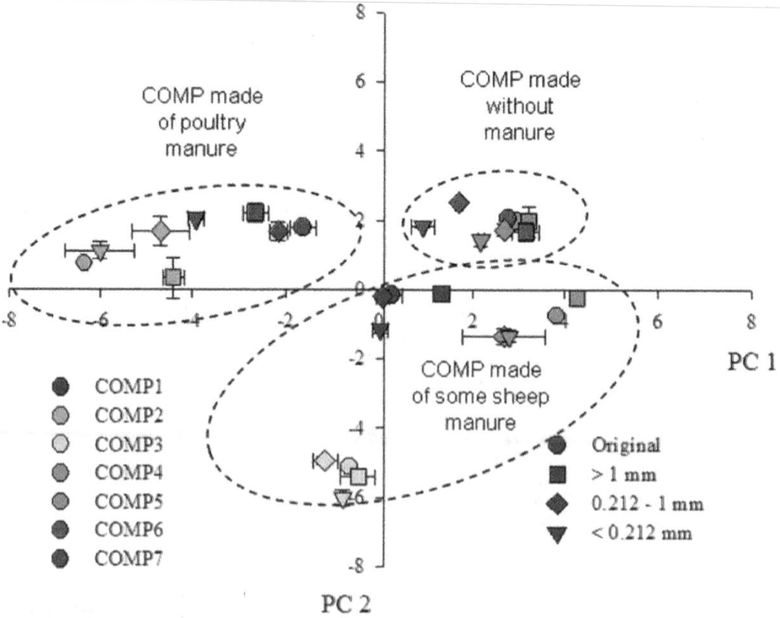

Figure 2. Scores of the whole and the particle size fractions of the COMPs in the space defined by PC1 and PC2.

3. Effects of composted olive mill pomace on nitrogen availability

Although it is expected that the application of COMP-organic N would also supply some available N for the plant, there is little information on the decomposition rate of COMP and the impact this may have on the N available for the growth and productivity of olive trees.

Previous work [49] showed a relatively low net N mineralisation in soil after application of compost obtained from olive mill wastewater taken from an evaporative pond and other agricultural by–products, suggesting that during decomposition of composted OMP, N immobilization could take place.

More detailed knowledge of the main soil N transformations and dynamics, following COMP application to soil, could be the key to regulating soil fertility through an improved capability to predict crop available N and organic matter sequestration and losses.

3.1. Nitrogen supply after the application of composted olive mill pomace

Available nitrogen supply throughout net N mineralisation (NM) and nitrification was studied for commercially produced COMPs differing in the raw materials co-composted with OMP (Table 1) in an aerobic incubation experiment under controlled conditions over two years. The highest rates of NM for the control soil and COMP samples were found during the first week of incubation. This positive net N mineralisation was probably from the N and C contained in the labile components of the COMP, which can be used directly by microbes as an energy source. After the initial stage of mineral N release, net N immobilization during COMP decomposition was found. Probably because the microorganisms began to decomposed the more recalcitrant compounds (with a higher C:N ratio) and this placed a demand on available soil N resulting in net N immobilization, which can ultimately be re–mineralised over time. Similar rates of mineralisation were found with mature composted materials [50,51]. This result was unexpected since, according to the TN contents, C:N ratios and lignin contents of the different COMPs, net N mineralisation would be expected according to the view that the addition of organic matter generally causes N mineralisation [52-54]. Raw materials co-composted with OMP had effects on net N mineralization. During the first year of incubation, N immobilization was found in all COMPs, except for COMPs made chiefly from OMP and OLM. After two years, decomposition of COMP resulted in a positive net N mineralization (i.e. net mineral N supply) for all COMPs (Figure 3), with values as high as 95% of that added as COMP-N. These results demonstrate that the organic N contained in the COMPs is not readily available over short to medium timescales, and suggests that most of the organic N may be in relatively recalcitrant forms. Indeed, during the composting process, N is immobilised in condensed aromatic compounds which might contribute to the reduced availability of N [55]. Therefore, the application of COMP may result in short–term declines in N availability in agricultural soils: it is therefore recommended that during the first year, the application of COMP should be combined with some other N-rich source of fertilizer. On the other hand, autumn COMP application could be a useful approach to lowering soil inorganic N levels, thereby reducing potential N losses by leaching.

3.2. Application of composted olive mill pomace to soil reduce nitrogen losses by leaching

Plant nitrogen use efficiency is usually lower than 50%, though highly dependent on crop type and environmental conditions. There are different ways by which the N is lost and no longer available for crop uptake, such as denitrification, ammonia volatilization and leaching, which might also be associated with environmental problems. N leaching can contaminate groundwater or surface waters through runoff. Indeed, although N contamination of groundwater arises from several sources, such as industrial waste, municipal landfills, mining, or septic systems, agricultural practices remain a major source [56-58]. About 15 – 55% of N applied to crops can be lost by leaching every year [59,60]. This is particularly important in olive oil groves as many olive crops are on areas vulnerable to nitrate pollution.

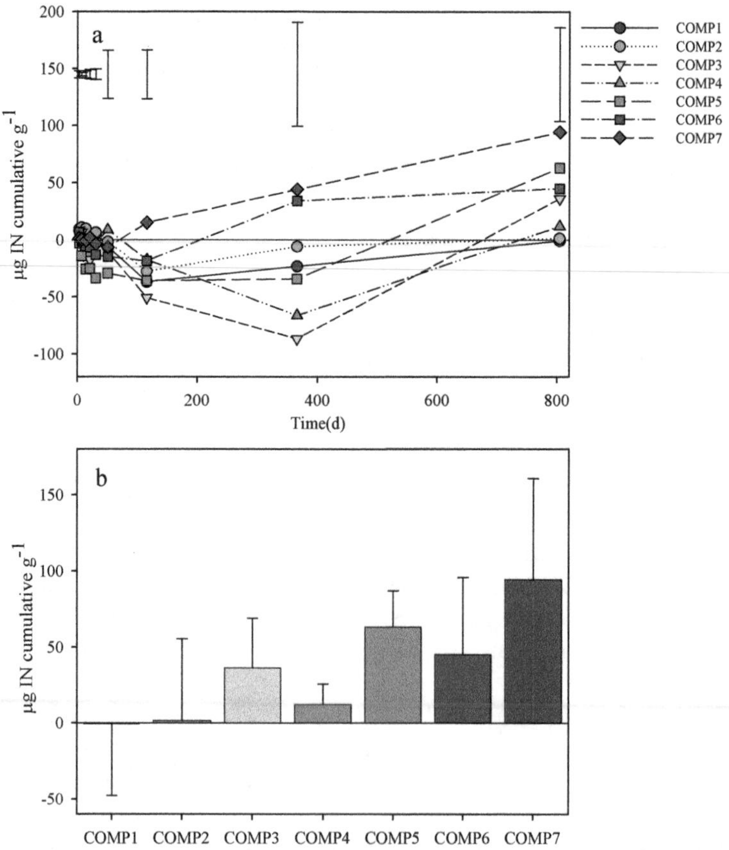

Figure 3. (a) Cumulative amounts of COMP–amended inorganic N (IN) for each sampling period and (b) at the end of incubation (365 d) for COMP amended soils. Bars for each sampling represent the mean of the standard deviations of the whole set of COMP–samples.

Taking into account that N contained in COMP is not readily available in the short-term, N lost by leaching is expected to be low after application of COMP to soils. Indeed, short-term (51 days) leaching losses of N after application of COMP were similar or even lower (between -5 to 10% of the added N) than the unamended control soils in an experiment carried out under laboratory conditions. These values were much lower than those found for commercial organic fertilizer (up to 30% of the added) and urea or sodium nitrate (up to 80% of the added) for the same amount of total N added (Figure 4).

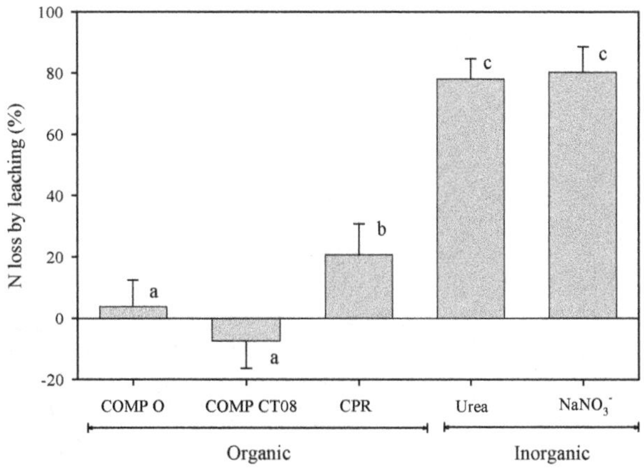

Figure 4. Percentage of N lost as nitrate by leaching of the total N added (equivalent to 100 μg N g^{-1}) after 51 d and three precipitation events (total precipitation equivalent to 80 l m^{-2}). Values are the mean of 4 replicates. Bar denotes standard deviation. Different letters denote significant differences between treatments ($P<0.05$).

Similar results were found in an experiment under outdoor conditions after one year. Nitrate leaching in soils amended with COMP was compared with soils amended with sheep manure (sheep M), commercial organic fertilizer (CPR), or inorganic fertiliser. Fertilisers were applied at two different rates (equivalent to 100 and 200 μg N g^{-1}) combined with two modes of application (on the soil surface or mixed with soil) and simulating autumn conditions. The lowest losses were for COMP amended soils (up to 7% of that applied) and the method of application had significant effects on mineral N leaching (Figure 5). In general, those soils which received COMP on the soil surface averaged negative mineral N losses (i.e. lower or similar losses to the control soil). N application rates had no effect on COMP IN leaching, regardless of the way the COMP was applied. Overall, inorganic N leaching for those treatments which received either Sheep M or CPR did not differ significantly, although leaching after surface application of CPR was higher than sheep M. Up to 37% of the fertiliser–derived N was leached for the 'double' surface application of CPR. No effects of rate or methods of application on IN leaching were found for CPR. However, for sheep M, leaching was higher at the double rate and lower for the 'surface', compared with the 'mixed in' application. The highest IN leaching reached 58% of the added N for inorganic fertiliser application (Figure 5). Results from the laboratory and outdoor experiment clearly showed the recalcitrant nature of the COMP-N (i.e. a high degree of N retention) despite the relatively low C:N ratio, confirming previous results on COMP-derived negative net N mineralisation and nitrification. Thus, application of COMP on olive oil farms distributed in the nitrate vulnerable areas might be a suitable strategy to diminish nitrate contaminated groundwater. This strategy agrees with [61] who reported that composted material can considerably lower the risk of groundwater contamination. Slow–release, controlled–release and stabilized fertilisers, such as composted OMP are recommended for several agro–environmental situations to improve nutrient use efficiency by reducing nutrient losses from the soil. Results from our studies clearly demonstrated the high efficiency of composted OMP to retain N and the beneficial reduction of nitrate leaching elicited by this compost.

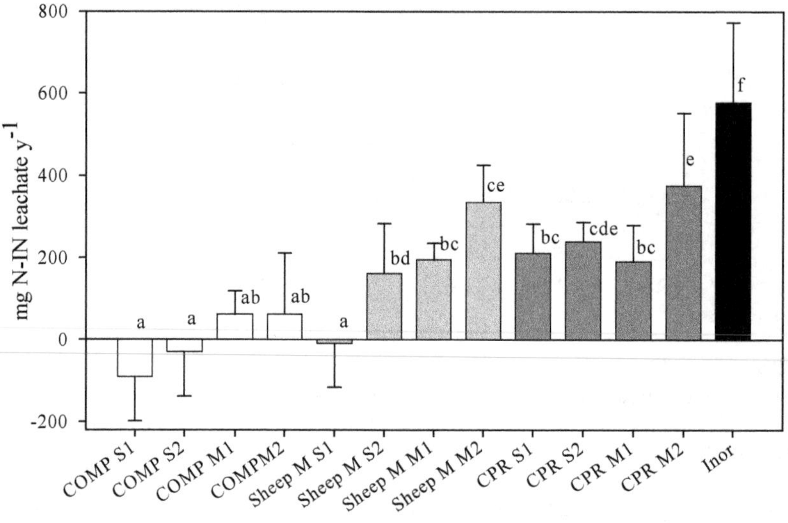

Figure 5. Cumulative fertiliser–derived mineral N (nitrate + ammonium) leaching after one year under natural rainfall and temperature in outdoor conditions for soils which received: i) composted olive mill pomace (COMP), ii) sheep manure (Sheep M), iii) commercial organic fertilisers (CPR), or iv) $NaNO_3$ (Inor) at 1 (250 μg N g^{-1}) or 2 (500 μg N g^{-1}) doses. M and S, stand for soil in which the fertilisers were mixed (M) with the soil or applied to the soil surface (S). Values are means of 4 replicates and bars denote standard deviations. Different letters denote significant differences (*P<0.05*).

3.3. N_2O emission derived of application of composted olive mill pomace

Agricultural soils are a significant source of atmospheric N_2O which is of concern because of the role this gas has in global warming [62]. Emissions of N_2O from soil have been shown to increase after the addition of plant residues [63] and organic fertilisers [64], the biochemical composition (or quality) being an important determinant of the magnitude of N_2O emissions [65].

N_2O emissions in soil amended with COMP (100 μg N g^{-1}) were determined under aerobic incubation over 51 days. Soil N_2O–N fluxes were constant and relatively low (2.71 ng N_2O–N g^{-1} d^{-1}), and did not vary significantly between COMP-amended and the control soils, except for the COMP which was made with moderate levels of sheep manure (COMP1) (Figure 6). The cumulative N-N_2O emissions after 51-days were only 0.23% of the added COMP-N. This result was not in agreement with the well documented increase in N_2O emissions after plant residue addition to soils as reported in other studies [63,65]. The very low increase in N_2O emissions after COMP application suggests that N losses via denitrification after the annual application of COMP to olive oil groves are expected to be very low, and thus application of COMP could provide organic N without increasing the emissions of N_2O

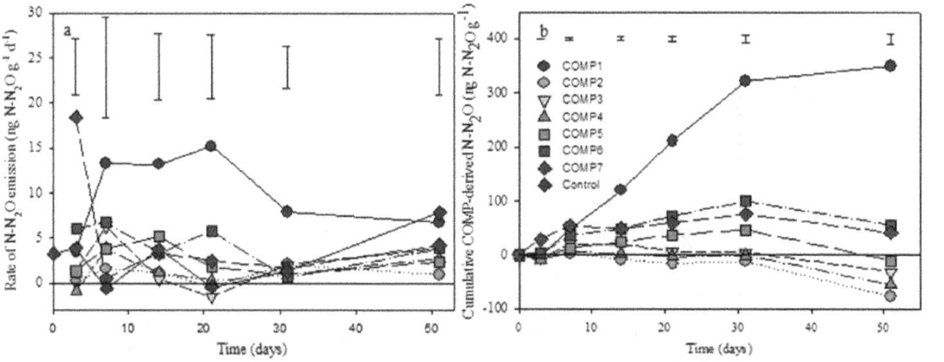

Figure 6. Time course of (a) N-N_2O emission rate and (b) cumulative N-N_2O emission in soil amended with COMP during 51 days of lab incubation under aerobic conditions. Bars denote the mean standard deviations for each treatment and sampling point.

4. COMP-carbon mineralization

As shown above, composted olive mill pomace (COMP) has a high content of total carbon, most of which of recalcitrant nature (e.g. expected low C decomposition) such as lignin or fibre. Therefore, COMP application in soil can be a good strategy to increase the organic matter content in soil, as well as the organic C. This practice could reduce CO_2 emission to the atmosphere increasing C storage in soil.

4.1. Carbon mineralization

Measurements of soil CO_2 emissions can provide useful insights into soil C cycling, and provide a basis for evaluating soil C dynamics and potential C sequestration in different agricultural systems [66], particularly in intensive production systems affected by the different cropping practices and residue management [67]. Soil C can only be assimilated, and recycled through the microbial biomass, or respired [68]. Soil respiration and mineralisation are generally thought to be related to the composition of the microbial biomass, which in soils, tends to vary both across substrate qualities [69] and in time [70]. Low C mineralisation measured as CO_2–C fluxes after mature COMP application have been shown [71]. However, the COMP used in this study was an experimentally produced COMP. The amount of COMP-C mineralized after 8 months of incubation after COMP application in eight soils differing in soil organic carbon (SOC) is shown in Figure 7. For all cases, the COMP-C mineralized (e.g. emitted as C-CO_2) was less than 10% of that added (4 ◎g COMP-C g^{-1}), and this was true, independently of soil texture or SOC contents. This result suggests that C compounds in COMP are complex and refractory, resulting in a high residence time in the soil. Moreover, the percentage of COMP-C mineralisation tended to be lower at higher rates of COMP application, which indicates a higher potential for COMP-C to accumulate in the soils at higher rates of application (Figure 8).

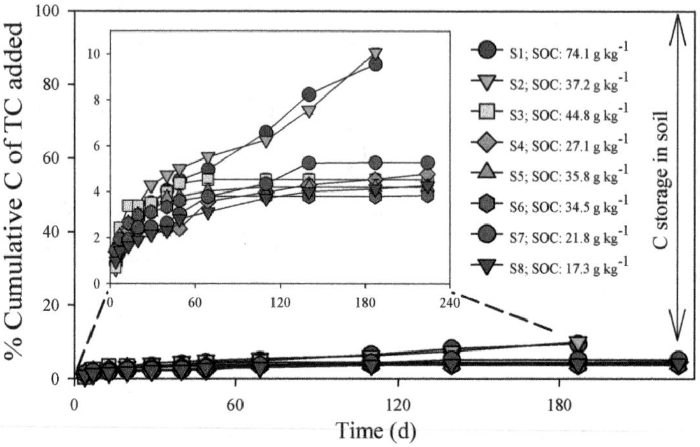

Figure 7. Percentage of C emitted as CO_2 during COMP decomposition for eight types of soil after 240 days of incubation. Bars denote standards deviations.

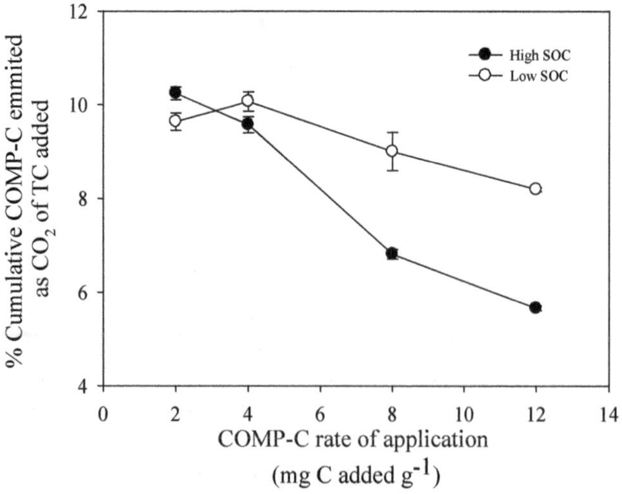

Figure 8. Effects of increasing the rate of COMP-C application to two soils differing in soil organic carbon (SOC) on cumulative COMP-C derived C-CO_2 emissions after 8 months of incubation. Bars denote standards deviations of three replicates.

5. Long-term effects of composted olive mill pomace application in olive groves

The majority of the olive grove soils of Andalusia are characterised by low levels of organic matter, and are exposed to progressive degradation processes. Thus, organic matter application is required to compensate for organic carbon deficiency and to improve soil fertility. As a consequence, the use of a transformed agro-industrial waste for agricultural use, such as composted olive mill pomace, might represent a realistic solution to overcome both the soil degradation of olive oil groves and a sustainable disposal of OMP, thus increasing the sustainability of olive groves. Regional authorities, have recognised the potential use of COMP in agriculture, and have promoted composting which has resulted in an exponential increase in the production of COMP during the last 5 years, with a production of about 70000 tonnes in the 2009-2010 harvest campaign [18]. However, the long-term effects of annual application of COMP on the soil fertility of olive oil groves should be evaluated to promote more olive mill pomace composting.

Soil samples were taken from four olive oil farms; Olvera (O), Reja (R), Tobazo (T) and Andújar (A), which annually received applications of COMP during 3, 4, 9 and 16 years, respectively (COMP olive oil farms, hereafter). Soil samples were also taken from comparable olive oil farms located in the vicinity (<20 m) of each of the COMP farms that never received composted olive mill pomace (NCOMP, hereafter). These NCOMP farms have similar environmental conditions, topography and soil texture to the COMP farms. Soil samples were characterized for physico-chemical and biological variables to determine the fertility and functionality of soil after COMP application.

In general, soil pH in the 4, 9 and 16 years COMP sites was 0.31, 0.23 and 1.47 units lower than the comparable NCOMP olive farming, respectively, whereas no differences were achieved in the site after 3 years of COMP application. Moreover, the soil water holding capacity (WHC) was significantly higher in the COMP treated soils, with increases of 2 – 4% with respect to NCOMP soils. The higher (WHC) in the soils under COMP farming was expected (Table 3), due to the relatively higher levels of organic matter in these soils: the effect of organic matter on the increased potential for soils to retain water was described previously for 77 soil profiles [72]. Moreover, soil water aggregate stability was also significantly higher after COMP application for all treated soils: up to 1.5 - 2.5 times greater for COMP compared with NCOMP treated soils (Table 3). The higher soil aggregate stability in the COMP soils agrees with the findings of [73], who showed that the application OMP to soils increased this soil property from 64 to 73% after 5 years. The cation exchange capacity of soils treated with COMP was also significantly higher than in the relative NCOMP soils, independently of the site (e.g. number of years since COMP was first applied) (Table 3).

Overall, COMP application significantly increased both organic matter and carbon contents, compared with the NCOMP soil, except at the site that received COMP for only three years. The soil organic matter content in soils treated for 4, 9 and 16 years with COMP (i.e. R, T and A sites), was from 2.1 to 8.5 times greater than the respective NCOMP treated soil (Table 3). Our data confirmed previous results which indicated that soil organic matter increased after amendment with OMP during the first weeks after application [74] or after several years [73,75]. Our results also showed that organic carbon was higher in the COMP treatment compared with the NCOMP treatment. The increase in organic carbon after COMP application indicates that the decomposition rates of COMP–C were relatively low, and lower than the rates of annual COMP–C application. These recalcitrant compounds might contribute to the reduced rate of C decomposition. Soil organic carbon was higher in the soils treated with COMP, being up to eight times higher in soil where COMP was applied during the last 16 years in comparison with NCOMP farming. When extrapolating to a hectare scale, these results show that soil organic C increased up to 30 tonnes per ha^{-1} y^{-1} in soil after COMP application. The low decomposition rate of COMP and high organic C content in COMP-amended soil indicates that application of COMP in olive oil groves is an appropriate strategy to sequester organic carbon into the soil and should be evaluated further.

Overall, the total N content was higher in the COMP than in the NCOMP soils although this depended on the site. At O and R sites, which received three and four years of COMP application respectively, differences with the comparable NCOMP soils were not significant. However, total N was 1.3 and 14.8 times higher in soil with 9 and 16 years of COMP application (sites T and A) than in the comparable NCOMP treated soils (Table 3). The higher TN in COMP soils was expected, since application of organic residues normally results in an increase in the soil content of N [49,73,76]. They showed a significant increase in the TN after two to three years of application of OMP or COMP, indicating that COMP–N is very resistant to mineralisation, and therefore is retained in the soil.

Generally, soil labile phosphorus (P) in the amended COMP soils was higher than the unamended soil. In the O, R and A sites, soil available P at the COMP soils was 13.9, 260.3 and 1607 % higher than the NCOMP soil, whereas no significant difference was found at T site (Table 3). At all sites, soil exchangeable potassium was significantly higher in the COMP treated soils than in the unamended soils.

Site		WHC (%)	SA (%)	CEC (meq100 g^{-1})	LOI (%)	TC (%)	TN (%)	Available P (µg P g^{-1})
O	COMP	26.3±0.81^{a}	51.8±2.0^{a}	22.2±1.22^{a}	3.95±0.95^{a}	2.29±0.39^{a}	0.25±0.04^{a}	10.6±0.04^{a}
	NCOMP	22.3±0.3^{b}	34.0±1.6^{b}	20.8±0.21^{b}	3.45±0.85^{a}	2.00±0.49^{a}	0.27±0.03^{a}	9.3±0.03^{a}
R	COMP	21.8±1.3^{a}	57.9±1.8^{a}	25.3±3.83^{a}	8.34±3.67^{a}	4.84±1.90^{a}	0.29±0.03^{a}	30.3±0.03^{a}
	NCOMP	20.5±1.7^{b}	37.1±0.9^{b}	18.6±0.27^{b}	3.96±1.09^{b}	2.30±0.57^{b}	0.23±0.02^{a}	11.5±0.02^{b}
T	COMP	23.6±0.5^{a}	56.7±0.2^{a}	21.1±3.46^{a}	6.31±2.22^{a}	3.66±1.15^{a}	0.25±0.08^{a}	6.9±0.01^{a}
	NCOMP	20.5±0.8^{b}	22.6±0.6^{b}	15.4±2.06^{b}	2.39±0.62^{b}	1.39±0.32^{b}	0.10±0.02^{b}	7.6±0.02^{a}
A	COMP	30.2±0.5^{a}	23.9±1.0^{a}	23.3±6.9^{a}	16.1±3.49^{a}	8.49±2.11^{a}	0.74±0.33^{a}	57.4±0.33^{a}
	NCOMP	28.7±1.3^{a}	23.7±1.5^{a}	10.6±1.2^{b}	1.88±0.41^{b}	1.09±0.21^{b}	0.05±0.02^{b}	3.57±0.02^{b}

Table 3. Water holding capacity (WHC), soil stables aggregates (SA), cation exchangeable capacity (CEC), organic matter (LOI), total C (TC), total N (TN) and available P of olive oil farming which received (COMP) or not (NCOMP) composted olive mill pomace at Olvera (O), Reja (R), Tobazo (T) and Andújar (A). Data are the mean of five replicates ± standard deviation. Different superscript letters for each site denote significant differences between COMP and NCOMP farming (one way ANOVA; $P< 0.05$).

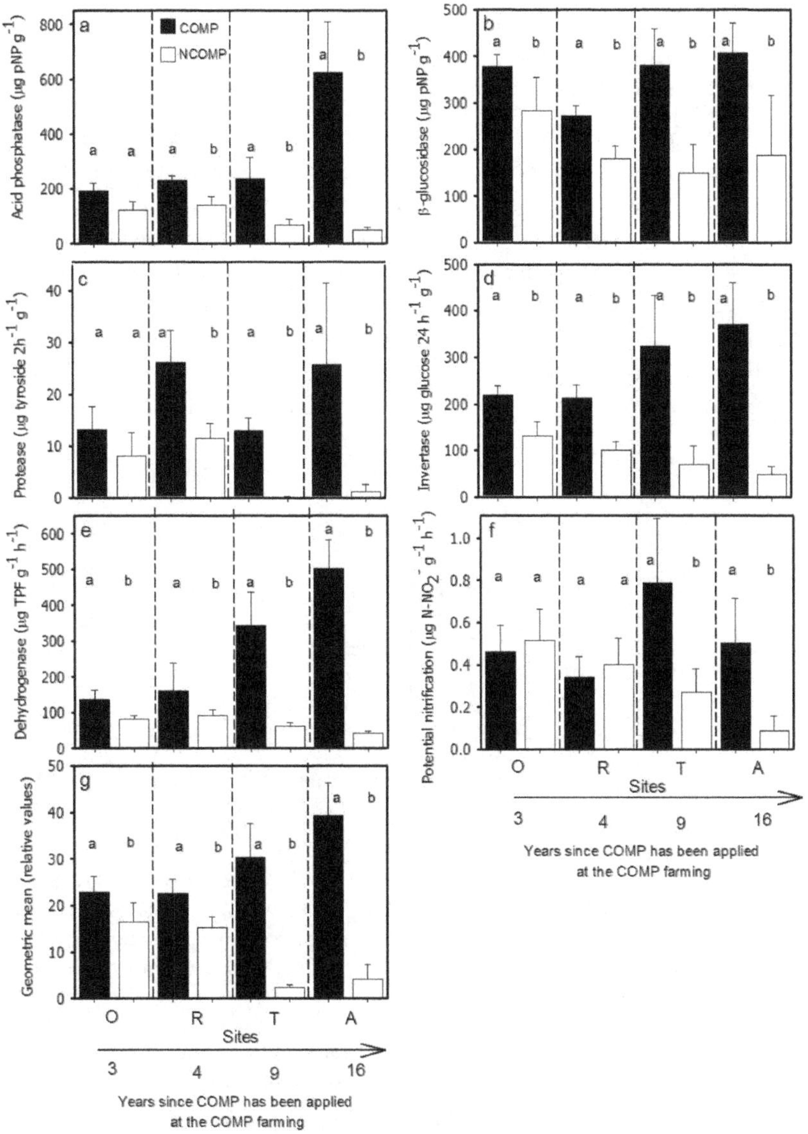

Figure 9. Soil acid phosphatase (a), β-glucosidase (b), protease (c), invertase (d), and dehydrogenase (e) activities, potential nitrification rate (f) and geometric mean of the assayed soil enzyme in olive oil farms which received (black) or not (white) COMP at Olvera (O), Reja (R), Tobazo (T) and Andújar (A). The number of years since composted olive mill pomace has been applied to COMP farming is also indicated. Different subscript letters stand for significant differences ($P < 0.05$) between COMP and NCOMP farming of each site.

Soil enzyme activities related to carbon, nitrogen and phosphorus cycling have been proposed as a tool to assess soil quality/health and functioning [77]. Enzymes activities related to the C, N and P cycles were all significantly higher in soils amended with COMP than in NCOMP soils (Figure 9). Others authors [78,79] have found similar results after application of manures of different origin.

When pooling all the analysed variables in a Principal Component Analysis it was found that the first principal component (PC1) was negatively correlated with soil organic matter, total N, available P, soil aggregate stability and soil enzyme activities and therefore, PC1 is strongly related to soil functioning. PC2, on the other hand, was positively correlated with clay content, WHC and cation exchange capacity. COMP-treated soils shifted upwards (higher WHC and CEC) and towards the left (higher soil functioning) with respect to their NCOMP soils in the PC1–PC2 space (Figure 10). Therefore PCA, which included all the analysed variables, clearly separated paired plots according to the COMP application, supporting the hypothesis that COMP application improves overall soil functioning.

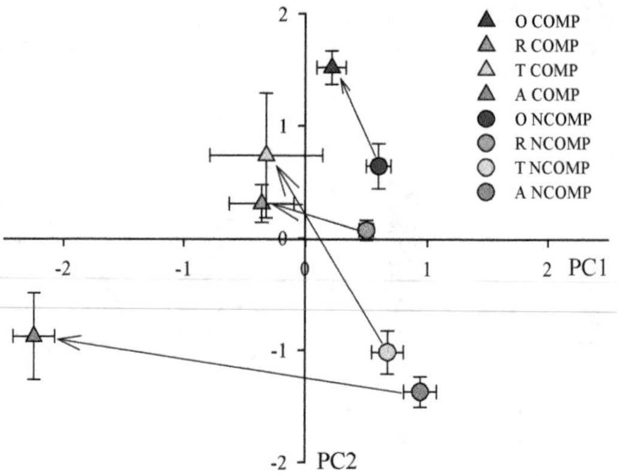

Figure 10. Ordination of the COMP and NCOMP farms at Olvera (O), Reja (R), Tobazo (T) and Andújar (A) in the space defined by the PC1 and PC2 axis resulting from the PCA analysis carried out with physico–chemical and biochemical soil properties. Coordinates are the means of t five replicates and bars represent the standard deviations of the mean. Arrows illustrate the differences in the position of the COMP and NCOMP farming at each site.

Finally, the vectorial distance between comparable COMP and NCOMP farms (e.g. differences in soil functioning) tended to increase with the period of COMP application, suggesting that there is potential for further increases in soil fertility and functioning after long-term application of COMP.

6. Ecological services associated with the use of composted olive mill pomace as organic fertiliser

There is a global trend towards developing agricultural production systems which are sustainable. This involves the more efficient utilisation of inputs and the reduction of waste products. Ideally, organic by-products should be transformed into useful products by processes such as nutrient recycling, replenishment of soil organic matter or generation of energy. In this context, the cost of waste disposal would be avoided and environmental pollution reduced. This fact has led y to a widespread increase in the value of by-products of the agricultural industry. This is also now the case for OMP.

Recycling of OMP through composting (relatively easy-to-use and low costing methodology) could be a sound strategy to provide some ecological services to olive oil groves (Figure 11). Firstly, composting OMP reduces most of the potential environmental pollution problems linked with the disposal of approximately 4 million of tonnes of OMP produced in Andalusia over a relatively short-time span (3 months). On the other hand, most of the nutrients (especially nitrogen, phosphorus and potassium) harvested with the yield, are contained in the OMP, and therefore after composting and application to olive oil groves helps to recycle these nutrients, reducing the need for chemical fertilisers. Our estimates show that between one to two-thirds of the Andalusian olive oil groves could be fertilised annually with the OMP produced in Andalusia after composting, with a subsequent reduction of about 25 – 60% in chemical fertilisers. In addition, the main beneficiary of the economic and environmental profits of composting OMP and application to olive oil groves is the farmer.

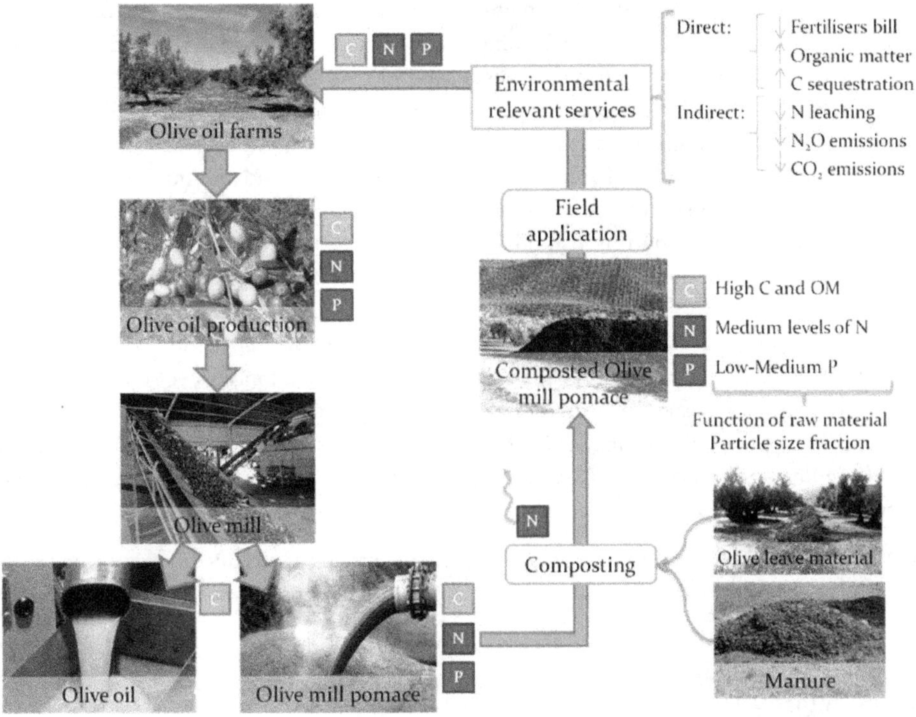

Figure 11. Fate of C, N and P of the fruit harvested in the olive oil farming when olive mill pomace (main by-product of the olive oil mill industry) is composted and applied to the olive oil groves. Some of the environmental services linked to the recycling of olive mill pomace throughout composting, are also indicated.

Some of the environmental services restored after application of COMP and shown in this chapter include: i) Increased soil organic matter and carbon. Indeed our data showed that both increased significantly after regular applications of COMP and there was a trend for increased differences in SOM and SOC between soils amended or unamended with COMP according to the number of years when COMP was applied. Other variables related directly or indirectly with the increase in SOM include i) an increase in soil microbial activity related to nutrient recycling (e.g. soil enzyme activities), cation exchange capacity, soil aggregate stability and available P and K, ii) increase in soil carbon sequestration. COMP-C derived mineralization is rather low (< 10% over one year at optimal conditions) and thus, regular applications of COMP could contribute to soil C sequestration and help alleviate the soil C-CO_2 emissions linked to olive oil cultivation, iii) reduction of the potential nitrate leaching and N_2O emissions. Our data on seven currently produced COMPs have demonstrated that during decomposition of OMP, soil mineral N is retained (i.e. net N immobilization) reducing not only nitrate leaching, but also N_2O emissions.

7. Conclusions

The recycling of nutrient and organic matter of the olive mill pomace after its composting and application to the soil of olive oil farming is a worthwhile strategy to avoid the potential environmental harm of olive mill waste disposal and could lead to increased soil fertility and functionality. The characteristics of the COMP currently produced are adequate for agricultural purposes (high organic matter and carbon, high level of potassium and from low to medium levels of N and P, and lack of phytotoxicity) and the quality was highly dependent on the proportion of manure co-composted with olive mill pomace. COMP-N is well humified and during decomposition soil mineral N can be immobilised depending on the proportion of raw materials co-composted, and therefore it is recommended to combine N rich fertiliser with COMP during the first years of COMP application. COMP-nitrate leaching (at a temporal scale of months to a year) and nitrous oxide emissions were negligible after COMP application to soils. COMP-C mineralization was very low (< 10% of that added after one year) and, therefore, COMP application to soils could enhance C sequestration in olive oil farming. Soil fertility and functioning was improved after three years of regular applications of COMP to soils of olive oil groves a there was a clear trend for a further increase over longer periods of application.

Overall, composted olive mill pomace is a worthwhile strategy to reduce the environmental problems associated with the disposal of OLM, and increases the sustainability and ecological services of olive oil cultivation.

Author details

Beatriz Gómez-Muñoz
Corresponding Author
Ecology Section, University of Jaén, Campus Las Lagunillas s/n, Jaén, Spain

David J. Hatch
Sustainable Soils and Grassland Systems Department, Rothamsted Research, North Wyke, Okehampton, Devon, UK

Roland Bol
Institute of Bio-and Geosciences, Agrosphere (IBG-3), Forschungszentrum Jülich GmbH, Jülich, Germany

Roberto García-Ruiz
Ecology Section, University of Jaén, Campus Las Lagunillas s/n, Jaén, Spain

Acknowledgement

This research was carried out in the framework of the "Olive grove project" of the General Secretary for Rural Development and organic production of the Junta de Andalucía, and with the economic help of the Minister of Science and Technology of Spain throughout the project referenced CGL200908303. We would like to thanks to Jose María Álvarez and the various olive oil farmers and olive oil mills from which soil and composted olive mill pomace samples were taken.

8. References

[1] Beaufoy G, Pienkowski M (2000) The environmental impact of olive oil production in the European Union: practical options for improving the environmental impact. European Forum on Nature Conservation and Pastoralism. European Commission, Brussels.
[2] Dios–Palomares R, Haros–Giménez D, Montes–Tubio F (2005) Estudio estructural del sector oleícola de Andalucía. Nivel de calidad y respecto medioambiental de las industrias almazaras. Comunicaciones EXPOLIVA.
[3] Moraetis D, Stamati FE, Nikolaidis NP, Kalogerakis N (2011) Olive mill wastewater irrigation of maize: impacts on soil and groundwater. Agr. Water Manage. 98: 1125-1132.
[4] Cabrera F, López R, Martinez-Bordiú A, Dupuy de Lome E, Murillo JM (1996) Land treatment of olive oil mill wastewater. Int. Biodeter. Biodegr. 38: 215-225.
[5] Kavvadias V, Doula MK, Komnitsas K, Liakopoulou N (2010) Disposal of olive oil mill wastes in evaporation ponds: effects on soil properties. J. Hazard. Mater. 182: 144-155.
[6] López-Piñeiro A, Cabrera D, Albarrán Á, Peña D (2011) Influence of two-phase olive mill waste application to soil on terbuthylazine behaviour and persistence under controlled and field conditions. J. Soil Sediment. 11: 771-782.
[7] Paredes MJ, Moreno E, Ramos–Cormenzana A, Martinez J (1987) Characterization of soil after pollution with waste waters from olive oil extraction plants. Chemosphere. 16: 1557–1564.
[8] Pérez J, de la Rubia T, Moreno J, Martinez J (1992) Phenolic content and antibacterial activity of olive mill wastewater. Environ. Toxicol. Chem. 11: 489–495.

[9] Linares A, Caba JM, Ligero F, De la Rubia T, Martínez J (2003) Detoxification of semisolid olive–mill wastes and pine–chip mixtures using Phanerochaete flavido–alba. Chemosphere. 51: 887–891.

[10] Sierra J, Marti E, Garau MA, Cruanas R (2007) Effects of agronomic use of olive oil mill wastewater – field experiment. Sci. Total Environ. 378: 90–94.

[11] Piotrowska A, Rao MA, Scotti R, Gianfreda L (2011) Changes in soil chemical and biochemical properties following amendment with crude and dephenolized olive mill waste water (OMW). Geoderma. 161: 8-17.

[12] Paredes C, Cegarra J, Roig A, Sánchez–Monedero MA, Bernal MP (1999) Characterization of olive mill wastewater (alpechin) and its sludge for agriculture purposes. Bioresource Technol. 67: 111–115.

[13] Alburquerque JA, Gonzálvez J, García D, Cegarra J (2004) Agrochemical characterisation of "alperujo", a solid by–product of the two–phase centrifugation method for olive oil extraction. Bioresource Technol. 91: 195–200.

[14] Roig A, Cayuela ML, Sánchez–Monedero MA (2006) An overview on olive mill wastes and their valorisation methods. Waste Manage. 26: 960–969.

[15] Cayuela ML, Millner P, Slovin J, Roig (2007) Duckweed (Lemna gibba) growth inhibition bioassay for evaluating the toxicity of olive mill wastes before and during composting. Chemosphere. 68: 1985–1991.

[16] Madejón E, Galli E, Tomati U (1998) Composting of wastes produced by low water consuming olive mill technology. Agrochimica. 42: 135–146.

[17] Cegarra J, Amor JB, Gonzálvez J, Bernal MP, Roig A (2000) Characteristics of a new solid olive–mill–by–product (''alperujo'') and its suitability for composting. In: Warman, P.R, Taylor, B.R. (Eds.), Proceedings of the International Composting Symposium ICS_99, 1. CBA Press Inc, pp. 124–140.

[18] Álvarez de la Puente JM, Jáuregui J, García–Ruiz R (2010) Composting olive mill pomace: The Andalusian experience. Biocycle. 31–32.

[19] García–Gómez A, Roig A, Bernal MP (2003) Composting of the solid fraction of olive mill wastewater with olive leaves: organic matter degradation and biological activity. Bioresource Technol. 86: 59–64.

[20] Alburquerque JA, Gonzálvez J, García D, Cegarra J (2006) Measuring detoxification and maturity in compost made from "alperujo", the solid by–product of extracting olive oil by the two–phase centrifugation system. Chemosphere. 64: 470–477.

[21] Canet R, Pomares F, Cabot B, Chaves C, Ferrer E, Ribó M, Albiach MR (2008) Composting olive mill pomace and other residues from rural southeastern Spain. Waste Manage. 28(12): 2585–2592.

[22] Alburquerque JA, Gonzálvez J, García D, Cegarra J (2007) Effects of a compost made from the solid by–product ("alperujo") of the two–phase centrifugation system for live oil extraction and cotton gin waste on growth and nutrient content of ryegrass (Lolium perenne L.). Bioresource Technol. 98: 940–945.

[23] Hachicha S, Sellami F, Medhioub K, Hachicha R, Ammar E (2008) Quality assessment of composts prepared with olive mill wastewater and agricultural wastes. Waste Manage. 28(12): 2593–603.

[24] Mustin M (1987) Le Compost: Gestion de la Matie`re Organique. Dubusc, F. (Ed.), Paris, pp. 117–263.

[25] Das KC (2007) Co–composting of alkaline tissue digester effluent with yard trimmings. Waste Manage. 28: 1785–1790.

[26] Iannotti DA, Grebus ME, Toth BL, Madden LV, Hoitik HAJ (1994) Oxygen respirometric method to assess stability and maturity of composted municipal solid waste. J. Environ. Qual. 23: 1177–1183.

[27] Moral R, Moreno–Caselles J, Pérez–Murcia MD, Pérez–Espinosa A, Rufete B, Paredes C (2005) Characterisation of the organic matter pool in manures. Bioresource Technol. 96(2): 153–158.

[28] Sánchez–Monedero MA, Roig A, Paredes C, Bernal MP (2001) Nitrogen transformation during organic waste composting by the Rutgers system and its effects on pH, EC and maturity of the composting mixtures. Bioresource Technol. 78: 301–308.

[29] Goyal S, Dhull SK, Kapoor KK (2005) Chemical and biological changes during composting of different organic wastes and assessment of compost maturity. Bioresource Technol. 96(14): 1584–1591.

[30] Pascual JA, Ayuso M, García C, Hernández T (1997) Characterization of urban wastes according to fertility and phytotoxicity parameters. Waste Manage. Res. 15: 103–112.

[31] Capasso R, Evidente AA, Schivo L, Orru, G, Marcialis, M.A, Cristinzio, G, 1995. Antibacterial polyphenols from olive oil mill waste waters. J. Appl. Bacteriol. 79: 393–398.

[32] Ait Baddi G, Alburquerque JA, Gonzalvez J, Cegarra J, Hafidi M (2004) Chemical and spectroscopic analyses of organic matter transformations during composting of olive mill wastes. Int. Biodeter. Biodegr. 54(1): 39–44.

[33] Palm CA, Gachengo CN, Delve RJ, Cadisch G, Giller KE (2001) Organic inputs for soil fertility management in tropical agro ecosystems: application of an organic resource database. Agr. Ecosyst. Environ. 83: 27–42.

[34] Fox RH, Myers RJK, Vallis I (1990) The nitrogen mineralization rate of legume residues in soil as influenced by their polyphenol, lignin and nitrogen contents. Plant Soil. 129: 251–259.

[35] Palm CA, Sánchez PA (1991) Nitrogen release from the leaves of some tropical legumes as affected by their lignin and polyphenolic contents. Soil Biol. Biochem. 23: 83–88.

[36] Melillo JM, Aber JD, Muratore JF (1982) Nitrogen and lignin control of hardwood leaf litter decomposition dynamics. Ecology, 63: 621–626.

[37] Constantinides M, Fownes JH (1994) Nitrogen mineralization from leaves and litter of tropical plants—relationship to nitrogen, lignin and soluble polyphenol concentrations. Soil Biol. Biochem. 26: 49–55.

[38] Palm CA, Rowland AP (1997) A minimum data set for characterization of plant quality for decomposition. In: G. Cadisch and K.E. Giller, Editors, Driven by Nature: Plant Litter Quality and Decomposition, CAB International, Wallingford. 379–392.

[39] Doublet J, Francou C, Pétraud JP, Dignac MF, Poitrenaud M, Houot S (2010) Distribution of C and N mineralization of a sludge compost within particle–size fractions. Bioresource Technol. 101(4): 1254–1262.

[40] Grilo J, Bol R, Dixon L, Chadwick D, Fangueiro D (2011) Long term release of carbon from grassland soil amended with different slurry particle size fractions: a laboratory incubation study. Rapid Commun. Mass Sp. 25: 1514–1520.

[41] Fangueiro D, Chadwick D, Dixon L, Bol R (2007) Quantification of priming and CO_2 respiration sources following the application of different slurry particle size fractions to a grassland soil. Soil Biol. Biochem. 39: 2608–2620.

[42] Robin P, Ablain F, Yulipriyanto H, Pourcher AM, Morvan T, Cluzeau D, Morand P (2008) Evolution of non–dissolved particulate organic matter during composting of sludge with straw. Bioresource Technol. 99: 7636–7643.

[43] Tester CF, Sikora LJ, Taylor JM, Parr JF (1979) Decomposition of sewage sludge compost in soil: III. Carbon, nitrogen, and phosphorus transformations in different sized fractions, J. Environ. Qual. 8: 79–82.

[44] Aoyama M (1985) Properties of fine and water–soluble fractions of several composts. Micromorphology, elemental composition and nitrogen mineralization of fractions. Soil Sci. Plant Nutr. 31: 189–198.

[45] Fangueiro D, Gusmão M, Grilo J, Porfírio G, Vasconcelos E, Cabral F (2010) Proportion, composition and potential N mineralisation of particle size fractions obtained by mechanical separation of animal slurry. Biosyst. Eng. 106: 333–337.

[46] Fangueiro D, Bol R, Chadwick D (2008a) Assessment of the potential N mineralization of six particle size fractions of two different cattle slurries. J. Plant Nutr. Soil Sc. 171: 313–315.

[47] Fangueiro D, Pereira J, Chadwick D, Coutinho J, Moreira N, Trindade H (2008b) Laboratory assessment of the effect of cattle slurry pre–treatment on organic N degradation after soil application and N_2O and N_2 emissions. Nutr. Cycl. Agroecosys. 80: 107–120.

[48] Beauchamp EG, Paul JW (1989) A simple model to predict manure N availability to crops in the field. In: J.A. Hansen and K. Henriksen, Editors, Nitrogen in Organic Wastes Applied to Soils, Academic Press, London.

[49] Cabrera F, Martín–Olmedo P, López R, Murillo JA (2005) Nitrogen mineralization in soils amended with composted olive mill sludge. Nutr. Cycl. Agroecosys. 71: 249–258.

[50] Parker CF, Sommer LE (1983) Mineralization of nitrogen in sewage sludges. J. Environ. Qual. 12: 150–156.

[51] Hérbert M, Karam A, Parent LE (1991) Mineralization of nitrogen and carbon in soils amended with composted manure. Biology, Agriculture and Horticulture, 7: 349–361.

[52] Chae YM, Tabatabai MA (1986) Mineralization in an aridisol mixed with various organic materials. J. Environ. Qual. 15: 193–198.

[53] Bernal MP, Kirchmann H (1992) Carbon and nitrogen mineralization and ammonia volatilization: from fresh, aerobically and anaerobically treated pig manure during incubation with soil. Biol. Fert. Soils, 13: 135–140.

[54] Hadas A, Portnoy R (1994) Nitrogen and carbon mineralization rates of composted manures incubated in soil. J. Environ. Qual. 23: 1184–1189.

[55] Senesi N, Plaza C, Brunetti G, Polo A (2007) A comparative survey of recent results on humic–like fractions in organic amendments and effects on native soil humic substances. Soil Biol. Biochem. 39: 1244–1252.

[56] Ferguson RB, Shapiro CA, Hergert GW, Kranz WL, Klocke NL, Krull DH (1991) Nitrogen and irrigation management practices to minimize nitrate leaching from irrigated corn. J. Prod. Agric. 4: 186–192.

[57] Hallberg GR (1985) Agricultural chemicals and groundwater quality in Iowa: Status Report, Cooperative Extension Service, CE–2158q, Iowa State Univ, Ames, IA, p. 11.

[58] Stanley EH, Short RA, Harrison JW, Hall R, Wiedenfeld RC (1990) Variation in a nutrient limitation of lotic and lentic algal communities in a Texas (USA) river. Hydrobiologia. 206: 61–71.

[59] Hallberg GR (1989) Nitrate in ground water in the United States. In R.F. Follett (ed.) Nitrogen management and ground water protection. Elsevier, New York, p. 35–74.

[60] Yadav SN (1997) Formulation and estimation of nitrate–nitrogen leaching from corn cultivation. J. Environ. Qual. 26: 808–814.

[61] Insam H, Merschak P (1997) Nitrogen leaching from forest soil cores after amending organic recycling products and fertilizers. Waste Management Research, 15: 277–292.

[62] IPCC (2007) Changes in Atmospheric Constituents and in Radiative Forcing, Cambridge University Press, UK and New York USA.

[63] García–Ruiz R, Baggs E (2007) ^{15}N–N_2O emissions from decomposition of plant residues in olive oil orchard. En: Towards a better efficiency in N use, pp. 250–253.

[64] Baggs EM, Stevenson M, Pihlatie M, Regar A, Cook H, Cadisch G (2003) Nitrous oxide emissions following application of residues and fertiliser under zero and conventional tillage. Plant Soil. 254: 361.

[65] Baggs EM, Rees RM, Smith KA, Vinten AJA (2000) Nitrous oxide emission from soils after incorporating crop residues. Soil Use Manage. 16: 82–87.

[66] Duiker SW, Lal R (2000) Carbon budget study using CO_2 flux measurements from a no till system in central Ohio. Soil Till. Res. 54: 21–30.

[67] Adiku SGK, Narh S, Jones JW, Laryea KB, Dowuona GN (2008) Short–term effects of crop rotation, residue management, and soil water on carbon mineralization in a tropical cropping system. Plant Soil, 311: 29–38.

[68] Manzoni S, Proporato A (2009) Soil carbon and nitrogen mineralization: Theory and models across scales. Soil Biol. Biochem. 41, 1355–1379.

[69] Cleveland CC, Liptzin D (2007) C:N: P stoichiometry in soil: is there a "Redfield ratio" for the microbial biomass? Biogeochemistry. 85: 235–252.

[70] Jensen LS, Mueller T, Magid J, Nielsen NE (1997) Temporal variation of C and N mineralization, microbial biomass and extractable organic pools in soil after oilseed rape straw incorporation in the field. Soil Biol. Biochem. 29: 1043–1055.

[71] Bernal MP, Sánchez–Monedero MA, Paredes C, Roig A (1998) Carbon mineralization from organic wastes at different composting stages during their incubation with soil. Agr. Ecosyst. Environ. 69: 175–189.

[72] Hollis JM, Jones RJA, Palmer RC (1977) The effect of organic matter and particle size on the water retention properties of some soils in the West Midlands of England. Geoderma. 17, 225– 238.

[73] López–Piñeiro A, Albarrán A, Nunes JM, Barreto C (2008) Short and medium–term effects of two–phase olive mill waste application on olive grove production and soil properties under semiarid Mediterranean conditions. Bioresource Technol. 99: 7982–7987.

[74] Piotrowska A, Iamarino G, Rao MA, Gianfred L (2006) Short–term effects of olive mill waste water (OMW) on chemical and biochemical properties of a semiarid Mediterranean soil. Soil Biol. Biochem. 38: 600–610.

[75] Montemurro F, Convertini G, Ferri D (2004) Mill wastewater and olive pomace compost as amendments for rye–grass. Agronomie. 24, 481–486.

[76] Benítez C, Tejada M, González JL (2003) Kinetics of mineralization of nitrogen in a pig slurry compost applied to soils. Compost Sci. Util. 11: 72–80.

[77] Dick RP (1994) Influence of long–term tillage and crop rotation combinations on soil enzyme activities. Soil Sci. Soc. Am. J. 56: 783–788.

[78] Mäder P, Fliessbacth A, Dubois D, Gunst L, Fried P, Niggli U (2002) Soil fertility and biodiversity in organic farming. Science. 296: 1694–1697.

[79] Melero S, Porras JC, Herencia JF, Madejón E (2006) Chemical and biochemical properties in a silty loam soil under conventional and organic management. Soil Till. Res. 90: 162–170.

Chapter 8

Social-Ecological Resilience and Maize Farming in Chiapas, Mexico

Nils Max McCune, Francisco Guevara-Hernández, Jose Nahed-Toral, Paula Mendoza-Nazar, Jesus Ovando-Cruz, Benigno Ruiz-Sesma and Leopoldo Medina-Sanson

Additional information is available at the end of the chapter

1. Introduction

Chiapas is considered one of the areas of origin for maize, and indigenous production systems remain a major element of maize farming and the food system of Mexico's southernmost state. These traditional systems often include a complex, long-term relationship between maize farming and the larger landscape (Medellín and Equihua, 1998). However, what often appear to be static, steady states of small farmer land use in fact are highly dynamic systems that are in the midst of major adaptations to new climatic and economic conditions. The globalization of the agribusiness model that came to dominate the U.S. landscape during the 20th century has put intense pressures on small farmers in Chiapas, adding to complications arising from shifting growing seasons and other effects of global climate change. However, history has shown that Mexican small farmers are never passive objects of their circumstances, and their responses to early 21st century marginalization take both centralized and local forms, in political and productive terms (Guevara-Hernández et al. 2011a). The balance of forces—capitalist agriculture, external input dependence, neoliberal national food policy on one hand; traditional knowledge, agroecological transitions, social movements on the other—in Mexican agriculture tends to distinguish two contending models for food systems, even as climate change, resource scarcity, and economic crises limit humanity's options.

Shifting cultivation and other traditional systems of food production in Chiapas are being highly influenced by the slow-motion arrival of industrial agriculture, which is based on maximizing short-term productivity through the use of synthetic inputs such as fertilizers and pesticides, as well as commercial seeds (Garcia-Barrios et al. 2010). Most farm inputs are supplied by Mexican subsidiaries of foreign-owned multinational corporations such as

Monsanto, Syngenta, and Dow. At the same time, an industrial food system means that large amounts of capital are invested in food sales, creating giant monopolies in processing and distribution, which significantly reduce the portion that farmers receive of the price paid by consumers. In other words, control over the food system—the set of activities that are built around capital flow and labor command in food production, shipping, transformation, consumption—is highly concentrated in the hands of input manufacturers and food processing, trading, and retail corporations, while the riskiest part of agriculture—the actual farming process—is still in the hands of hundreds of thousands of small farmers (Magdoff et al. 2000).

In Chiapas, these farmers live in hills and valleys, in forested land and former forests, in dry shrub lands and in lush jungles. Their communities and farmlands compose a peasant landscape, in which patches of forest are interspersed in a complex mosaic of farms, backyards, homes, schools, rivers, roads, and towns. Depending on the type of agriculture practiced, soil may be highly degraded or intact. Chiapas is a center for biodiversity; many endemic species live in and around forest patches and agroecosystems (Ramírez-Marcial et al. 2001). In this setting, the demands of rural social movements such as *¡Sin Maíz No Hay País!* (Without Corn There is No Country!) increasingly refer to the goal of food sovereignty. Food sovereignty means a fundamental emphasis on local and domestic production, based on land access for small farmers and ecological production practices. It rejects food as a commodity to be included in free trade agreements or dumping schemes meant to undermine countries' domestic production capacity. As a political proposal, food sovereignty implies a radical democratization and decentralization of the agriculture-food system, including the destruction of corporate power over food. On a more cultural level, food sovereignty is an affirmation of rural community, local knowledge, and gender equality.

Both the agribusiness model and the food sovereignty model are highly complex, integrated systems that involve the relationship of society and nature. The stability of each system depends on distinct factors that combine social and ecological drivers, while the capacity of each to respond to shock or disturbance will depend on unique intrinsic qualities. In the case of industrial or export-focused agriculture, we have a system that has been shown to be destructive to peoples and ecosystems, due to its focus on short-term profits through maximizing monoculture productivity. Here we draw from theoretical contributions from restoration ecology that use models of alternative stable states to study change in complex systems (Suding et al. 2004). We argue that monoculture/capitalist agribusiness represents one pull of attraction, or alternative state, indeed a food system unique to late-stage global capitalism. On the other hand, we propose that the agroecology/food sovereignty framework may in fact represent another alternative state, far more promising for building resilient food systems in the 21st century. In order to develop this line of inquiry, we start by examining more closely the resilience paradigm and the two proposed stable states at an abstract/global level. Then we describe historical, agroecological, and political elements of food system resilience in the case of a maize-growing community in Chiapas.

2. The resilience paradigm

Research into sustainable agriculture has increasing come to embrace the conceptual approach of food systems, as these reflect the interface of alimentation, human activities, public policies, cultural norms and social well-being, along with land, farms, ecosystems, and economies. The complex interactions between these processes at distinct scales, and involving various institutional and economic actors, may produce the outcome of food security. The food system approach may be useful for developing cohesive strategies across policy sectors, including agrarian and land access sectors, natural resource and environmental management, agriculture, trade, economy, industry, science and technology, health, and education, among others (Ericksen et al. 2010). Efforts to achieve food system sustainability in the midst of global environmental and economic changes are beginning to coalesce around certain concepts that help determine the most significant problems in food systems and identify management strategies at several levels of analysis (farm, community, national, international) to increase social, ecological, and economic sustainability (Pretty et al. 2011).

The management of complex, adaptive systems has become a dynamic field of new trans-disciplinary theory, especially with regard to life supporting systems of human activities, such as agriculture, in sensitive ecological contexts. Social-ecological systems (Berkes and Folke, 1998), or coupled human and natural systems (Liu et al. 2007), have become a central concept to allow greater understanding of the interdependencies and feedbacks between social and ecological systems. The contributions of systems ecology are applied in order to understand the complex internal dynamics and adaptability of these coupled systems. Many of the concepts that inform such studies of systems originate from ecology, for two reasons: one, its emphasis on qualities that emerge from a set of relationships between elements, rather than the reductionist focus on elements in isolation; and two, the growing academic and popular concern for the relationship between humanity and the biosphere that tenderly exists on the surface of the Earth's crust (Lang, 2009).

In the efforts to understand the intrinsic qualities of social-ecological systems, researchers from several different disciplinary backgrounds have approached the concept of resilience (Shattuck, 2012). The resilience principle stems from systems ecology theory (Hooper, 1973) that suggested that instead of static, unchanging climax communities, natural ecosystems could evolve between several alternative stable states, with biotic and abiotic feedback mechanisms accelerating or preventing system change. Disturbances began to be seen as an integral part of ecosystem function, and resilience as an emergent system capacity to absorb a certain magnitude of shock and maintain key system functions before reaching a critical threshold and switching to an alternative stable equilibrium with new system properties (Holling, 1973; Noy-Meir, 1975). Noy-Meir (1975) used the analogy of a mechanical ball-in-container (figure 1) to describe alternative steady-states. The original steady-state is stable to fluctuations within a certain range, but too hard a push in one direction will send it over the turning point and toward a new steady-state. The major concern in light of global environmental change is that ecosystems will be pushed beyond their limits, into new steady-states that provide less ecological services (Walker et al. 2004).

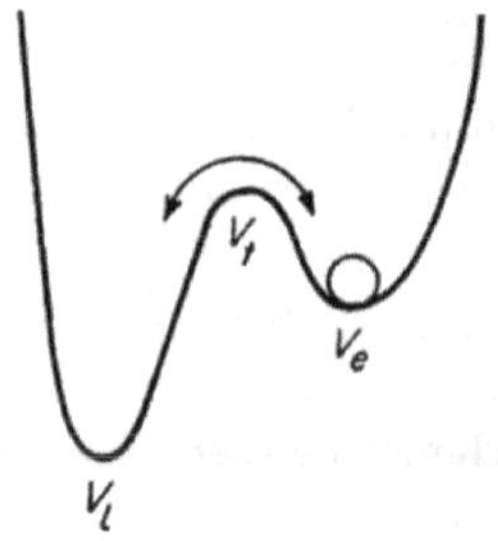

Figure 1. A physical model of the two-steady-states situation (from Noy-Meir, 1975).

The two-steady-state model is a very simple illustration of a key concept in resilience studies: the threshold. The high point of the center curve in figure one is the threshold, or point of no return, for the original system. Resilience systems may absorb strong shocks below this point; one tiny push above it will result in what could be irreversible and accelerated change. One of the major objectives of resilience research is to identify and characterize system thresholds, in order to understand what makes systems able to absorb some shocks without changing overall function, what makes some changes temporary and others permanent, and how to shift thresholds through system adaptations. Social and ecological dynamics in industrial food states can be very different from dynamics in traditional or sovereign food states. Efforts to promote transition to a sovereign food state need to better understand feedbacks and constraints of the industrial food state. Alternative state models, used in restoration ecology to focus on internally reinforced states and recovery thresholds (Suding et al. 2004), may help guide historic conversions to sovereign food systems.

The recent attention paid in academic literature toward the concepts of risk, robustness and resilience in social and economic systems is surely related to such troubling events as the global financial crisis of 2008; the Great Recession that is still very much limiting employment and well-being among most nations in the world; the so-called Arab Spring that has produced protest movements for systemic change in diverse countries like Egypt, Tunisia, Yemen, Libya, Syria, Spain, Greece, Portugal, France, and the United States; the global food price crises of 2007 and 2011; the impending scarcity of hydrological and energetic resources; and finally, the arrival of such damning evidence as increased incidence of extreme weather events (e.g. hurricanes, droughts, floods) and long-term changes associated with excessive greenhouse gases in the Earth's atmosphere.

The sudden growth in use of the term "resilience" has been studied elsewhere and associated with a shift toward the understanding of complex systems as being more dynamic and less tied to any one climax or stable state. The Resilience Alliance and the Stockholm Resilience Centre, high-profile scientific platforms that bring together orthodox neoliberal economists with systems ecologists (Walker and Cooper, 2011), emphasize an adaptive cycle within complex systems that includes phases of growth, decadence, self-destruction and renewal. Curiously, this self-organization that complex systems are seen to have is used to argue against inserting planning mechanisms and regulation on economic systems. Indeed, the free market is argued to be a complex system capable of self-regulation. The same analysis considers the shock of deregulation and structural adjustment programs to be a healthy opportunity for renewal. A skeptic might insert here the example of New Orleans or Iraq, both places where disaster (to be located somewhere on a gradient between "natural" and human-made) created an opportunity to rebuild in a new image and for the benefit of a distinct group, in both cases resulting in massive profits for contractors that, incidentally, worked under U.S. government contact (i.e., the planned economy). Essentially, orthodox neoliberals create conditions for huge profit by monopoly capital, rather than constructing any sort of self-organized or resilient world state. On the contrary, we argue that the resilience of a local food system is often inversely related to its integration into the world capitalist economy.

Shattuck (2012) proposes a framework for studying resilience in food systems, in order to effectively prioritize goals of human well-being and biodiversity conservation. This author combines literature on biophysical restraints, adaptive capacity, and political economy to develop a food system resilience framework that considers ecological processes and underlying macroeconomic causes of livelihood vulnerability. Here we presuppose that as an emergent property, resilience of a given food system-state is always subject to influence by resilience of systems that operate at larger or smaller scales. We have built our analysis around a resilience assessment in a rural community of Chiapas, but have sought to contribute to a much-needed debate on the global scale. In the study community, we evaluated aspects of economic, social and ecological resilience, based on interviews, surveys, and data sampling that were carried out over the course of two years of fieldwork in the Fraylesca region of Chiapas during 2010-12.

3. Resilience of the industrial food system

Allenby and Fink (2005) define resilience as the "capability of a system to maintain its functions and structure in the face of internal and external change and to degrade gracefully when it must." Accepting this definition, we should define the functions of the capitalist industrial food system and describe its structure, before looking for the critical thresholds that it may not cross without converting into a qualitatively different system. As with all activities driven by capital, the paramount function of the capitalist food system is to reproduce capital in greater quantities. This function has lead to a rationalization of economies surrounding food, based on the principle of maximizing the difference between costs and revenues in the application of capital to production, processing, distribution, and sales. In the productive sphere, rationalization means the maximization of commodity production. In order to maximize revenues, capitalist food systems have developed enormous structures for food processing in order to add to commodity value, as well as advertising to boost revenues. As capital has been slow to penetrate the actual farming process itself (Levins, 2007), due to its risky nature and bio-physical limitations, it has instead reduced the on-farm value added to commodities (i.e. cut into the farmer's income) by creating an industry of costly farm inputs, which are generally accepted as a part of the modern, monoculture form of growing food.

To think of the industrial agriculture model as a stable state on a food system continuum requires identifying its major components, defining the limits and time scale of the system, determining the values, peoples, and natural resources involved, analyzing the political economy and legal character of industrial agriculture, and recognizing the cross-scale interactions that all have an impact on system resilience (Kinzig et al. 2007). The essential components of a mature industrial agriculture model include the integration of food into free trade agreements, government support for agribusiness, market control over land and water resources, external input-intensive production models, monoculture and specialization, as well as an agricultural research establishment that focuses on developing profitable technologies. These conditions are largely met in Mexico as a whole, except for the millions of small-scale producers who continue to meet food needs at the local and national level using few external inputs and some form of communal land rights. The contradictory proliferation—and hybridization—of this model in socially and ecologically adverse circumstances deserves further attention.

The industrial food system represents the technological and organizational apex of a model first put into place in lands colonized by European powers. The monoculture is an invention of colonial economies, which treated dominated nations only as sources for cheap raw materials (including cheap food for a growing industrial working class). In Ireland, the monoculture potato production system was enforced by British colonial law that prevented the Irish from planting other crops. When a common pathogenic fungus destroyed the potato harvest, millions of Irish were killed by the ensuing famine. This is a classic example of the risk of the food system built on monoculture. Despite many such examples of spectacular failure, the industrial food system remains deeply committed to monoculture production systems around the world. The development of a global food system based on the increasing excursion of capital into farming has been a complex process, by which capital completely surrounded farming by taking over farm input and post-harvest economies while only slowly moving into the actual farming itself. Early stabs at industrial agriculture included the guano boat and phosphorus mining fertilizer industries in the late 1800s. In California, industrial agriculture and land takeovers were always linked (Walker, 2004), as wheat farming prospecting triggered a new "gold fever" and led to a bonanza period of often-falsified speculation on real estate.

A great new era began for industrial agriculture after the Second World War. Many countries were in ruins, and baby booms gave impetus to the US war materials industry to "convert swords to plowshares" and sell them to reconstruction programs. Factories that produced nitrogen-based explosives already had the entire infrastructure necessary to produce nitrogen fertilizers, tank assembly lines could easily be converted to create tractors, and many of the nastier chemicals used in war efforts were found to have satisfyingly lethal effects on insects and unwanted plants in agriculture. Even more importantly, the call for technical solutions to hunger was seen as an antidote to the more radical demands from structural change and wealth redistribution in order to combat poverty-caused hunger in the hopeful post-war and post-colonial world. Thus it was that the "green revolution," a broad program of agricultural research and technology development mostly focused on producing new high-yield varieties, was developed as production-focused solution to hunger in Asia, Africa, and Latin America.

Industrial agriculture is based on the intensive use of external inputs—such as improved seeds, fertilizers, pesticides, and irrigated water—to maximize yields. In the United States, this led to more overall production, causing the prices that farmers receive for their production to fall. As input costs rose and farm prices fell, small farms were squeezed by the low per-unit return on farming. The large farms that could produce at such a scale as to be profitable—despite the narrow margin between input costs and sales prices—started to swallow a much greater share of U.S. farm income. In 1969, the 1.2 percent of U.S. farms with the greatest annual income earned 16 percent of net farm income; by the end of the 1980s, they earned nearly 40 percent (Rosset, 1998). It is not necessarily the technologies of industrial agriculture that cause this concentration of agricultural income, but their application in societies where the advantage is already with wealthier growers and large agribusiness corporations. In such a social context, green revolution technologies tend to accelerate the concentration of food system resources, such as land and capital, in the hands of a few large players.

At a landscape level, the agribusiness model becomes best consolidated in conditions of potential ecological homogeneity and market control over economic resources (Perfecto et al. 2010). For this reason, its arrival to the Fraylesca region of Chiapas has been uneven—dominating the landscape in the large valleys, and barely felt in the most remote *ejidos* and family farms. The *ejido* system, as a form of collective property embedded in the national food system, has been a buffering element that kept local food systems viable in much of the Mexican countryside, especially in the southern states. Meanwhile, the large population of northern Mexico has created a more drastic contrast between delicate, rain-fed systems, and industrial farms built on fossil water.

This partially explains the highly disproportionate amount of private and public investment in agriculture in the northern states, while most support for farmers in southern states take the form of social welfare programs (Fox and Haight, 2010). The capitalist agribusiness model has been consolidated in northern Mexico in the generations since the end of the Second World War. In southern Mexico, it has arrived in waves—the most tidal of which was the destruction of state-owned grain warehouses and price regulations—which have yet to completely break the small farmer food system that retains a large geographic and nutritional importance, despite the dismantling of the economic structure that had been built around it since the Revolution.

Resilience in the industrial food system depends on two major objective factors: avoiding ecological destruction that would affect profit margins, and continued growth into new markets to prevent negative effects of overproduction. As global economies become more integrated than ever before and resource scarcity on a global level seems imminent, agribusiness corporations have moved into biotechnologies as a way to absorb huge sums of capital and—they hope—create vast new seed, energy, and pharmaceuticals markets. As a global system, agribusiness can easily leave behind devastated ecosystems and farm communities once degradation has reached the point that farming is no longer profitable. This has especially been the case in areas where long-term irrigation and synthetic fertilizer use have increased salt content of soils beyond thresholds of productivity, or on deforested land where original soil fertility is quickly exhausted to abysmal levels. While this kind of ecological and economic destruction is not threatening to the agribusiness model as a whole, it does threaten to create a social blowback, in the form of rural social movements and consumer groups, strong enough to threaten the future of the agribusiness model. In this sense, widespread social rejection of industrial agriculture is a subjective factor (i.e. dependent upon people's consciousness) that deeply influences the resilience of the system.

4. Resilience of the sovereign food system

Many efforts have been made to define food sovereignty (Patel, 2009). As an evolving concept, it has also been subject to growing social and academic interest, giving its full meaning an emergent quality in the historical conditions of 21st century social struggle (García-Linera, 2011). Nonetheless, we present a non-exhaustive list of components for economic, social and ecological resilience in sovereign food systems. While definitions are still being agreed upon, economic resilience among rural peoples may consist, at this particular historical moment and in many parts of the world, of several interacting components: 1) land access and unalienable rights to produce; 2) capacity to produce an abundance and variety of food necessary to meet most local food needs, essentially the potential to subsist with local production; 3) minimal dependency upon external inputs (e.g. hybrid seeds, pesticides) the availability and price of which are controlled by monopolies or foreign corporations; 4) maximum capacity to use local and renewable sources for energy and material needs (e.g. water, light, soil nutrition, farm labor); 5) use of diversified land-use and production systems, that may include extraction, agriculture, animal production, and small-scale processing in order to appropriate the value added by labor; 6) diverse income sources and form, which may include local products and off-farm employment, local and regional markets, direct contact with consumers, or in-kind payments; 7) real participation in the planning, design, and implementation of economic activities, through grassroots organizations or through governmental planning processes; and 8) the capacity to adapt and transform economic systems to better suit ecological and social necessities. Economic resilience allows systems of economic activities to withstand climatic shock, sudden scarcity or loss of markets, or long-term disturbance.

Social resilience, perhaps more difficult to define, includes at least the following components: 1) free and universal access to education and culture; 2) methods for sharing information and ideas vertically and horizontally in a way that combines theory and practice; 3) strong social organizations organized with democratic principles; 4) access to a common identity that admits and is strengthened by diversity; 5) access to universal and affordable health care; and 6) respect for social and economic human rights, such as the human right to food. This is clearly not a static situation, but rather a dynamic learning and adaptation process subject to conflict and contestation, as well as consensus. Ecological resilience could consist of: 1) ecosystems based on material cycles (e.g. hydrological, carbon, nitrogen), energy flow through trophic webs, and ecological interactions between biotic agents (e.g. competition, predation, mutualism, commensalism, parasitism); 2) productive patterns that maintain the possibility for indefinite temporal continuity without degradation to material cycles; 3) diversity of function in the ecosystem, with the greatest possible number of niches filled; and 4) redundancy of function, so that the potential loss of some species can be compensated by the activity of others.

In order to think of food sovereignty as a stable state or domain of attraction, it becomes necessary to define the "pull" that is capable of directing a food system transition, once the thresholds of the industrial food system have been reached. The direction and strength of this force of attraction almost certainly depends on which thresholds have been reached in the industrial food system-state. Sometimes in order to better understand the resilience pull effect, we can ask ourselves, in any given context, what the easiest option is. In the contradictory dominant order of a globalized, capitalist food system, where are there activities that, when certain thresholds have been passed, become easier to do by conforming to the logic of food sovereignty rather than the logic of capital accumulation? Clearly, the access to conventional farm inputs is a defining pull toward the capital-influenced agricultural model. Thus maybe one starting point for a regime shift could be the end of access to conventional, yield-intensifying chemicals and seeds. This was indeed the case in Cuba, the world's greatest example yet of a national-level transition from an industrial agriculture model to an organic, diversified, low-external input agricultural model. When conventional inputs, petroleum, and imported food all become unavailable due to the fall of the Soviet Union and the US trade embargo, Cuba's small farmers, scientific community, and government teamed up to direct a national inward-looking agricultural effort, based on organic urban gardens, agroecological small farms, and the breaking-up of unproductive state farms into cooperatives more directly controlled by workers (Rosset and Benjamin, 1994). In this case, the perturbation was an acute food crisis; the response was a rapid, dramatic regime shift toward the food sovereignty framework.

In other cases, the defining pull that defines food state transitions could be the social demand for land, as was the case in those that accompanied the Mexican Revolution. There is also the eternal drive for greater social justice and equality, which has been a major component of food sovereignty-themed social movements in countries such as the United States and Brazil. In the case study below, the driving pull toward a food sovereignty state of the food system is the concern for human health.

5. The focal system

By focusing on one maize-growing community in rural Chiapas, we set out to understand the resilience of a maize production system to external disturbances and internal contradictions that jeopardize its natural resource base and the health of its inhabitants. That is to say, we are interested in the factors of social-ecological resilience in a small farmer agroecosystem at the community/landscape level over the next couple decades. In Mexico, land reform resulted in the creation of *ejidos,* or agrarian communities of small producers with internal political structures. The *ejido* system is itself a complex adaptive system that has survived decades of neoliberal food and land policy at the national level. Within one *ejido,* our study focuses especially on the food production, distribution, and consumption surrounding what is known as the *milpa* system, or the fields where maize is grown, in 25 hectares within the limits of the *ejido.* The overall shift toward the agribusiness state in Mexican maize farming is an uneven, long-term trend which takes place over the course of decades and has a series of social and ecological feedbacks.

The critical components of the current system include farmer families, land access, synthetic fertilizers, hybrid seeds, and chemical herbicides, as well as maize purchasers. Critical components of the restored agroecosystem will include local knowledge and farmer identity, community interest in health and nutrition, soil fertility, functional agrobiodiversity, farmer-to-farmer knowledge exchanges, strong local organizations, traditional seed varieties, organic fertilizers, and crop rotation (Milestad et al. 2010). Important natural resources in the focal system include biodiversity, clean water, fertile soil, forest carbon, and knowledge in the form of traditional maize varieties. Key people include the producer families, including elders, women, men, and children, while critical values include the relative interest in short-term income versus long-term economic sustainability, the capacity to coalesce around the concept of health, and the impacts of belonging to the small farmer social class (Guevara-Hernández et al. 2011b).

The *ejido* system is a form of local governance, not only for land issues but also for other social issues such as health. Property rights reflect a mix of collective and private land-holdings, with complex informal arrangements used for producing food on the land that is closest to the community, regardless of who is the legal owner. There is a definite lack of strong rural organizations in the region, leading the *ejido* structure, the elementary school, and the local church to hold a monopoly over collective action in the community. Maize farming communities in the region, as in most of Mexico, are highly dependent on government anti-poverty programs, as these have come to replace most productive subsidies and credit mechanisms. Obviously, multinational farm input corporations are untouched by democratic institutions that might be used to control the use of toxic chemicals in the community or promote local seeds (Bakan, 2004). Indeed, scale factors deeply influence the capacity to characterize the resilience of the focal system, because its resilience is intertwined with that of the agribusiness model at the international level, as shown below in table 1.

In order to assess social-ecological resilience, at the focal scale of one maize-growing community in Chiapas during the recent past and near future (10-30 years), we will look for the cross-scale interactions between components in table 1 and describe their feedback mechanisms (Buchmann, 2010). In the next section, we give a context for understanding the local food system in the study community.

Scales of variables	*Food system components*
Spatial Scales	
Microbiological	soil ecology, nutrition assimilation, chemical exposure to organisms
Field	net primary productivity, cost of production, maize yield, planned agrobiodiversity, diversity of associated plant and insect species, soil health, use of trees, use of organic fertilizers, pest and plant disease
Community/landscape	**attitudes, knowledge, organization, experimentation, crop diversity, out-migration, conflict, human health, nutrition, economic necessity, access to social assets, access to land, landscape matrix quality**
Subregional	health indicators, land use, public policies, level of influence of agribusiness, growing seasons, farm prices, social equality, farmer organizations, political economy and actors, capacity for self-reliance in key crops
National	food policy, environmental policy, trade policy, agrarian policy, education policy, health care policy, popular participation in democracy, adaptive governance, level of influence of transnational corporations
Global	climate change, global social movements, capitalism
Temporal Scales	
Hourly, Daily or Weekly	food security, physical activity, household labors, planting dates, weed control, soil biological processes
Seasonal	production cycles, climate factors, weed and insect communities, off-farm income, farm prices, training programs, implementation of government supports
Annual	farm productivity, learning-by-experimentation, community demographics, government policies, farm prices
Decade-level	**soil erosion and compaction, landscape mosaic quality, land use changes, trade policy, market influences, climate change, crop suitability**
Longer-term	land use, population factors, national sovereignty, world economic system

Table 1. Spatial and temporal scales in food systems. Focal scale is in bold.

6. Regional characteristics and historical context of the study community

The Fraylesca region is a hot and dry tropical zone that comprises the Central Valleys of Chiapas. Its major city, Villaflores, is located about two hours' drive south of the state capitol of Tuxtla Gutierrez, but the region continues another 100 km to the southeast. The Fraylesca traces its name back to the monks who habited the zone during the early colonial period. It is one of the regions of Chiapas with the least presence of indigenous language-speaking groups, probably due to the productivity of its lands and resulting displacement of the indigenous population during the colonial period. During the 1960s and 1970s, the Fraylesca region was known as the maize equivalent of the breadbasket (*granero*) for southern Mexico, where flat, alluvial valley floors gave typical maize harvests of 5-8 metric tons per hectare. Green Revolution technology, introduced through concentrated efforts to modernize maize farming systems in the flatlands, trickled upstream into the hills as population growth and limited land access pushed families upwards. By the 1990s, the vast majority of traditional maize varieties had been lost in the region, due to adoption of hybrid varieties in government seed programs and corporate advertising. The traditional shifting *milpa* agriculture system was generally replaced by a system of permanent fields in which maize is planted in monoculture during the first rainy season from June to July. Chemical laden maize fields came to dominate the landscape, while posters advertising agricultural chemicals and hybrid seeds are pinned to trees along the highways.

Recent decades have produced change in the Fraylesca region. Increased costs of maize farm inputs, together with soil degradation and low farm prices, appears to be putting the commercial maize farming system in economic jeopardy. Cattle-ranching has been increasingly embraced by hillside farmers, who graze beef cattle on maize stalks during the dry season and set them into the forest during the rainy season. More commercial lowland cattle operations buy the chicken manure from massive chicken farms in the region and feed it to their cattle, in order to produce greater volumes of milk and beef. This practice is generally disliked by the population, but that has not prevented it from becoming the conventional practice adopted by ranchers and dairy farmers. Intertwined with the growth of cattle-raising and the uncertainty of maize cultivation, the changing climate has added to the insecurity of social-ecological systems in the Fraylesca region. Growing seasons have shifted as annual precipitation has begun to concentrate in the second rainy season of the year from September to November, increasing the risk of cob-rot fungal disease. Rainfall has become scarcer during the long dry season from December to May, leading farmers to concentrate their cropping activities between the months of June and October.

Land tenure in the Fraylesca region is subject to similar social tensions to those that have characterized Chiapas as a whole during the last 50 years. The agrarian reform of the Mexican Revolution was slow in arriving to Chiapas, and the *finca* system of large landlord estates remained intact into the 1920s. The first *ejidos*, or agricultural communities created and protected by Mexico's agrarian reform laws, were created in the region as a result of social struggle in the 1920s in valley floors, and currently resemble small towns of paved streets, parks, and residential neighborhoods. As population pressure increased, peasant families have challenged landlord estates across the landscape, building makeshift communities in remote hills and asking for government recognition. Until the constitutional counter-reforms of 1992, Mexican land policy included a legal process for recognizing land claims through the Secretariat of Agrarian Reform (SRA) and reimbursing landowners for the forfeiture of unused land to new agrarian communities, which in turn could become *ejidos*. These second- and third-generation settlements differ in many respects from the older *ejidos*, in that they have much less access to health and education services, markets, transportation, and government supports. In newer *ejidos*, maize farming and cattle grazing take place on slopes that are much more vulnerable to erosion than the alluvial flatlands of the valley floors. As the Zapatista rebellion and federal military occupation of much of the Los Altos region took place in 1994-95, indigenous communities displaced by the violence began to look for land far into the hills of the Fraylesca region. These communities, fleeing from bloodshed, occupied land belonging to large and small landholders alike and sought federal recognition. While many such communities have obtained a certain level of land security by gaining *ejido* status, others remain in situations of precarious land tenure despite more than a decade of waiting for governmental recognition, even while cultivating the landscape and constructing homes.

7. Characterizing small farmer maize production in 24 de Febrero

The *ejido* of 24 de Febrero is about 45 minutes' drive south of Villaflores, in the municipality of Villa Corzo. The community is situated at 16°06′30′′ north and 93°22′33′′ west, at an altitude of 900 meters above sea level (Figure 2). The climate is considered subhumid tropics, with an annual precipitation of 1,248mm, concentrated in five months from June to October, and a mean temperature of 24° Celsius. Of the 1,240 hectares that belong to the *ejido*, 650 hectares are considered forestland and areas of important habitat for rare species of mammals, birds, reptiles, and amphibians. No more than 40 hectares are dedicated to maize cultivation in any given year, and often on fields with between 5 and 20 years of continuous maize monoculture production. Meanwhile, the area dedicated to cattle production shifts across the landscape, from pastures to former maize fields to forestland.

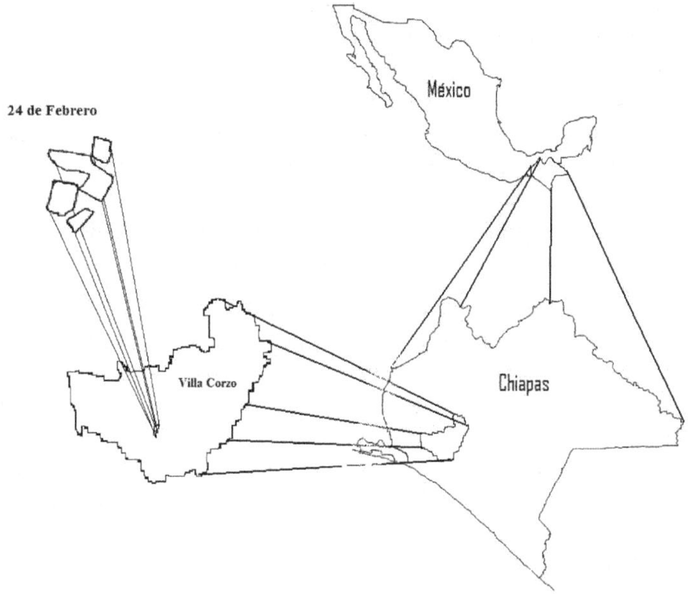

Figure 2. Geographic location of the *ejido* 24 de Febrero, in Villa Corzo municipality, state of Chiapas. Source: Rural Development Studies Network (2011).

The community of 24 de Febrero is primarily made up of one extended family of blood relatives and in-laws. None of the residents of the community speak an indigenous language. In contrast to many *ejidos*, the founders of the settlement were small farmers in nearby lands between the current community and the valley floor. In the mid 1980s, a group of peasants from a different part of the Fraylesca region organized with a lawyer to occupy lands belonging to one of these small farmers. In response, a large part of the extended family organized to create an *ejido* on the contested lands, and thus avoided a land conflict between peasant groups. So it was that the *ejido* 24 de Febrero was founded in 1986 with 5 homes. In subsequent years, the settlement has grown to include over 50 homes, with most new construction by sons and daughters of the community's founders. In the entire community, there are only 38 maize farmers, due to cost/benefit pressures that have taken out of production several former fields that are now considered to be too far to walk from the community.

Transition from the original subsistence system

OG plus no fires, OG plus fertilizante, etc.

Risks in the degraded state: soil erosion and cob-rot disease

We need to evaluate soil erosion in the 18 fields, as well as damage due to the cob-rot disease.

Farmers in 24 de Febrero plant both purchased maize seed and several native or mixed varieties that have been used by local farmers for generations. Planting usually takes place in the months of June and July, when torrential rains soften land. Maize fields are typically about 15 to 45 minutes' walk from the population center of the community, and are generally on sloped hillsides that are deeply eroded. Some farmers mix their seeds with chemical pesticides in order to limit damage from ants, while others use local herbs such as *epozote* to the same effect. Planting is carried out using hollow gourds to contain seeds, and a wooden stick with a metal tip to open up small holes in the untilled soil. Two or three seeds are tossed into each hole, which are made every 40 cm in rows of 80 cm width. Some farmers still follow the traditional practice of mixing squash seeds in with their maize seeds, in order to plant squash every 3m or so throughout the field.

Generally farmers clear fields for planting by using a systemic herbicide such as glyphosate, applied soon after the first rains in May. A few farmers still burn fields before planting maize, although only in areas that haven't been planted with maize in several years. Around 4 days after planting, most farmers apply a contact herbicide, such as 2,4D amina or paraquat. At 15-20 days, farmers apply a dose of nitrogen or phosphate fertilizer. Another herbicide treatment is carried out at 40 days with 2,4D amina and paraquat. At 45-50 days, a second fertilization is carried out. Maize plants are bent over below the ears only in fields where beans are planted in between rows, during the months of September and October. Sweet maize is harvested in September and October for family consumption, while the vast majority of maize ears are left to dry in the fields and harvested from December to March (table 2).

<table>
<tr><td>Jan</td><td>Feb</td><td>Mar</td><td>Apr</td><td>May</td><td>June</td><td>July</td><td>Aug</td><td>Sep</td><td>Oct</td><td>Nov</td><td>Dec</td></tr>
<tr><td colspan="2">Maize harvest</td><td colspan="3">Home gardens, sugar cane and off-farm activities</td><td colspan="2">• Plant maize
• Plant squash</td><td rowspan="2">40 days of heat without rain</td><td colspan="2">• Plant beans and harvest sweet maize</td><td colspan="2">•Bean and maize harvest begins</td></tr>
<tr><td colspan="5">Dry season</td><td colspan="2">First rainy season</td><td colspan="3">Second rainy season</td><td colspan="2">Dry season</td></tr>
</table>

Table 2. A seasonal calendar for cropping activities in 24 de Febrero.

During harvest, farmers remove only the corn ear, and leave all crop residues in the field. Soon afterwards, cattle are generally moved into harvested fields to graze upon the maize stalks. Cattle manure and pulverized maize stalks encompass the major sources of soil organic matter. Crop rotation is a very rare practice in the community, as maize is the only commercial crop and additional food crops such as sugar cane, banana, chili peppers and yucca are grown in small, separate batches. The practice of leaving fields in fallow, or several years of woody bush and chaparral tree regrowth in between cycles of maize production, was largely abandoned with the adoption of synthetic fertilizers by the community. Crop association, however, remains as a traditional practice in several fields, where farmers plant squash seeds along with their maize and beans in the latter rainy season.

7.1. Conventional and alternative practices

Figure 2 shows the ecological quality of several chemical practices in the community. A value scale from 0 to 3 was applied to four indicators of chemical usage, following a method for quantifying ecological quality of management practices (McCune et al. 2011). A value of 0 denotes practices with no benefit and with harmful impacts to ecological processes, 1 represents practices with no benefit but with a minimum of harmful effects, 2 denotes practices with minimal or insufficient benefits to ecological processes, and 3 represents practices with broad ecosystem benefits and that are applied with ecological criteria.

For example, in the case of chemical inputs as shown in Figure 2, the value attributed to each product is essentially the inverse of usage intensification; e.g. the higher value for 2,4D amina shows that this product is less widely used than paraquat. The indicator practices were chosen to be sensitive to changes from the conventional practices found in the community, in order to indicate where processes of innovation may be entering into maize farming. Combinations of organic with conventional fertilization practices remain very rare in the community, as do responses that indicate that farmers believe that it is possible to produce without conventional inputs. Incipient processes of innovation were found in the use of composts for fertilization, and in substitution for paraquat, a contact herbicide that is also a respiratory toxin. In 2011, four farmers were experimenting with liquid mixes made from the leaves of two common trees (*Ficus* spp. and *Byrsonima crassifolia*), along with fine salt and one-tenth the normal dosage of paraquat. Results were encouraging, although the next steps for expanding the usage of these homemade liquids are unclear. Despite the farmer experimentation taking place, chemical fertilizers and paraquat-based herbicides were most intensively used of the four indicators of agrochemical use among the 18 farmers surveyed.

The information in Figure 3 can be useful for determining the kinds of dependence produced within small farmer communities in Chiapas. The extreme dependence on synthetic fertilizers is an indicator of the level of soil erosion present in agricultural fields. Aside from highly unusual, small-scale efforts at growing organic maize for specialty markets or home use, the only maize-growing systems in Chiapas without this dependency

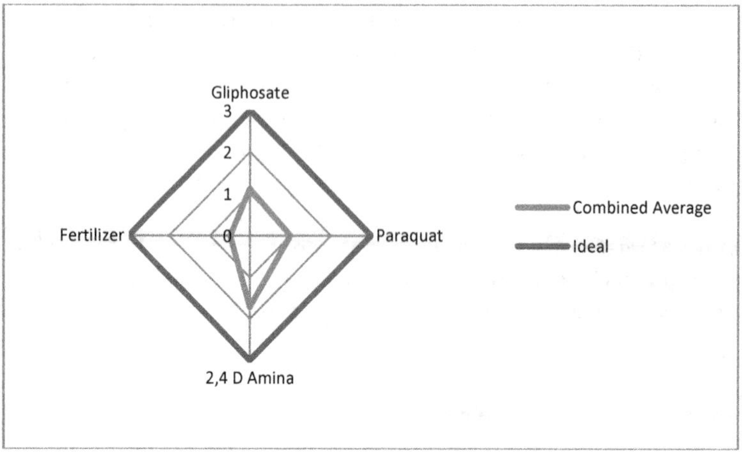

Figure 3. Ecological quality of chemical input use for maize production in the ejido 24 de Febrero, using an indicator system on a scale of 0 to 3. The value of zero represents maximum chemical application, while the value of three represents no usage.

on chemical fertilizers are the traditional shifting agriculture systems in which the forest re-growth of long-fallowed fields is cut and burned, and seeds planted into the rich layer of ash.

It is useful to identify the constraints and feedbacks of soil fertility management in Chiapas maize production. Fertilizers represent the greatest cost of maize farmers, with typical costs in 24 de Febrero reaching well over $1,000 US per hectare. Yet substitution of synthetic fertilizers with organic soil amendments is difficult, because hillside soils are so badly eroded that existing soil has almost no nutritional content, and up to 40 tons of organic matter per hectare would need to be applied in order to satisfy nutritional requirements of maize. The production and transportation of this volume of organic fertilizers would require large inputs of labor, difficult for farmers to provide as livelihood diversification strategies have left less time for maize-production activities than before. Thus the availability of time and labor (or cash in the case of purchased organic fertilizers) is a limiting factor for the efforts to break the dependency on synthetic fertilizers. The sloped maize fields represent an additional restraint, in that erosion is likely to undo most soil amendment applications until a massive labor effort goes into erosion-reduction practices, such as stone or stick terracing.

Feedbacks between management and ecological factors also complicate efforts to reduce fertilizer dependency. For example, populations of soil organisms that could improve soil structure and nutrition over time are likely to be negatively affected by the application of chemical fertilizers. In addition, cattle grazing in maize fields during the dry season can exacerbate erosion and also cause soil compaction. Compared to other crops, maize is hardy to degraded soil structure as long as sufficient nutrients are present. Indeed, farmers feel that its capacity to adjust to poor soils is an aspect of maize's centrality as "the" subsistence crop. Thus the economic need to produce every year is combined with the fact that maize is the only crop that can be produced under such marginal conditions, to create a system of monoculture maize production year after year in the same fields. Legal prohibitions on burning are meant to limit landscape degradation, but they also effectively end shifting agriculture, since fire is the typical way to open sloped fields to agriculture. Without shifting fields or rotations, soil degradation is accelerated to alarming levels.

The use of herbicide cocktails is a characteristic of maize farming in the Fraylesca region. The major restraints on reducing herbicide dependence are related to the labor opportunity cost, as manual weed removal takes a great deal of time that can be otherwise used productively by households. New ideas such as the organic weed retardant may represent the most promising directions for reducing herbicide dependence. Management feedbacks also exist with regard to herbicide use. After herbicide disturbance to fields, pioneer species such as aggressive weeds are the first plants to take advantage of nutrients and light in the newly opened spaces. The practice of herbicide application tends to increase the relative abundance of plants species that establish competitive relationships with maize crops. Fields with frequent use of herbicides tend to have major problems with a few key weeds, whereas fields without herbicide use have a greater diversity of associated plant diversity, including beneficial and non-competitive species.

A final interesting aspect of farmers' chemical use is the high dependence on herbicides and very infrequent of insecticides in the maize cropping system. In general, insect pests are not considered to be more than a nuisance, while farmers identify over 10 beneficial insects, mostly insect predators and parasitoid wasps (see below).

Over the course of two or three years, the community has progressively given more importance to ecological considerations, partly as a result of a training course in environmental education offered by the National Forestry Commission and carried out in the *ejido* by a local NGO in 2010. As a result of this course and follow-up activities by a participatory research team from the Autonomous University of Chiapas, several farmers have engaged in communication and experimentation with the purpose of substituting organic and traditional inputs for chemical inputs in agricultural activities. These and other alternative activities, such as a promotion of herbal medicine, are being adopted explicitly out of concerns for human health that emerged in monthly *ejido* assemblies, where the residents attributed poor health to chemical usage and the diminishing quality of diets. A first practice of seed saving using such traditional materials as ash, lime, and several kinds of herbs was carried out with the participation of 18 farmers. Of these, five attempted to grow fields of organic maize in 2011 using compost and organic fumigants. Figure 4 shows indicators for the appropriation of agroecological practices by the same 18 maize producers in the *ejido*. A similar value scale as with chemical use was applied to four indicators of alternative productive activities: nutrient cycling, crop rotation, crop association (intercropping), and seed management.

Our results showed a much greater appropriation of nutrient recycling practices than alternative seed management and crop association activities, basically due to the local customs of leaving crop residues in the field, moving cattle into former crop areas to eat the maize stalks, and the total absence of plowing. Crop rotation was shown to be a major

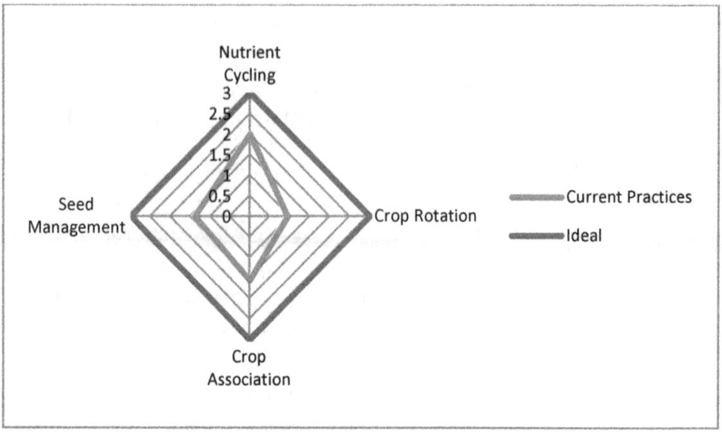

Figure 4. Ecological quality of alternative practices for maize production in the ejido 24 de Febrero, using an indicator system on a scale of 0 to 3. The value of zero represents total absence of alternative practices, while the value of three represents widespread use of ecological practices for reasons that farmers understand.

problem in the community, as annual crops of maize dominate the agricultural landscape during the single growing season. Maize is the preferred crop due to its importance as a food crop, its durability as a commercial crop, and its response to fertilizers even in highly eroded soils, providing a lightly positive cost-benefit balance to farmers for years even as soil quality declines. Cultural preference for maize makes diversification of the productive landscape a complex and sensitive process.

7.2. Biological interactions in maize production systems in 24 de Febrero

Weed and insect communities within the maize fields show that even under existing conditions and technological patterns, the small farmer landscape is capable of supporting a rich diversity of species and functional groups. This is an especially important finding, given that agriculture and biodiversity conservation are often considered in neoliberal theory to be mutually exclusive, even competitive uses for land in the tropics (Grau and Aide, 2008). The idea of contradictory agricultural and conservation goals, and the necessary segregation of the two, has led neoliberal resource economists to support wilderness reserves in some parts of the rural tropical landscape and industrial agriculture in the rest (Aide and Grau, 2004). The problem, as pointed out by rural organizations, is that small farmers are essentially excluded from both parts of the landscape, and conservation policy then becomes a tool for the dispossession of family farmers and rural communities. In addition to small farmer objections, the neoliberal model of biodiversity conservation has been challenged on ecological grounds, as recent decades of theory on metapopulations has shown the importance of migration between habitat patches for species survival. According to this conservation paradigm, also known as the convergent model of mixed land-use (Miki et al. date unknown), agricultural systems that retain elements of the original ecosystem can promote successful migration between patches of wilderness. This may make the ecological quality of agroecosystems even more important than the conservation of habitat patches to biodiversity conservation in the tropics.

In 18 fields of 24 de Febrero, weeds and crops were measured monthly for the percentage of area they covered within three 50cm x 50cm quadrants placed using a random block design in each field, for a total of 54 quadrants, during the six-month maize growing season. In the total studied area of 13.5 square meters, over 30 species of weeds were found, reflecting high overall richness despite the use of herbicides. Tree cover and type varied among fields, but crucially, trees were present in all fields. The uses for weeds and trees were various; eight weeds were considered to be edible, and 11 were identified as medicinal plants. Among trees, several were nitrogen-fixing legumes and others were fruit-bearing, with the remainder having social use as building material, medicine, or firewood.

We identified 29 families of insect herbivores, seven families of secondary consumers or predators, 12 families of parasitoids, and two families of pollinators in five samplings across the eighteen study fields during the growing season of 2011. The insect community within maize fields reflects a high level of biodiversity and is likely to have the net result of stabilizing yields, creating an "ecological homeostasis" through complex networks of trophic, life-cycle, and density-dependent interactions (Vandermeer et al. 2010). The critical interactions within such an autonomous agroecological service such as pest control may be highly complex and occur on various spatial and temporal scales.

In 24 de Febrero, maize fields also bear beans, squash, tomato, edible herbs, several medicines, and several use categories of trees. This multi-use aspect of agricultural fields may lend itself to system resilience, since any detrimental impact on maize production is partially offset by the other functions of the same land. Land-use diversity is an element of system resilience that is pronouncedly strong in small farmer settings (Altieri, 2010). While the use of the farm landscape in 24 de Febrero retains an important level of diversity, it is also useful to ask why it doesn't have even more, especially given the supposition that small farmers maintain diverse productive systems. To understand the drivers of land-use change and agricultural intensification, it becomes necessary to examine the social and economic vulnerability of small farmers in Chiapas.

7.3. Characterizing social resilience

Livelihood is an important concept for understanding risks in social-ecological systems. Both vulnerability and livelihood trace their conceptual roots in the search by Sen for adequate measures of well-being (1993). Livelihood has to do with the relationship between households and the conditions of their production and reproduction as an economic unit, including housing, employment, income, access to basic necessities and to consumer goods, transportation, health, and education. It is generally used to define baseline measures of human well-being, and as such applied to small scale rural producers and the rural or urban classes without property. While livelihood studies generally examine immediate aspects of economic life at a household level, vulnerability studies tend to focus on structural factors, such as legal, political, cultural, ecological or economic factors that threaten livelihoods.

In order to make an initial characterization of livelihood resilience factors in the community of 24 de Febrero, an indicator system was created to include among its variables: food access, health care access, access to credit, access to public programs, access to markets, access to alternative technologies, and access to education (Figure 5). These access indicators are significant to everyday life under normal circumstances, but they are also indicative of social risks that could become urgent under changing conditions.

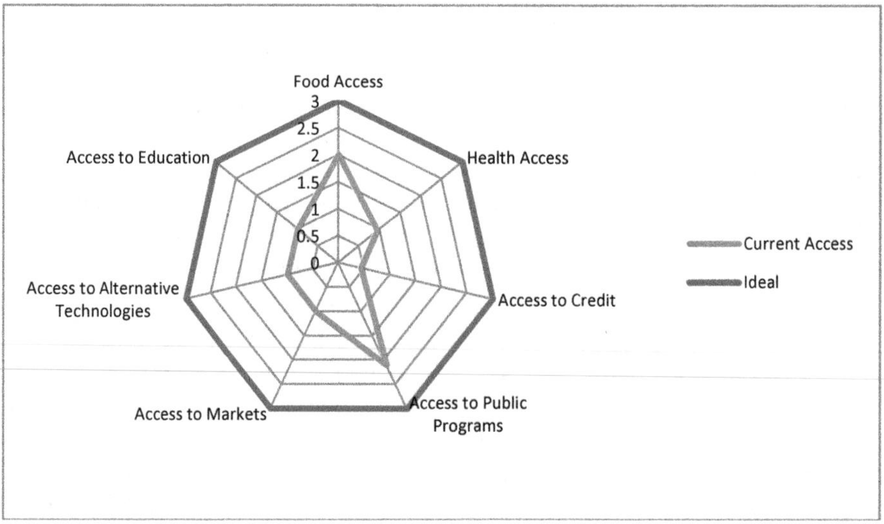

Figure 5. Factors of livelihood resilience in the ejido of 24 de Febrero, using an indicator system on a scale of 0 to 3. The value of zero represents abandoned or systematically denied access rights, while the value of three represents free and universal access as well as participation within planning or implementation processes.

Results confirm the existence of several types of social vulnerability in the community. Indicator values for access to food and public programs were substantially greater than those for other variables. Values were notably low for access to education and access to credit, two indicators related to opportunity. With regard to education, the attained value of one means that the average response to interview questions was that beyond elementary school, monetary costs associated with education made it inaccessible. The very little access to credit for farmers in the maize growing regions of Chiapas is a matter of considerable importance for the community of 24 de Febrero, and contributes to migration by young people to the United States in pursuit of sufficient cash to construct homes or purchase fertilizers. At the same time, the conservation of certain traditional practices is often attributed to the lack of farm credits, which would enable farmers to pursue a more technified production strategy.

8. Food sovereignty and the nation-state

While we have characterized many aspects of the maize production system in 24 de Febrero, the question remains: where is the community located on a gradient between agribusiness and the food sovereignty model? Clearly, the food system of the community is somewhere in between the two states that we have described. Its combinations of traditional and conventional technologies, cash- and subsistence-oriented agriculture, monoculture and multi-use systems, indeed, environmental stewardship and degradation, give the local food system a character highly compatible with the sovereignty system state. However, the community is within a nation that has been subjected to the full formula of capitalist agribusiness.

Here we come to a fundamental issue of scale: the local food system in 24 de Febrero can only be understood in its larger context, as part of Mexico's food system. On one hand, it is an adaptation on the traditional *milpa* system of maize production on collectively held lands that has characterized Mexican food systems for millennia. On the other hand, it is the result of the extension of capital logic and conventional technologies to far corners of the Mexican countryside, bringing junk food and chemical input dependencies to the rural household. The ambiguity is a signal of the importance of scale, and it may well be possible that systems within the gravitational pull of one steady state could also exist within another. For example, the Procede land certification policy enacted during the 1990s was thought to be the end of collective landholding in Mexico, as it partitioned private titles for *ejido* lands and legalized land sales (De Ita, 2000). However, the internal resilience of the *ejido* system, based on social and political feedback mechanisms, was strong enough that certification did not have the same effect that it has in other parts of the world, such as Africa and the Middle East.

At a larger historical scale, collective resource-use regimes such as the *ejido* system may be momentarily compatible with both capital-driven and socially-planned economies. In this sense, a valid comparison can be made between *ejidos* of Mexico and agricultural production cooperatives in Cuba, both of which are based on profound land reform and collective agrarian property governed by local assembly. Such institutions can exist within countries dominated by the industrial food model, but the dynamic of the overall food system will determine how long they last and how they change. The *ejido* was created as a compromise between the radicalized peasantry of the Mexican Revolution and conservative groups of power that were interested in limiting resource redistribution. By giving *ejidos* to peasants, militant rural organizations could be demobilized and wages could be kept low during industrialization, since industrial workers' wages were supplemented by their access to productive land. Essentially, the *ejido* was a major tool for the consolidation of a new bourgeois regime after the Mexican Revolution. It was the eventual reorientation of the national economy toward global capital and away from nation-building in the late 1970s that brought the *ejido* system into conflict with the emerging neoliberal resource management regime. The temporary compatibility of agrarian systems that have a food sovereignty character, such as the Mexican *ejido*, within industrial food systems that are in the process of consolidation, is a matter of scale and historical contingency. Components embedded in one kind of food system can reflect a distinct qualitative character, as long as they are so limited in scale and impact as not to push the larger system to a threshold.

The dependence of the maize production system in 24 de Febrero on foreign multinational corporations and the chemicals they sell is a sign of its integration into the international capitalist development model that dominates Mexico from outside. As maize sellers, they are limited to the white maize varieties sought by middlemen buyers en route to large-scale tortilla production (mixed first with maize from the state of Sinaloa or, increasingly, the United States) or exportation to Guatamala. Surrounded by an agricultural landscape of maize fields with hybrid seeds, the integrity of their traditional varieties is at risk of contamination. It would seem that in neoliberal Mexico, the community is within the outer reach of the industrial food model and as such, subject to its contradictions. Nonetheless, on a household level, there remains a level of resistance to the industrial food system, which takes the form of self-sufficiency in basic grains, conservation of landrace maize and bean varieties, use of home gardens, traditional labor-sharing arrangements and artisan food processing. These are all components of the local food system consistent with a food sovereignty framework, but they are gradually disappearing from the landscape. What is the role of the Mexican State in this transformation?

8.1. Devolutionary governance

Since the Mexican Revolution of 1910, national policy toward agriculture has reflected a struggle between peasant groups that have fought for land access and favorable policies toward small farmers, and a combination of business and political elite from within and outside of Mexico that have sought to develop a capitalist, export agriculture model. Since the dawn of the twentieth century, capital investments began creating two basic tracks for Mexican agriculture: the capital-intensive irrigated, specialized farms in Central and Northern valleys and plains, and subsistence agriculture in most other non-urban land in Mexico. This split in land use reinforced the nation's conception of monoculture as "modern," and diversified, low-input farming as "backward."

The agricultural research and technology program that came to be known as the Green Revolution was largely a US Cold War-era effort to resolve issues of hunger and poverty in Mexico with technical solutions, rather than new social and economic policy (Perfecto et al. 2010). The proliferation of new, "modern" seed varieties, as well as irrigation infrastructure, synthetic fertilizers and pesticides during the 1960s, 1970s, and 1980s, was a highly uneven process that reflected the compromise between corporatist governmental policy beneficial to large agricultural interests, and the commitment to

Since the 1980s, Mexico's government has opted for a free trade strategy in all productive spheres, as a result of changes in the economic ideology of the ruling party, as well as external pressure from international lenders and the United States. For Mexican industry, this meant the final and unequivocal abandonment of the import-substitution industrialization strategy. In commerce, it eventually led to the signing of the North

American Free Trade Agreement in 1993, which opened up Mexican markets to a flood of cheap products from the United States. In agriculture, the adoption of the free trade model meant a shift in strategy from the goal of self-sufficiency that had characterized agricultural and land policy since the revolution. Guaranteed farm prices for basic grains such as maize disappeared, as the state reduced its presence in the countryside and international private actors stepped into the void. Cheap grain from subsidized farmers in the United States began to flood Mexican markets, adding to the economic insecurity of millions of Mexican maize farmers. Meanwhile, the price of tortilla, the basic and essential form of maize in the Mexican diet, has more than tripled for consumers since NAFTA was signed, as a result of concentration of the maize storage and processing sectors by several transnational corporations.

One of the most controversial issues in contemporary Mexico is the entrance of genetically modified maize into the country, almost universally from the United States, as seed, feed, or food. In 2001, Mexican and U.S. researchers accidently found traces of genetically modified maize in landrace varieties of rural Oaxaca (Quist and Chapela, 2001, Nature 414), and subsequent studies have confirmed the contamination of maize landraces by modified genes across Mexico. Given the extraordinary cultural and alimentary importance of maize in Mexico, the loss of traditional agrobiodiversity in this crop represents a loss of national patrimony and sovereignty. In 2007, President Felipe Calderon created by decree a federal program to support *in situ* conservation of landrace maize varieties by farmers. However, the Secretary for Agriculture, Livestock, Fish and Food (SAGARPA for its initials in Spanish) was cold to the proposal, as it went against the productivity focus of its programs. Thus it fell to the Secretary for Natural Resources and the Environment (SEMARNAT) to take on the maize biodiversity program. SEMARNAT, in turn, sent the new law to its National Commission for Natural Protected Areas (CONANP), which began to apply the program, its implementation having now been reduced to agricultural areas within nature reserves.

CONANP's Program for *in situ* Conservation of Landrace Maize, or PROMAC (*Programa de Maíz Criollo*) as it is more commonly known, is still a new fish in a very complex pond of federal and state programs that combine agriculture and natural resource conservation. It pays about $100 US per year to farmers who have been growing landrace maize varieties in nature reserves to continue growing them, and advocates the conservation of the traditional *milpa* productive system (maize in association with squash, beans, and other edible plants). PROMAC funds are used based on the discretion of each nature reserve, and can be used to hold seed exchange fairs, conduct capacity-building trainings for farmers, build seed banks and even create maize-based cultural centers. While this program clearly has the potential to strengthen the peasant maize production system, its capacity to help small farmers and protect landrace maize varieties depends on how it is implemented in each nature reserve. In interviews, many nature reserve officials compare PROMAC to PROCAMPO: a program created during the administration of Carlos Salinas (1988-1994) to buffer his free trade economic shocks and which pays annual subsidies to all citizens who show documents proving that they grow crops or raise livestock. Despite its populist appeal, PROCAMPO is a notoriously inefficient program, as its implementation provides ample opportunity for fraudulent payment claims and requires no participation in training programs or production plans. Many Mexican politicians at the national level oppose support of any kind for maize production, because they see it as a marginal subsistence activity that is outside of the free market agricultural strategy of specialized crop exports and basic grains imports. Thus from its origins, PROMAC has been born into a hostile and disjointed institutional atmosphere, in which some nature reserves have ignored the program while others have encouraged farmers to enter into it.

While PROMAC is among the more important federal programs for maize farmers in Chiapas, due to the significant amount of farmland within protected areas, on the state level there is a program called Solidarity Maize that is closely connected with the governor's office. This program ostensibly gives payments in the form of agricultural inputs to all maize farmers in the state. Given that fertilizers represent the greatest expense in maize farming in many regions of Chiapas, small farmers are generally in favor of Solidarity Maize and eager to participate. Unfortunately, the program reaches a relatively small portion of the actual maize farmers in Chiapas, while creating a massive informal market for sacks of fertilizers that are often exchanged for political allegiance long before reaching farmers.

In 2008, a group of farmers and advocates formed the Landrace Maize Network (*Red de Maíz Criollo*) in Chiapas in order to stem the loss of traditional peasant varieties of maize and defend the *milpa* production system. This group protested the fact that the supports from the state government through the Solidarity Maize program set small farmers on a course toward conventional, chemical-laden agricultural practices. The Landrace Maize Network achieved a commitment by the state government to offer organic fertilizers to those producers who request them, setting a new precedent for governmental support for alternative agriculture. Unfortunately, to date very few farmers know that they have the option to request organic farm inputs.

The farmers of 24 de Febrero have yet to receive support from PROMAC, despite their long-term commitment to growing traditional maize varieties, and they have not received support from the Solidarity Maize program either. In fact, PROCAMPO is the only government support that they receive. Despite being relatively close to population centers, and following the requirements to be considered in state and federal programs, they have been left out of the little support for small-scale agriculture that exists in Mexico.

9. Conclusions

The future is unknown, social and ecological drivers of change are linked, and periodic, qualitative change is part of life. That is one view of the world, carefully developed in resilience theory since Holling's (1973) seminal essay. Here we have posited two contrasting (but not exhaustive) food system possibilities, in part to demonstrate the openness of history. We do not see evidence for necessary evolution toward stable equilibrium in either social or ecological systems. Rather, the last several hundred years have shown that history is full of surprises, and theory of stages of development is often more hindering than helpful. In its insistence on the existence of the unknown, orientation toward emergent properties, and focus on feedback mechanisms, resilience theory is of extraordinary usefulness in social science.

Despite such impressive strides in systems thinking, we find the brave new world embraced by resilience science to itself be exclusionary in terms of possible outcomes for humanity. Far too often, the capital system is naturalized into the feedbacks of the social-ecological framework, rather than understood as a historical system that is liable to the same phases of growth, decadence, collapse and renewal as something qualitatively different. Indeed, in the present conditions of global financial crisis and historical highs of economic inequality, it would be quite blind not to accept the collapse of global capitalism as a historical possibility. With natural resource exhaustion and an exploitative human-nature relationship increasingly understood as inevitable contradictions of this economic system (O'Conner, 2001), the alternative stable states model is highly relevant for the testing of alternatives at distinct scales.

In *ejido* assemblies, the residents of 24 de Febrero have identified the directions that they would like the community to take. The local vision is of organic agriculture as a response to what appear to be increasing health problems, such as high blood pressure, diabetes, and cancer. The community has made collective decisions to begin a process of experimentation to combine what people remember of the practices used by past generations with technical advice from a research team of the Faculty of Agronomic Sciences at the Autonomous University of Chiapas. The role of local innovation is being filled (Milestad et al. 2010). But is this enough to pull the system toward a food sovereignty stable state?

The resilience literature identifies the need for adaptive governance (Allen and Holling, 2010; Folk et al. 2005). This can be interpreted in several ways, but the creation of public-private partnerships is often suggested, albeit in terms of "bridging organizations" or multi-stakeholder groups. In sum, governance for resilience is understood to take place in what Walker and colleagues (2002) call with refreshing honesty "more-or-less democratic, pluralistic, capitalist" societies. The resilience principle, as applied to social-ecological systems, has been applauded and feared because it normalizes—and absorbs all critique to—the neoliberal development model (Walker and Cooper, 2011). When adaptive governance is understood to mean increased private influence over formerly public spheres, especially natural resource management, then the objectives of its research agenda may well include building resilience to "shocks and disturbances" like market crashes, critical social movements, and dissent.

Unfortunately, history may show that the degraded system is global, and that what is "too big to fail" in the international economic system has indeed already failed, in terms of its social and ecological impacts. The destruction of global agrobiodiversity—a fundamental component of food system resilience—that took place during the last half-century has not been accompanied by a solution to world hunger, as more than one billion undernourished people make plain. If resilience is the "capability of a system to maintain its functions and structure in the face of internal and external change and to degrade gracefully when it must,

(Allenby and Fink, 2005)" then perhaps the most prudent option would be to begin a wide debate about the graceful exit of the agribusiness model. The development of sovereign food systems at the local level requires cross-scalar interactions that should not be limited to farmer innovations, but include farmer-to-farmer networks, strong rural organizations, and redistributive public policy. If these requirements cannot be met within the neoliberal development model, then the next step is to ask what kind of transformation is necessary at the global level, beyond the narrow market valuation of Nature.

Author details

Nils McCune
Ghent University, Belgium
Red de Estudios para el Desarrollo Rural AC, Mexico

Francisco Guevara-Hernandez
Facultad de Ciencias Agronomicas, Universidad Autonoma de Chiapas, Mexico

Jose Nahed-Toral,
El Colegio de la Frontera Sur, San Cristobal de las Casas, Mexico

Paula Mendoza-Nazar
Facultad de Medicina, Veterinaria y Zootecnia, Universidad Autonoma de Chiapas, Mexico

Jesus Ovando-Cruz
Red de Estudios para el Desarrollo Rural A.C., Mexico

Benigno Ruiz-Sesma
Facultad de Medicina, Veterinaria y Zootecnia, Universidad Autonoma de Chiapas, Mexico

Leopoldo Medina-Sanson
Idem

10. References

Allen, Craig, and C.S. Holling. 2010. Novelty, adaptive capacity, and resilience. *Ecology and Society* 15(3): 24-39.

Allenby, Brad, and Jonathan Fink. 2005. Toward inherently secure and resilient societies. *Science* 309: 1034-1036.

Bakan, Joel. 2004. *The Corporation: the pathological pursuit of profit and power.* New York: Free Press.

Buchmann, Christine. 2010. Farming System Dynamics: The quest for a methodology to measure social-ecological resilience in subsistence agriculture. Presentation in: *Adapative management in subsistence agriculture.* 9th European IFSA symposium. 4-7 July 2010, Vienna (Austria).

García-Linera, Álvaro. 2011. *La Potencia Plebeya: acción colectiva e identidades indígenas, obreras y populares en Bolivia.* Havana: Fondo Editorial Casa de las Américas.

Ericksen, Polly, Beth Stewart, Jane Dixon, David Barling, Philip Loring, Molly Anderson and John Ingram. 2010. The Value of a Food System Approach. In: *Food Security and Global Environmental Change*. London: Earthscan.

Folk, Carl, Thomas Hahn, Per Olsson, Jon Norberg. 2005. Adaptive governance of social-ecological systems. *Annual Review of Environmental Resources* 30: 441-473.

Fox, Jonathan and Libby Haight. 2010. Mexican agricultural policy: Multiple goals and conflicting interests. In: *Subsidizing Inequality: Mexican Corn Policy Since NAFTA*. Woodrow Wilson International Center for Scholars.

García-Barrios, Luis, Yankuic M. Galván-Miyoshi, Ingrid Abril Valdivieso-Pérez, Omar R. Masera, Gerardo Bocco, and John Vandermeer. 2010. Neotropical forest conservation, agricultural intensification, and rural out-migration: the Mexican experience. *BioScience* 59(10): 863-873.

González, Humberto and Alejandro Macías. 2007. Nutritional Vulnerability and Mexico's Agro-Alimentary Policy. *Descatos* 6: 14-31.

Grau, H. Ricardo and Mitchell Aide. 2008. Globalization and Land-Use Transitions in Latin America. *Ecology and Society* 13(2): 16-28.

Guevara-Hernández, Francisco, Nils McCune, Heriberto Gómez-Castro, René Pinto-Ruíz, Francisco Medina-Jonapá, Adalberto Hernández-López, and Carlos Tejada-Cruz. 2011a. Conflicting regulatory systems for natural resource management in Southern Mexico: An ethnographic case study. *International Journal of Technology and Development Studies* 2(1): 30-62.

Guevara-Hernández, Francisco, Nils McCune, Luis Rodríguez-Larramendi, and Gillian Newell. 2011b. Who's who? Power mapping, decision making and development concerns in an indigenous community of Oaxaca, Mexico. *Journal of Human Ecology* 36(2): 131-144.

Holling, C.L. 1973. Resilience and stability of ecological systems. *Annual Review of Ecological Systems* 4: 1-23.

Kinzig, Ann, Lance Gunderson, Allyson Quinlan, and Brian Walker. 2007. *Assessing and managing resilience in social-ecological systems: a practitioner's workbook.* Resilience Alliance: Stockholm.

Levins, Richard. 2007. The Maturing of Capitalist Agriculture: Farmer as Proletarian. In: Lewontin, Richard, and Richard Levins, *Biology Under the Influence: Dialectical Essays on Ecology, Agriculture, and Health.* New York: Monthly Review Press, pp. 329-341.

Lin, Brenda. 2011. Resilience in Agriculture through Crop Diversification: Adaptive Management for Environmental Change. *BioScience* 61: 183-193.

Magdoff, Fred, John Bellamy Foster, and Frederick H. Buttel. 2000. *Hungry for Profit: the agribusiness threat to farmers, food, and the environment.* New York: Monthly Review Press.

McCune, Nils, Yanetsy Ruíz González, Edith Aguila Alcántara, Osvaldo Fernández Martínez, Calixto Onelio Fundora, Niria Castillo Arzola, Pedro Cairo Cairo, Marijke D'Haese, Stefaan DeNeve, and Francisco Guevara Hernández. 2011. Global questions, local answers: soil management and sustainable intensification in diverse socio-economic contexts of Cuba. *Journal of Sustainable Agriculture* 35(6): 650-670.

Medellín, Rodrigo and Miguel Equihua. 1998. Mammal species richness and habitat use in rainforest and abandoned agricultural fields in Chiapas, Mexico. *Journal of Applied Ecology* 35: 13-23.

Milestad, Rebecka, Susanne Kummer, and Christian Vogl. 2010. Building farm resilience through farmers' experimentation. Presentation in: *Knowledge, innovations and social learning in organic farming*. 9th European IFSA symposium. 4-7 July 2010, Vienna (Austria).

Miki, Rosa, Mariana Zarazua, Adriana Aguilar, Lizzette Luna, Juan Carlos Hernández, Martín Cadena, and Mariano Torres. Date Unknown. Divergent vs. Convergent Development Models. Sustainability Science Seminars.

O'Conner, James. 1998. *Natural causes: essays in ecological Marxism*. New York: Guilford Press.

Patel, Raj. 2009. What does food sovereignty look like? *Journal of Peasant Studies* 36(3): 663-673.

Perfecto, Ivette, John Vandermeer, and Angus Wright. 2010. *Nature's Matrix*. Earthscan: London.

Ramírez-Marcial, Neptalí, Mario González-Espinosa, and Guadalupe Williams-Linera. 2001. Anthropogenic disturbance and tree diversity in Montane Rain Forests in Chiapas, Mexico. *Forest Ecology and Management* 154: 311-326.

Rosset, Peter. 1998. Do we need new technology to end hunger? In: Lappé, Francis Moore, Joseph Collins, Peter Rosset and Luis Esparza, *World Hunger: 12 Myths*. Grove Press/Earthscan.

Rosset, Peter, and Medea Benjamin. 1994. *Cuba: the greening of the Revolution*. Food First Press: Oakland.

Sen, Amartya. 1992. Capacidad y Bienestar. In: Nussbaum, Martha, and Amartya Sen (Eds.), *La Calidad de Vida*. Oxford: Oxford University Press, the United Nations University, pp. 54-80.

Shattuck, 2011. Resilience and Vulnerability in Agriculture: A Cross-Disciplinary Approach. Forthcoming.

Suding, Katharine, Katherine Gross, and Gregory Houseman. 2004. Alternative states and positive feedbacks in restoration ecology. *Trends in Ecology and Evolution* 19(1): 46-53.

Vandermeer, John, Ivette Perfecto, and Stacy Philpott. 2010. Ecological Complexity and Pest Control in Organic Coffee Production: Uncovering an Autonomous Ecosystem Service. *BioScience* 60: 527-537.

Walker, Brian, Stephan Carpenter, John Anderies, Nick Abel, Graeme Cumming, Marco Janssen, Louis Lebel, Jon Norberg, Garry Peterson, and Rusty Pritchard. 2002. Resilience management in social-ecological systems: a working hypothesis for a participatory approach. *Conservation Ecology* 6(1): 14-30.

Walker, Jeremy, and Melinda Cooper. 2011. Genealogies of Resilience: From Systems Ecology to the Political Economy of Crisis Adaptation. *Security Dialogue* 14(2): 1-28.

Walker, Richard. 2004. *The Harvest of Bread: 150 years of agribusiness in California*. Berkeley: University of California Press.

Chapter 9

Land Degradation, Community Perceptions and Environmental Management Implications in the Drylands of Central Tanzania

Richard Y.M. Kangalawe

Additional information is available at the end of the chapter

1. Introduction

Land degradation particularly through soil erosion is an important concern in many parts of the world including semiarid areas of central Tanzania. One of the issues that have gained importance is the concern on implications of local perceptions in resource management. The Irangi Hills (Figure 1) are severely affected by soil degradation; hence they provide a vivid example for studying environmental degradation, local perceptions, and land-management strategies that the local population use to cope with the degradation problems [1]. The Irangi Hills have been influenced by various land management interventions, for example, the implementation of various soil-conservation measures that have contributed to considerable changes in the spatial and temporal land-use patterns during the 20th century, and particularly over the last four decades. The Irangi Hills, located in Kondoa District in semiarid, central Tanzania (Figure 1), constitute about 10% (c. 1256 km^2) of the District area that has been particularly affected by sheet and gully soil erosion [2, 3 ,4, 5, 6,7]. In many places soil erosion has reduced the agricultural potential of the land by the physical removal of topsoil, sand deposition on lower slopes and valley floors, and gullying and incipient badlands development [7]. The sub-humid, north-eastern parts of the Irangi Hills are more severely degraded than the semiarid south and south-west.

A combination of factors makes soil erosion a particularly serious problem in these semiarid areas of Tanzania. The problem has often been associated with local mismanagement of the land resources through among others, overgrazing, over-cultivation, burning of grasslands and woodlands, resulting in over-exploitation and consequent soil erosion [8]. However, it has been shown in recent studies in the Irangi Hills that the problems and causes may be more complex than was earlier presumed, including factors such as tectonic activities and historical changes in climatic conditions [9, 10, 11, 12].

In southern Africa, where many researchers are engaged in land management and conservation issues, many have included the aspect of perception and their influence on resource management by various groups of the society [13, 14, 15]. Local perceptions can be established by interviewing local people about how they viewed various resource conditions [16]. The importance of acknowledging the socio-economic environment of the various decision making groups involved is discussed in [17]. Also Dahlberg [18] accorded considerable importance on local views in the study of environmental change and degradation in Botswana. Understanding the local people's perceptions on environmental issues is thus a prerequisite in making successful and sustainable resource management strategies.

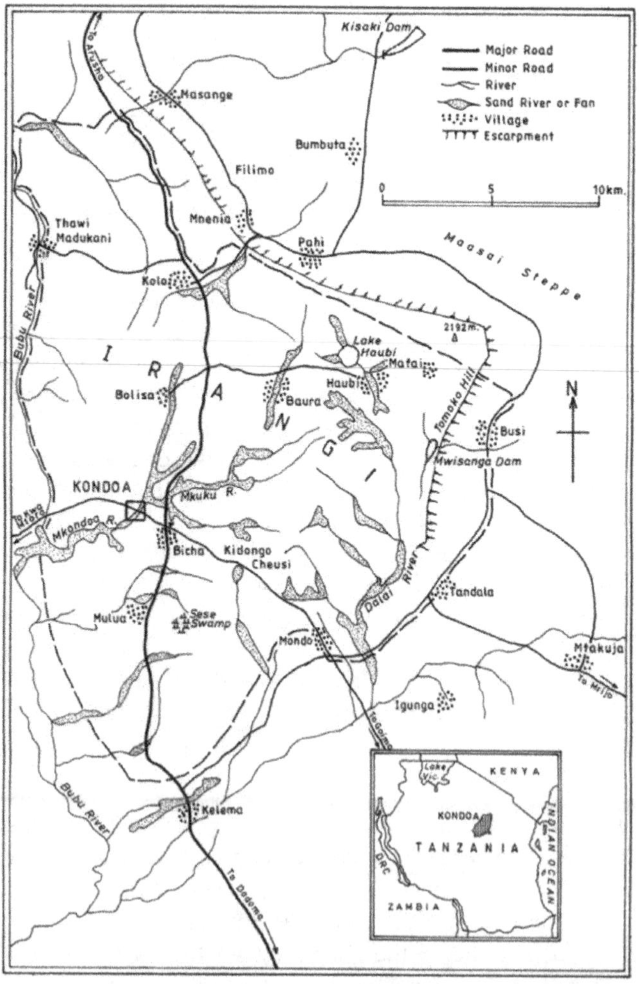

Figure 1. Map of the Irangi Hills showing the Kondoa Eroded Area and the surveyed villages of Mafai, Baura and Bolisa. The insert is a map of Tanzania showing the location of Kondoa.

Local people's perceptions of environmental issues can be looked at from three perspectives. First, people will perceive land degradation on the basis of their socio-economic interests. In this case, farmers will be more aware and concerned about environmental changes and damages that affect agricultural productivity such as soil erosion. Secondly, when these people understand that their physical environment is deteriorating they will attempt to control some of their activities leading to degradation [19], thereby be more willing to support land management programmes if they are aware that their actions are harmful to the environment [20]. Various social, economic, technological and ecological often exist, changing degrading practices especially where communities are aware of the negative impacts of current land management practices. Communities in the Irangi Hills have demonstrated their willingness to participate in conservation initiatives as indicated by their adoption of various conservation measures reported later in the chapter.

The third perspective from attitude survey shows that a large majority of farmers are concerned about soil and/or land degradation as a general community problem, disregarding the fact that their own holdings are likely to be also at risk [21]. Under such circumstances then no actions may be taken although such people hold positive attitudes towards conservation. However, it is believed that when the landowners themselves have been involved in fact-finding on their own land they become instrumental in implementing planned courses of action [22]. Thus basing on the local people's perceptions of environmental resources and knowledge then it is possible to develop methods which can allow the people themselves to provide the solutions to their environmental problems [19, 22, 23]. Generally, planning for and implementation of effective soil and water conservation measures in a site require, among other things, a detailed understanding of the extent, risk and spatial distribution of the problem [24, 25], including local concerns.

While sustainable management of resources leads to sustainable development, the deep-rooted poverty leads to overdependence on natural resources for livelihood which in some instances has undermined the capacity of the communities to manage their resources sustainably. This problem is more critical in developing countries where rapid population growth leads to the invasion of marginal lands and unsustainable land use practices that in turn encourage environmental degradation and perpetuate poverty.

This chapter is based on a study undertaken in the Irangi Hills to examine the farming community perceptions and awareness on environmental degradation. The main objective was to examine on methods used in the area to improve agricultural productivity and control soil degradation. The specific objectives were: to assess farmers' awareness on soil degradation and conservation, and to study the farming system and resource use in the area. It also examined the mechanisms that are taken by the local communities to address land resource management issues, including approaches used in soil conservation and how such approaches help to improve agricultural productivity and local livelihoods in general. It further examined the factors limiting effective community involvement in land/soil conservation initiatives.

2. The study area

The Irangi Hills (Figure 1) are located in the severely eroded area of Kondoa District known as the Kondoa Eroded Area (KEA). Kondoa District is located in the northern part of Dodoma Region at latitudes 4º10' - 5º44' South and longitudes 34º54' - 36º28' East. The land area of the Kondoa District is approximately 13,210 km^2, out of which the Kondoa Eroded Area covers 1256 km^2. The Irangi Hills forms the largest part of KEA [5]. The altitude of the Irangi Hills ranges from 1200 to 2000m above sea level. The climate of Kondoa District is semiarid, characterised by an average annual rainfall of between 600 and 800 mm, with a long-term average of 640 mm per year [5]. However, in the more elevated parts of the Irangi Hills up to 900 mm of rain have been recorded [26]. The rainfall season is usually between November and April/May, with a dry spell in February. The period between May and October is usually dry. The rainfall pattern in this area is highly variable and most of the precipitation occurs in short duration storms. The district is characterised by high evapotranspiration rates that double the amount of precipitation [27].

The majority of soils in the semiarid areas of central Tanzania originate from granitic, gneissic and schistic parent material. These soils are of low fertility, base-exchange capacity, bulk density and water-retention capacity [6]. These soils also have low organic matter content, a condition that makes them extremely erodible [7]. The Irangi Hills are severely affected by soil erosion. Studies on soil erosion in the area indicate rates of between 27 and 37t ha^{-1} yr^{-1} [28], in the sub-humid and semiarid parts of the hills respectively. The soils in Irangi Hills are generalised as coarse loamy to sandy loams in texture, being sandiest in the surface horizon. This implies the need for proper management in order to sustain agricultural productivity. Different strategies used by farmers in KEA, for instance, in coping with poor soil fertility have been described by Kangalawe [1, 29].

Soil conservation initiatives started in Kondoa district since the colonial administration in the 1930s. During the 1940s to 1950s soil conservation involved measures such as reduction of livestock numbers, ridge cultivation, contour bunding of uncultivated land, rotational grazing and gully erosion control [30]. Farmers were also required to plant sisal around farmlands to save the arable land from further destruction.

Soil conservation measures during that period were associated with colonial force, where some of the activities were assigned to people as punishment for disobedience of local rulers and tax aversion. They were thus considered as an interference with local traditions and became quite unpopular [4]. In 1973 the government of Tanzania started a state-run soil conservation project in Dodoma region, popularly known as HADO (Hifadhi Ardhi Dodoma). This was a deliberate attempt to come to grips with the menace of soil erosion and degradation in the region [5]. Three kinds of approaches were undertaken to enhance soil conservation, including mechanical, biological and administrative measures [4]. *Mechanical measures* involved barriers such as terraces and earth banks across the slopes that were built to slow down surface runoff. *Biological measures* involved earth binding with plantations of different kinds, such as grass strips across the slope, planting grass in sandy rivers and in gully bottoms, rotation of crops, and the spread of residue on the fields. The *administrative measures* encompassed tree planting demonstration plots, organising grazing and farming techniques such that the land is protected as much as possible. These measures were complemented by evacuation of all domestic animals in 1979 [5]. Expulsion of livestock from the Irangi Hills in 1979 as part of soil conservation approaches instituted in the KEA resulted in ecological transformation from the heavily browsed shrubs and scattered trees to impressive regeneration of herbaceous, woodland and woody shrubs, as well as grassland vegetation cover. Economically, however, this agro-pastoral society was no longer able to depend on livestock as a form of capital, or as a security against the harsh climatic conditions of the region.

3. Methodology

3.1. Sampling procedure

The Kondoa Eroded Area (KEA) covers twenty-eight villages. A list of these villages was made and three of them were selected at random, namely Mafai, Baura and Bolisa. The selected villages had 370, 350, and 750 households respectively. The selected villages are subdivided into sub-villages for administrative purposes. To allow for adequate representation, 10% of households from each sub-village was randomly selected for inclusion in the sample. A total sample of 147 households was selected for interviews. A random sampling procedure was employed in selecting the sample households from lists of household heads that were made for each of these villages. Sampling is a common practice in research. The random sample of 10% of the villages and households selected for this study is considered to be representative enough for statistical analysis [31]. Under certain circumstances, such as resource constraints, even a smaller sample of 5% is regarded as being representative enough [32].

3.2. Data collection and analysis

Household interviews, using a standard questionnaire, were the major means used to collect both qualitative and quantitative information. The questionnaire survey was complemented by informal surveys that involved discussions with key informants, including village leaders, extension workers, district agricultural officials and HADO staff. These informal surveys were conducted in order to get some general overview on soil degradation, community perceptions and agricultural performance in general. These surveys also provided a means and direction in crosschecking the responses from formal interviews. The key informants were found in the respective villages and/or at district level. Information from key informant interviews was analysed by triangulation with all other sources. To determine the level of awareness of soil degradation three exploratory questions were asked. Firstly, whether the study community perceived land/soil degradation as a problem in their villages. Secondly, what criteria are used by this community to determine the quality of land/soil in general. Thirdly, whether they associated land/soil degradation with crop cultivation or livestock management systems of the area. These aspects are addressed in the following sections.

The surveys were complemented by field observations in farmers' fields. Field visits involved observations of various land degradation features, such as soil erosion and sedimentation, surface runoff, sandiness of soils, crop vigour, presence of indicator-plant species; and agricultural practices, including among others, types of crops grown, cropping patterns and on-farm soil conservation measures. Field observations also included sampling soils from selected transects for subsequent laboratory analysis of soil nutrients. Three farmer-led transect walks were undertaken with small groups of farmers in each village and soil samples were collected from representative sites of major soil groups as identified by farmers. Soil profile pits were dug to a depth of one meter and samples taken at 20 cm intervals [33]. A total of eighty samples were collected for analysis. Nutrient content of the soil was determined to provide an estimate of the inherent soil fertility status and for soil nutrient balance assessment. The data was analysed using statistical measures of central tendency (means), and frequency distribution (percentages) [34, 35]. The frequency distribution data was cross-tabulated into contingency tables.

4. Results and discussion

4.1. Community' perceptions on land/soil degradation

4.1.1. Local perceptions of soil degradation

Response to the inquiry on whether the study community perceived soil degradation as a problem in their villages have shown that 58% of the respondents considered soil degradation as being a serious problem in their vicinities. This perceptions may be influenced by differences in socio-economic characteristics inherent among the local people. Socio-economic characteristics such as endowment of livelihood assets by households determine the ability of a household to use, for example, agricultural inputs like fertilisers or manure as a way of improving soil productivity. In the Irangi hills, for instance, wealthy farmers who could afford using fertilisers and/or manure did not perceive soil fertility as a major issue. Those who perceived soil degradation as a problem mentioned the generally low but declining soil fertility of the Kondoa soils, soil erosion and runoff, sandiness of soils and sedimentation as key indicators of soil degradation in their villages. Figure 2 presents the proportions of responses on indicators of farmers' awareness of soil degradation processes.

The small percentage of respondents mentioning soil erosion in Mafai village could be attributed to that this village is surrounded by a protected catchment forest and is generally less degraded compared to the other two villages. The presence of these indicators seem to show that rural people are aware of their environment and its related problems, and particularly so with those which affect the farm productivity and/or those that resulted into more visible landscape changes such as soil erosion. However, the fact that less than half of the respondents indicated that soils are inherently infertile suggests that productivity has declined significantly within living memory and that people were unaware that their yields were probably rather low from the outset. It is explained by Tosi et al. [36] that the low inherent soil

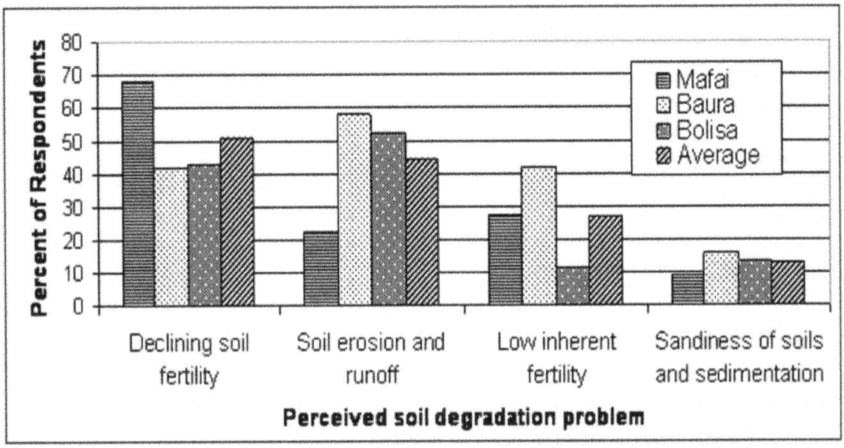

Figure 2. Percent responses on community awareness of soil degradation problems.

fertility is attributed to parent material factors. Soils in this area are reported to have formed from metamorphic and igneous rocks which are poor in plant nutrients. The dry climatic condition of the study area also limits high productivity of organic matter, resulting in poor surface cover and low incorporation into the soil as binding agent and for fertility enhancement [37].

Declining soil fertility was perceived as the major indicator of soil degradation in the studied villages. A majority of the farmers (80%) attributed such decline to continuous cultivation without resting the fields, whereas 20% ascribed it to inadequate application of manure and/or fertilisers. One explanation to continuous cultivation was the increasing land shortage that has led to intensified crop cultivation and short or no fallow periods. Studies conducted in neighbouring villages and in other parts of the Irangi Hills also revealed that most farms are cultivated every season without fallow and are thus subjected to continuous loss of soil fertility [1, 38].

The general assessments of soil fertility in Kondoa District that soils in the area are of low fertility [1, 7, 39, 40]. Results from laboratory analyses for soil nutrients (Table 1) confirm the low levels of soil nutrients in all the three villages studied. These results also indicate that Mafai soils had more advantage in terms of nitrogen content than both Baura and Bolisa. These soils had about four times as much total nitrogen as Baura and Bolisa, whereas Baura had soils richer in available phosphorus than the other two villages. The variations in nutrient contents may be attributed to, among other factors, the severity of soil degradation that characterises the studied villages and different management practices of the farms. Many soil conservation structures like contour bunds and planted trees were also observed in Mafai compared to the other two villages. Soils in Mafia village are less eroded, hence have more nutrients in the sampled surface layer attributable to presence of higher organic matter content. A large part of the village is covered by a protected catchment forest and generally less severely degraded compared to Baura and Bolisa.

Soil characteristic	Mafai	Baura	Bolisa
pH (H_2O)	4.80	5.20	5.40
Total Nitrogen (%)	0.29	0.07	0.08
Available P (ppm)	4.20	7.20	3.20
CEC me/100g	17.12	3.84	13.89
Exchangeable K (me/100g)	0.61	0.61	0.52
Exchangeable Ca (me/100g)	4.39	1.83	2.74
Exchangeable Mg (me/100g)	1.61	0.40	1.20
Exchangeable Na (me/100g)	1.86	0.92	0.80
Organic carbon (%)	2.62	0.31	0.54
Particle size			
% Sand	54.83	80.27	69.20
% Silt	14.57	10.60	17.63
% Clay	30.60	9.13	13.17

Table 1. Some characteristics of soils from the study area (measured at 0-20 cm depth)

Soil erosion and surface runoff featured as indicators of soil degradation as indicated by about 44% of respondent farmers (Figure 2). Awareness of soil erosion as a soil degrading process featured more prominently among Baura and Bolisa respondents. Visual observation of the landscape in these villages confirms the local people's response. Both Baura and Bolisa have landscapes dissected by more pronounced gullies (Figure 3) compared to Mafai village (Figure 4). Discussions with key informants in these villages indicated that historically the two villages had large numbers of livestock prior to destocking in 1979 that rendered many places devoid of vegetation because of overgrazing. This situation exposed the land surface to agents of soil erosion, such as runoff. The extensive gullies seen today in these and many other villages in the Irangi Hills are said to have formed along former cattle tracks aligned down the slope [6].

Figure 3. Gully erosion in the studied villages. This is a common feature of the landscape and in many parts of the Kondoa Eroded Area

Figure 4. An uneroded part of the landscape wihin the Kondoa Eroded Area.

Sedimentation and sandiness of the soil was perceived as a problem by only a few farmers (see Figure 2). This response was particularly obtained from farmers whose fields laid in stabilising sandfans that have soils with very low organic matter levels, low moisture holding capacity and poor fertility status. Such soil characteristics are also common in other parts of the Irangi Hills, such as in Haubi and Mulua [1, 7]. Sedimentation was reported to take place in depositional footslopes and valley bottoms where the eroded materials from hillslopes accumulate. In many places sedimentation of sandy materials buried the former fertile clayey topsoil [1, 7]. One would expect this indicator to be mentioned by most respondents, however since farmers have had their settlements and fields in that kind of environment for generations they do not often mention it as a major concern. The explanation to this situation would be that soil erosion and sedimentation in this area dates long in history to the extent that very few benchmark areas remain that could show the earlier landforms not affected by sedimentation [7].

4.1.2. Assessing land and soil quality

Findings from this study showed that there are several traditional ways communities use to evaluate and to explain the quality of the land and the soils they are cultivating. Three categories of responses appeared to be most prominent, namely crop vigour and crop yields, presence of indicator-plant species and density of vegetation under fallow [29].

A healthy and vigorous crop growth, reflected by a good crop stand in the field, was used as an important indicator that the soil is fertile enough, if moisture and other factors are not limiting [29]. Under such circumstances, even if the weather conditions worsen during the growing season such that final yields are poor, the farmer would have realised the potential fertility of a certain piece of land. A stunted crop with less vigorous growth in the field when other factors such as moisture are considered not limiting was locally perceived to indicate a high probability that soils on which the crop is growing are of low quality and infertile.

Majority of respondents (95%) considered crop yields as the best measure to comprehend land/soil quality. It was noted that low or declining crop productivity could be a clear indicator of declining soil fertility, and hence soil degradation. The use of this indicator by the local farmers in evaluating land/soil quality is also appreciated by experts in land degradation, where crop output decline is regarded as a proxy indicator of soil degradation in farmlands [1, 40, 41]. It is particularly important because it affects people directly in terms of food availability and security. However, this factor alone is not sufficient to establish that degradation is taking place since cropping conditions vary considerably between years and between individual farmers. The influence of other factors such as crop pests and diseases and climate variability may affect crop yields [42, 43]. In the Irangi Hills most of the respondents indicated also that low crop yields could be due to low and/or erratic rainfall. This aspect needs to be investigated further to establish detailed linkages between climatic patterns and crop yield trends in the area. Nevertheless crop yields are an important indicator of proximate soil conditions if other factors are not constraining.

4.2. Locally perceived association between farming systems and soil degradation

Table 2 presents the locally perceived relationships that were cited by the respondents as being the contribution of the farming practises to the observed land/soil degradation in the study areas. About 52% of the respondents associated soil degradation to continuous cropping while 23% considered inadequate manure application to be responsible for the diminishing soil quality. The overuse of the soil in continuous tillage without fertiliser supplementation, coupled by grazing on plant residues, weeds and crop stubble, has deprived the soils of both nutrients and organic matter [1, 29].

Perceived Relationship	Mafai (n=28)	Baura (n=26)	Bolisa (n=42)	Total (n=96)
Continuous cropping	57.1	30.7	61.9	52.1
Lack or inadequate availability of manure	10.7	53.8	11.9	22.9
Overgrazing in the past decades	14.3	11.5	11.9	12.5
Cultivation on steep slopes	14.3	4.0	4.8	7.3
Lack of on-farm conservation measures	3.6	0	9.5	5.2

Table 2. Perceived relationships between cropping/livestock management systems and soil degradation in the Irangi Hills (%)

Overgrazing was pointed out to be one of the processes that facilitate degradation by 12.5% of respondents. The low figure is explained by the fact that to the Rangi community, which is traditionally agro-pastoralist, having big herds of livestock is just part of their culture thus locally overgrazing is rarely seen as a major problem. A similar explanation regarding perceptions on overgrazing has been reported recently among communities in the Sukumaland and in the Iramba area [42, 44]. Overgrazing of livestock had similar effects on the soils of steep slopes and on shallow and stony soils, where continuous cultivation has not been practised [36]. Lack of on-farm conservation measures, especially before HADO started its activities, significantly contributed to the degradation features witnessed in the present days [1, 7]. Thus elaborate extension services are probably needed regarding various mechanisms that may contribute to sustainable farm production, such as on-farm erosion control, agroforestry practices and proper residue management. Proper farmer education would inculcate the culture of conservation among communities. Other associations presented in Table 2 did not feature as important concerns among farmers, but because of their role in soil degradation they are worth some attention.

While incorporation of crop residue and manure were meant to improve organic matter content and replenish soil fertility in the farm, contour ridges were constructed to check runoff and control erosion, and as such prevent further loss of soil fertility through nutrients washed away in eroded soil material. Only limited quantities of manure were however applied per unit area, especially since 1979 when livestock were evicted from the KEA. The limited supplies of manure and the high fertiliser prices are responsible for their low usage as adaptive mechanisms in fighting against soil degradation at farm level [29]. The only reliable way of replenishing soil fertility has been through crop residue incorporation into the soil. However, during dry seasons the residue provides valuable feed for livestock, whereas considerable proportions are burnt when preparing the land for a succeeding crop hence not much residue is left for incorporation into the soil. This may have negative consequences on subsequent crop productivity. Similar experiences are reported for other parts of the Irangi Hills, for example in Haubi and Mulua villages [1].

4.3. Community involvement in land resource management

4.3.1. Community participation in soil conservation

The pressure on land has practically increased all over the country particularly during the 20th century as a result of population growth. This has, in many instances resulted in unsustainable cultivation techniques including shortened fallow periods [38, 45] that consequently impoverish the soil. Enhanced long-term productivity and sustainability of the land resource thus require sound soil conservation measures in the farming systems that enhance maintenance and/or improvement of soil and land quality in general. This is an important consideration as it influences agricultural productivity and local livelihoods.

In many instances environmental degradation has stimulated a variety of responses and adaptation mechanisms by local communities. This study made an enquiry on whether farmers had undertaken any deliberate efforts to protect their land holdings from soil degradation. Majority of respondents (95%) indicated to have used one or more conservation techniques in their farms as a means of adjusting and adapting to soil degradation processes. Table 3 presents the various soil conservation approaches as mentioned by the interviewed farmers. The first three combinations of approaches, that is, contour ridges; tree, sisal and grass planting; manure application and incorporation of crop residue in soils, appeared to be the most prominent conservation strategies adopted by majority of farmers who were practising conservation, accounting for 97% of farmers in Mafai, 85% in Baura and 81% in Bolisa village. The generally sloping terrain in Mafai village partly explains the reported increase in the use of contour ridges, trees, sisal and grass planting to protect the soil from erosion.

Measures taken	Mafai (n = 35)	Baura (n = 34)	Bolisa (n = 70)	Total (N = 139)
Contour ridges; tree, sisal and grass planting; Manure application and Incorporation of crop residues in soils	80	55.9	68.5	68.1
Contour ridges and tree planting	17.1	29.4	12.9	19.8
Tree planting only	0	5.9	12.9	6.3
Stall-feeding cattle	2.9	5.9	4.3	4.4
Crop rotation	0	2.9	1.4	1.4
Total	100	100	100	100

Table 3. Soil conservation measures undertaken by respondent farmers (in %)

A majority of farmers have also planted trees as one of the soil conservation practices advocated by HADO. Table 4 presents a list of tree species that are planted in the study area and associated uses. As for crop rotation, the low response reported in Table 3 was mainly attributed to the small farm holdings that necessitate farmers to practice continuous cultivation of same fields.

Name	Scientific name	Purpose[1]	Mafai (n = 31)	Baura (n = 31)	Bolisa (n = 67)	Total (N = 129)
Silk oak	*Grevillea robusta*	Ct, Fw, Fi	87.1	90.3	95.5	91.0
Guava	*Psidium guajava*	Fr	51.6	51.6	49.3	50.8
Pawpaw	*Carica papaya*	Fr	29.0	29.0	76.1	44.7
Lemon	*Citrus limon*	Fr	19.4	22.6	17.9	20.0
Orange	*Citrus sinensis*	Fr	25.8	0	26.9	17.6
Eucalyptus	*Eucalyptus sp.*	Fw, Ct	35.5	3.2	1.5	13.4
Mango	*Mangifera indica*	Fr, Sh	0	16.1	14.9	10.3
Pomegranate	*Punicum granatum*	Fr	3.2	3.2	11.9	6.1
Leucaena	*Leucaena leucocephala*	Fd, Fi	3.2	3.2	1.5	2.6
Cypress	*Cuppressus sp.*	Fi, Ct	3.2	3.2	0	2.1
Iron wood	*Senna siamea*	Ct, Fw	0	3.2	1.5	1.6

[1]Ct = construction material/timber, Fr = fruits, Fd = fodder, Fi = soil fertility improvement, Fw = fuelwood, Sh = shade.

Table 4. Tree species planted in the study area, ranked according to preferences and percent of respondent farmers that have planted them

Findings from this study are in agreement with studies elsewhere that farmers often attempt to adjust to environmental degradation by using various measures and strategies [19, 46]. The measures taken may be different depending on the natural environment and sociocultural backgrounds of the area concerned. According to Nsiah-Gyabaah [19], farmers' adjustment to their environment can generally be effective when they are able to predict the short term inter-annual variability patterns (e.g. in soils and weather). Under such situation they may be able to successfully use available innovations and local expertise to maximise the benefits of both soils and weather. However, where changes are unpredictable farmers may face difficulties in establishing adaptation mechanisms.

4.3.2. *Challenges to community participation in soil conservation practices*

Community participation in conservation practices is of great importance as it seeks to guarantee access and control over resources by the communities living in them, but who depend on these resources to satisfy their various needs (ecological, economic, social, cultural and spiritual needs). Community participation ensures more commitment in ensuring that resources are more sustainably managed, where apart from communities depending on these resources for a living and conserving them, they at the same time become their guardians [47, 48]. The active participation of various stakeholders in decision-making is crucial for ensuring the long term sustainability of community-based resource management initiatives. In several occasions however, soil conservation has not received the expected involvement of local communities. Some of the reasons that have influenced the local people's attitudes towards land/soil conservation efforts in the Irangi Hills are discussed here.

One of the reasons put forward was the tendency to underestimate the seriousness of the soil erosion problems by many people in the area. Where the tenure system is not elaborate individual farmers may not be concerned with problems of land degradation regardless of their holdings being at risk as such land degradation is considered as a general community problem. Such attitudes may result in no action being taken against land degradation even when there are no clear hindrances. The implication of the foregoing is that effective conservation is likely to be achieved when land tenure systems are properly articulated [1]. Thus efforts are needed to ensure integrated community-level planning that could promote individual farmers efforts without undermining community interests. Recent experiences from studies in Kishapu and Kahama Districts in Shinyanga Region, and in Nyarugusu area in Geita District Mwanza region indicated that many people practice soil and water conservation mainly in their own lands/farms [42], including setting aside private conservation areas such as Ngitili [49]. Adoption and/or practicing certain conservation measures are much influenced by the farmer's economic situation, including resource endowments. For instance, farmers with sufficient land holdings can afford to conserve some of it under the Ngitiri system, while land constrained farmers may not. Similar experiences would be the case for other conservation measures that require heavy investment by the farmer, for example making of soil erosion control structures that may need additional labour, and using fertilisers and/or manure.

Apart from the underestimation of the land degradation problem and inarticulate tenure system, it is also evident that the Rangi people themselves are traditionally not conservationists. When the pressure on land increases and harvests dwindle, some villagers will leave to take up new land elsewhere while others remain. The Rangi as a group can be said to practice an expansionist permanent agriculture [2]. Their principal solution to problems of soil erosion, for example, has thus been to move temporarily or permanently to areas with better conditions, while also retaining a foot in the Rangi core areas [2, 50, 51].

Experiences from the Irangi Hills and elsewhere in East Africa indicate that resource management is closely influenced by the prevailing socio-political environment [52]. It was reported, for instance, that some people in the Irangi Hills were of the opinion that the low adoption of soil conservation measures was based on negative attitudes that were inculcated among people during the colonial era, when such activities were basically coercive [1, 29]. During the colonial period soil conservation was seen as a form of oppression [50, 53]. In Kenya, for example, soil conservation was made compulsory under colonial administration and forced labour was used for community conservation programmes [22]. Although some of the soil conservation techniques employed were effective (e.g. in Machakos district), the coercive methods used were very unpopular [4, 22, 53] resulting into some local communities withholding their own conservation initiatives. This was also experienced in the Irangi Hills and among other communities in Kondoa District [50].

In the Kondoa case, the opposition to conservation during the 1940s was closely linked to a wide discontent within the Rangi community with the way the colonial government appointed local leaders. Thus a more explicit political discourse (struggle against colonialism) became linked to soil conservation policies [50]. Consequently, during the nationalist struggle in the late 1950s it was no longer possible to enforce communal turnout and most conservation work came to a halt [54]. The Kondoa Chief of the time came to side with the Rangi underground opposition against regulations on land-use. It is not surprising then that even independent governments found it difficult to reverse the previous nationalist attitudes and support conservation measures [22, 53]. Land degradation was thus left to continue. These experiences reflect to the need for local community participation in developing and implementing resource management strategies.

4.4. Land-use dynamics and environmental degradation

4.4.1. Land availability over time

In general, there have been changes in land-use use types in the study area due to several factors. Such changes can be considered to be adaptive mechanisms to population growth, economic development and to changing climate. Population growth is usually associated with increased exploitation of natural resources. According to local knowledge in the study area, prior to Tanzania's independency in 1961, arable land for agriculture and livestock was relatively plenty and used for different purposes, including cultivation and livestock grazing. The increase in population has led to expansion of agricultural and livestock activities. This has contributed significantly to changes in land-use intensity and cover types.

An assessment was made on perceptions regarding land availability over three time periods, from post independence, villagisation and current period. Experiences from the selected villages indicated that during post independence and villagisation period agricultural land was fairly easily available at the areas close to homesteads. It was reported that, land availability is currently very difficult than ever before, the main cause being increase in population pressure, which has led to increased demand for land by villagers. Eventually this has created land shortage in most of the villages. The pattern of land availability over years is also experienced in other parts of Kondoa District [55]. Farmers reported that currently the places with easily available land are those located far from homesteads, which was considered to be a major limitation for people who are not able to manage distant farms.

4.4.2. Decreasing farm sizes

For arable land use, there has been a general decline in the farm sizes. The local people attributed the decline in the farm sizes to several other factors (see Table 5). Thus there is a concern that land is increasingly being insufficient. The big proportion of the people reporting to experience land shortage reflects to that the problem of land shortage is much bigger than currently envisaged, and may worsen particularly with the fast growing population.

Perceived cause declining farm sizes	Percent
High population growth/increase	80.3
Expansion of settlements	6.9
Soil erosion and gullies	4.0
Expanding livestock keeping (more livestock)	2.9
Expansion of family sizes	2.0
Low soil fertility - land is tired/exhausted	2.0
Tree conservation	0.9
Intensive rainstorms (more erosion)	0.9
Total	100

Table 5. Different factors considered by the local community to cause declining farm sizes

The decreasing farm size is one of the causes of household food insecurity in the area. The small farm sizes limit the possibilities to practice fallow rotation. Consequently continuous cultivation culminates into declining soil fertility and reduced crop productivity. The problem may as well be aggravated by loss of land as a result of soil erosion, particularly gully erosion.

4.4.3. Chancing land use patterns and local adaptive mechanisms

Experiences on land use and the way rural land users interact with environmental resources show that communities have increasingly interacted with their local environment by faltering land use practices. However, continued land degradation has had severe environmental and social-economic consequences resulting in poor agricultural productivity, perpetuating food insecurity and poverty among the concerned communities. This necessitates the analysis of land-use dynamics, land degradation and their inter-linkages with livelihoods and poverty alleviation strategies adopted by the respective communities. Generally, there have been changes in land use patterns, which in many instances has involved increasing land use intensity as adaptive mechanisms to increased population growth, economic development and to changing climate [25]. Increase in population has led to expansion of agricultural and livestock activities, contributing significantly to changes in land use intensity, decline in farm sizes, land fragmentation and land degradation. Sustainable management of natural resources can thus be achieved by having integrated land use practices, including developing elaborate village land use plans.

Livelihood diversification into non-agricultural enterprises was also reported to have been a means of adapting to the changing environments and for poverty alleviation. Diversification as applied in the rural context is a process whereby rural households construct an increasingly diverse portfolio of activities and assets in order to survive [56]. Diversification can be viewed from different angles. It may imply a matter of survival associated with harsh local environmental conditions, or alternatively, it may be considered a matter of opportunity involving pro-active household strategies employed to improve living standards [57]. Generally, however, a diversity of livelihood sources is considered to be one of the ways through which households may develop security against agrarian environments, particularly in semiarid areas.

5. Conclusions

This chapter has examined how perceptions and knowledge facilitate and/or act as barriers to sustainable land management. It has been realised that farmers in the Irangi Hills are aware of land/soil degradation and its various processes, with levels of perception varying between villages and among respondents depending on the severity of the land degradation problem. Declining soil fertility and soil erosion, demonstrated by existence of spectacular gullies and extensive depositional sand fans, have been found to contribute significantly to the general understanding of land/soil degradation problems.

Awareness of land degradation was also reflected by the various criteria that the local people use in assessing the potentials and constraints that farmlands and the landscape in general are facing. Low crop productivity has been identified as one of the important constraints; attributed mainly to declining soil fertility, unreliable rainfall, and to a lesser extent, soil erosion. Farmers seemed to be quite aware of the association between cropping and/or livestock management systems and land/soil degradation. An extension service focusing on the various mechanisms that may contribute to sustainable farm production, such as on-farm erosion control, agroforestry practices and proper residue management is necessary. This is particularly important in a situation where continuous cultivation has become the norm because of increasing land shortage. Thus the various adaptations that farmers are already exposed to with regard to improving agricultural productivity in the erosion-stricken farmlands need to be promoted and developed further.

Limiting factors to local participation in soil conservation initiatives would probably be successfully corrected through proper farmer education that inculcates the culture of conservation among communities. It is important therefore that steps are taken to address the diversity of the intricate attitudes and socio-political environments among rural communities. It is recommended that further comparative studies be conducted so as to come up with sound strategies that will motivate the local people to participate more in issues related to sustainable land resource management. Where relevant such strategies and approaches to extension work need to be tailored to individual communities rather than using large homogenous programmes.

Increase in population has led to expansion of agricultural and livestock activities, contributing significantly to changes in land use intensity, decline in farm sizes, land fragmentation and land degradation. Since agricultural production (crops and livestock) is the major means of poverty alleviation in these semiarid areas, it calls for enhanced environmental conservation so that agricultural productivity can be sustained for the betterment of the community livelihoods. This is particularly important now given the additional factor of climate change. Thus, sustainable management of land and other natural resources can only be achieved by having integrated land use practices, including developing elaborate land use plans. Poverty alleviation strategies and other policies also need to ensure sustainable development with minimal impacts on the environment. The diversification of livelihood activities (farm and non-farm) recorded in the area indicates that local communities are struggling towards poverty alleviation and as a way of adaptation and coping to environmental degradation.

Author details

Richard Y.M. Kangalawe
Institute of Resource Assessment, University of Dar es Salaam, Dar es Salaam, Tanzania

Acknowledgement

This study was undertaken with a financial support from NORAD through its Management of Natural Resources and Sustainable Agriculture fellowship programme. Many thanks are extended to the farmers in Bolisa, Baura and Mafai villages who enthusiastically participated in this study and for their inspirations that paved a way towards completion of this work.

6. References

[1] Kangalawe RYM. (2001) *Changing land-use patterns in the Irangi Hills, central Tanzania: A study of soil degradation and adaptive farming strategies*. PhD Dissertation No. 22.

Department of Physical Geography and Quaternary Geology, Stockholm University: Stockholm.

[2] Östberg W (1986) *The Kondoa Transformation: Coming into Grips with Soil Erosion*. Research Report No. 76. Scandinavian Institute of African Studies: Uppsala.

[3] Christiansson C, Kikula IS, [51] W (1991) Man-land interrelations in semiarid Tanzania: A multidisciplinary research programme. Ambio 20 (8): 357-361

[4] Christiansson C, Mbegu A, Yrgård A (1993). The Hand of Man: Soil conservation in Kondoa Eroded area, Tanzania. Nairobi: Regional Soil Conservation Unit.

[5] Mbegu AC (1988). The HADO Project: What, Where, Why, How? Forestry and Beekeeping Division, Ministry of Lands, Natural Resources and Tourism: Dar es Salaam.

[6] Payton RW, Christiansson C, Shishira EK, Yanda P, Eriksson MG (1992) Landform, Soils and Erosion in the north-eastern Irangi Hills, Kondoa, Tanzania. Geografiska Annaler 74 (A 241): 65-79

[7] Payton RW, Shishira EK (1994) Effects of soil erosion and sedimentation on land quality: Defining pedogenetic baselines in the Kondoa District of Tanzania. In: Syers JK, Rimmers DL. Editors. Soil Science and sustainable land management in the tropics. England: Wallingford CAB International 88-119.

[8] Darkoh MB (1986) Experiences of arid land development in Tanzania: In: Arid land development and the combat against desertification: An integrated Approach. UNEP: Moscow.

[9] Eriksson MG (1998) Landscape and soil erosion history in central Tanzania. A study based on lacustrine, colluvial and alluvial deposits. Dissertation No.12. Department of Physical Geography, Stockholm University: Stockholm.

[10] Eriksson MG, Olley JM, Payton RW (1999) Late Pleistocene colluvial deposits in central Tanzania, erosional response to climate change. GFF 121:198-201

[11] Eriksson MG, Olley JM, Payton RW, (2000).. Soil erosion history in central Tanzania based on OSL dating of colluvial and alluvial hillslope deposits. Geomorphology 36: 107-128

[12] Ngana JO (1996) Climate and hydrology of the Kondoa Eroded Area. In: Christiansson C., Kikula IS. Editors. Changing Environments: Research on Man-Land Interrelations in Semiarid Tanzania. Report No. 13. Nairobi: Regional Soil Conservation Unit:.

[13] Whyte AVT (1977) *Guidelines for field studies in environmental perception*. MAN Technical Notes No 5. Paris: UNESCO.

[14] Hackel JD (1990) Conservation attitudes in southern Africa: A comparison between KwaZulu and Swaziland. Human Ecology 18 (2): 203-209

[15] Hunter Jr ML, Hitchcock RK, Wyckoff-Baird B [1990] Women and wildlife in the southern Africa. Conservation Biology 4(4): 448-451

[16] Showers KB, Malahleha GM (1992) Historical environmental impact assessment: A tool of for analysis of past interventions in landscapes. Working Paper No. 8. The Project of

Agriculture of the Joint Committee on African Studies, ISAS: National University of Lesotho.

[17] Rydgren B (1993) Environmental impacts of soil erosion and soil conservation. A Lesotho case study. PhD Thesis, UNGI Report No 85. Upsalla: Uppsala University.

[18] Dahlberg AC (1996). Interpretation of environmental change and diversity: A study from North-eastern Botswana. Dissertation Series No. 7. Department of Physical Geography, Stockholm University: Stockholm.

[19] Nsiah-Gyabaah K (1994) *Environmental degradation and desertification in Ghana. A study of the Upper West Region.* Aldershol & Brookfield: Avebury Studies in Green Research..

[20] Herberlein TA (1972) The land ethic realised: Some social psychological explanation for changing environmental attitudes. Journal of Social Issues 28: 78-82

[21] Pitt MW, Yapp TP (1992) Perceptions of land degradation and awareness of conservation programmes in North-eastern New South Wales. In: Haskins PG, Murphy BM. Editors. People protecting their land, Proceedings of the 7th ISCO conference: Sydney. pp. 115-124

[22] Critchley W (1991) Looking after our land. Soil and water conservation in dryland Africa. Oxfam.

[23] Toulmin C, Chambers R (1990) Farmer-First: Achieving sustainable dryland development in Africa. IIED Paper No. 19. London: IIED.

[24] Bewket W, Teferi E (2009) Assessment of soil erosion hazard and prioritization for treatment at the watershed level: Case study in the Chemoga Watershed, Blue Nile Basin, Ethiopia. Land Degradation & Development 20: 609–622, DOI: 10.1002/ldr.944

[25] Kangalawe RYM (2009) Land use /cover changes and their implications on rural livelihoods in the degraded environments of central Tanzania. African Journal of Ecology 47 (Suppl. 1): 135–141

[26] Ngana JO (1990). *Modelling for periodic features in seasonal rainfall and its implications to water resources and agricultural planning*. Research Report No. 27. Dar es Salaam: Institute of Resource Assessment, University of Dar es Salaam.

[27] Christiansson C (1981) Soil erosion and sedimentation in semi-arid Tanzania. Studies on environmental changes and ecological imbalance. Uppsala: Scandinavian Institute of African Studies.

[28] Eriksson MG (1999) Influence of crustal movements on landforms, erosion and sediment deposition in the Irangi Hills, central Tanzania. In: Smith BJ, Whalley WB, Warke PA. Editors. *Uplift, erosion and stability: Perspectives on long-term landscape development*. London: Geological Society of London, pp.157-168

[29] Kangalawe RYM (1995) *Fighting soil degradation in the Kondoa eroded Area, Kondoa District, Tanzania: Socio-economic attributes, farmer perceptions and nutrient balance assessment.* M.Sc. Thesis. Aas: Agricultural University of Norway.

[30] Mbegu AC, Mlenge WC (1983) Ten Years of HADO. Dar es Salaam: Forest Division, Ministry of Natural Resources and Tourism.

[31] Clarke R (1986) The handbook of ecological monitoring. Oxford: Clarendon Press. 298p

[32] Boyd HK, Westfall R, Stasch F (1981). Marketing Research: Tests and Cases. Illinois: Richard D. Inc.

[33] Hodgson JM (1985) Soil survey field handbook: Describing and sampling soil profiles. Technical Monograph No. 5. Harpenden Herts: Soil Survey, Rothamsted Experimental Station. 99 pp

[34] Johnson RA, Bhattacharyya GK (1992). Statistics: Principles and methods. Second edition. John Wiley & Sons, Inc.

[35] Ryan BF, Joiner BL, Ryan Jr TA (1992) Minitab Handbook. Second edition. : Belmont California: Duxhury Press.

[36] Tosi JA, Hartshorn GS, Quesada CA (1982) *HADO Project Development Study and Status of the Catchment Forestry*. A Report to the Ministry of Natural Resources and Tourism, Tanzania. San Jones: Tropical Science Centre.

[37] Smith JL, Elliott LF (1990). Tillage and residue management effects on soil organic matter dynamics in semi-arid regions. Advances in Soil Science 13: 69-88

[38] Mohamed SA (1996) Farming systems and land tenure in Haubi Village of the Kondoa Eroded area. In: Christiansson C, Kikula IS. Editors. Changing environments: Research on Man-Land Interrelations in Semi-Arid Tanzania. Report No. 13. Nairobi; RSCU/SIDA. pp.77-83

[39] Mowo JG, Floor J, Kaihura FBS, Magoggo JP (1993) Review of fertiliser recommendations in Tanzania. Part 2 - Revised fertiliser recommendations for Tanzania. Soil fertility report No. F6. ARI Mlingano-Tanga: National Soil Service.

[40] Dejene A, Shishira EK, Yanda PZ, Johnsen FH (1997) Land degradation in Tanzania: Perceptions from the village. World Bank Technical Paper NO. 370. Washington DC: World Bank.

[41] Kikula IS (1989) Possible implications of the similarities and differences in perception of land degradation between planners and the planned. Research Paper No. 19. Dar es Salaam: Institute of Resource Assessment, University of Dar es Salaam.

[42] Kangalawe, RYM, Liwenga, ET, Majule, AE (2007) The Dynamics of Poverty Alleviation Strategies in the Changing Environments of the Semiarid Areas of Sukumaland, Tanzania. Research Report submitted to REPOA, Dar es Salaam.

[43] Kangalawe RYM, Lyimo JG (2009). Climate Change and Rural Livelihoods in Semiarid Tanzania. Proceedings of the 10th WaterNet/WARFSA/GWP-SA Symposium: "Integrated Water Resources Management within the context of Environmental Sustainability, Climate Change and Livelihoods". Entebbe: 28 – 30 October 2009.

[44] Kangalawe RYM., Majule AE, Shishira EK (2005) Land-use dynamics and land degradation in Iramba District, central Tanzania. Addis Ababa: Organisation for Social Science Research in Eastern and Southern Africa.

[45] Madulu NF, Mbonile MJ, Kiwia HDY (1993) Environmental impacts of migration in rural Tanzania. In: *Population, Environment and Development*. New York: United Nations, pp.73-91

[46] Cummingham OR, Jenkins QAL (1982) Natural disasters and farmers: A neglected area of research by rural sociologists. The Rural Sociologist 2(5): 325-330

[47] Nyega N (2008) Assessment of the beach management units strategy on the Lake Victoria fisheries resources in Ilemela District, Tanzania. MSc Dissertation, University of Dar es Salaam.

[48] Shadrack S (2009). Effectiveness of community based forest management approach in transforming local use of forest resources in Rufiji District. MSc Dissertation, University of Dar es Salaam.

[49] Mlenge W (2004) Ngitili: An indigenous natural resources management system in Shinyanga. Nairobi: Arid Lands Information Network - East Africa.

[50] Mung'ong'o CG (1995) *Social processes and ecology in the Kondoa Irangi Hills, Central Tanzania*. Meddelanden Series B 93. Stockholm: Department of Human Geography, Stockholm University.

[51] Östberg W (1995) *Land is Coming Up: The Burunge of Central Tanzania and their environments*. Stockholm Studies in Social Anthropology, 34. Stockholm: Department of Social Anthropology, Stockholm University.

[52] Mung'ong'o CG, Kikula IS, Mwalyosi RBB (2004). Geophysical and socio-political dynamics of environmental conservation in Kondoa District. Dar es Salaam: Dar es Salaam University Press.

[53] Kauzeni AS, Kikula IS, Shishira EK (1987) Developments in Soil Conservation in Tanzania. In *History of Soil Conservation in the SADCC region*. Research Report No. 8. SADCC Soil and Water Conservation and Land Utilisation Programme, Co-ordinating Unit: Maseru.

[54] Berry L, Townshend J (1973) Soil Erosion Conservation Policies in the Semi-Arid regions of Tanzania, A Historical Perspective. In: Rapp A, Berry L, Temple P. Editors. *Studies on Soil Erosion and Sedimentation in Tanzania,* BRALUP Research Monograph No. 1. BRALUP: Dar es Salaam; 241-253.

[55] Liwenga ET, Kangalawe RYM, Masao, CA (2007) The Implications of Agricultural Commercialization on Agro-diversity Management and Food Security in the Drylands of Central Tanzania. Research Report submitted to the Research Programme on Sustainable Use of Dryland Biodiversity (RPSUD), University of Dar es Salaam.

[56] Scoones, I. (1998). Sustainable rural livelihoods: A framework for analysis. IDS Working Paper No. 72.

[57] Ellis, F. (1998). Household strategies and rural livelihood diversification. Journal of Development Studies 35 (1): 1-38.

Chapter 10

Methodology for the Regional Landfill Site Selection

Boško Josimović and Igor Marić

Additional information is available at the end of the chapter

1. Introduction

One of the most important causes of environmental pollution is certainly an inadequate waste management. The three factors that have primarily influenced this problem area are: ever increasing amount of municipal solid waste (which causes increasingly pronounced space occupation), increasing amount and types of hazardous waste, as well as lack of awareness on the importance of acting promptly in this field. Particular problems in waste management occur in developing countries, where the awareness of the importance of environmental protection has not yet achieved a satisfactory level and where, out of economic or political reasons, professional guidelines associated with waste management are not observed. Problems emerge either due to a lack of legislation, or obsolete legislation, lack of trained personnel, inadequate waste management infrastructure, financial constraints in the establishment of a modern waste management system, population lacking the awareness about solid waste management, impossibility of selecting appropriate space for landfill development, lack of standards, etc.

Great problems in waste management in Serbia are caused by increasing amount of waste, lack of sanitary landfills built under international standards (which is inefficient and ecologically acceptable), as well as by the fact that the principle of hierarchy in waste management is not observed at all. Problems emerging in the field of environmental pollution and the manner of responding to pollution through the planning documentation, only increase the importance of this problem area.

Waste management is a complex process which implies a control of the entire waste management system (from waste generation, through collection and transportation of waste, to waste treatment and disposal), along with the support of legislation and appropriate institutional organization. The accent in the present paper is placed on spatial planning as an inevitable instrument for strategic waste management. The paper also points out the importance of spatial aspect in the waste management planning process.

2. Disposal of solid waste

The final functional element in the waste management system is waste disposal. Waste disposal is a final fate of all types of waste, either municipal solid waste, collected and transported directly to landfills, or industrial waste or other materials from waste treatment facilities which are of no use-value any longer [1].

Landfill forms the basis of every waste management plan, because there will always be waste to be disposed of.

Sanitary landfills are sites selected for waste disposal, such as natural or artificial (excavated) depressions, engineered facilities, where the waste is, through appropriate technological processes, compacted as densely as practicable to minimize its volume and covered with a layer of soil or some other material in a systematic and sanitary manner. Before proceeding with such work, a terrain to be used must be selected, surveyed and prepared [2].

Sanitary landfills are necessary in any combination, even for some other form of solid waste treatment, because there will always be waste to be disposed of. Uncontrolled dumps must be closed along with necessary sanitation. This requires knowledge of a series of notions, processes and activities which should enable proper landfill planning, design, construction, exploitation and funding, as well as control of landfill environmental impacts [3].

3. Spatial aspect of waste management

Considering that waste management system is realized in space, it is quite clear that characteristics of space greatly determine the choice of an adequate management system, i.e. its spatial organization. This primarily refers to the selection of sites having physical elements of the system, such as, primarily, sanitary landfills, transfer stations, recycling centers, etc. In this context, physical-geographical and anthropogenetic characteristics of space are of great importance. Relative to these characteristics, conceptual solutions to the waste management system are defined, and landfill site selection process is carried out for elements of the waste management system.

3.1. Requirements for the regional landfill site selection

In the waste disposal process, a controlled disposal procedure is unavoidable, either for the disposal of genuine waste or materials that remain after the treatment process, or, as necessary, if the main process cannot be carried out in certain period because of interruption, defect, overhaul, or out of other reasons. Sanitary landfills are necessary in any chosen waste management option, because there will always be waste to be disposed of on landfills. In this sense, locating potential landfill sites, as the most commonly used process through which a huge amount of collected waste is treated, should be given great attention in the waste management process, i.e. in spatial planning process. This is a very delicate and very important process from the viewpoint of the protection of key environmental factors (land, water and air), landscape values, as well as the protection of population health [4]. Out of this reason, it is also necessary to dedicate great attention to the investigation of a character, as well as potential and real landfill impact on the environment. This enables the elaboration and implementation of measures to eliminate or minimize negative effects.

Sanitary landfill planning and construction is only a part of a complex solid waste management process which encompasses the treatment of waste from its generation, through minimization of its amount, selection, recycling, collection, transport and disposal, to landfill recultivation and bringing of land to new use. However, although sanitary landfills are only a part of a wider waste management process, this activity is characterized by a very complex and long-term process which must take into account natural and anthropogenetic characteristics of space.

Sanitary landfill is available land for solid waste disposal at which engineering methods of waste disposal are used in a manner in which threats to the environment are minimized. The landfill site selection and technology of devices and equipment for sanitary waste treatment and disposal should be in the function of the protection and rational use of space.

Sanitary landfill development implies activities in several phases where certain sequence must be obeyed. The process is usually carried out in four phases:

- landfill site selection (site investigation process),
- identification of a landfill site (through the planning documentation) and elaboration of conditions for bringing it to the intended use,
- elaboration of construction (technical) documentation,
- landfill construction.

3.1.1. Preconditions for sanitary landfill construction

It is necessary to consider the following requirements and requirements for sanitary landfill construction:

- Spatial and urban planning requirements
- Spatial and regional requirements
- Landfill site selection
- Required land area
- Transportation distances
- Local site conditions
- Topography
- Climate conditions
- Hydrogeological conditions
- Geological conditions

- Geo-mechanical conditions
- Environmental protection

Landfill site selection.- In planning, landfill site selection occupies extremely important place. In the widest sense, the natural, social, political, economic and technical factors have an important role in landfill site selection, thus it follows that the selection is to be made by a multidisciplinary team of experts. Given that landfill construction is considered as a non-economic activity, a special task is to select landfill site from the aspect of the use of buildable land and its price, as well as other natural or urban values which have an important role in relation to rationality and planned landfill remediation or its rehabilitation, i.e. reconstruction [5]. From technical and technological aspects, for the planning, design, construction and exploitation needs, it is necessary that on each landfill site the following is ensured:

- Complete sanitary security for people living in the surrounding residential areas, as well as personnel working at landfills
- Protection of land, air, ground and surface water from pollution
- Rational use of land, as well as save land (increased levels of waste compaction using special machines, as well as a deposition height)
- Maximum number of machines and equipment for all types of works

3.2. Implementation of GIS tools in waste management

Information system is an arranged set of information on things and facts in surroundings, with the aim to get acquainted with a system. Right decision making in planning and space organization depends to a great extent on knowledge, i.e. the quality and importance of information available to decision makers.

The GIS is a powerful set of computer tools for collecting, storing, searching as necessary, transformation and display of real-world data for various purposes [6].

As one of the most complex information systems that cover all spatial problems, the GIS has many advantages out of which the most significant are:

- It covers all elements of geo-space and ecological elements
- It includes natural and social elements of space

The use of GIS is appropriate for:

- Spatial planning
- Mapping for various purposes
- Traffic planning
- **Waste management planning**
- Natural resource inventory and management
- Computer mapping of population and entering of census data
- Creating hazard maps and programs of procedures in such cases, etc.

Essentially, GIS contains data on:

- Air
- Earth's surface
- Water areas
- Lithosphere
- Soil
- Biocenosis
- Anthropogenetic spatial elements, etc.

The GIS, as already mentioned, consists of spatial data and descriptive data.

Spatial data refer to locations, i.e. spatial relationships between phenomena and objects. They are obtained based on literature, maps, aerial photos taken from aircrafts, etc., and they are useful only if they can be converted into maps.

Descriptive data are linked to localities, polygon line or body and are the system accompanying content.

The GIS key features are:

- Possibility of spatial search of phenomena
- Possibility of overlapping contents and combining individual contents into a new quality
- Logical operations with spatial and descriptive data

Geographic information systems are most frequently compatible with most of related systems (geodetic, agricultural, geologic, mining, water resources management, forestry, urban planning systems, etc.), but also with census databases, statistical information systems, technological databases, databases associated with health, education, science, etc. Using GIS mapped data, we carry out precisely what an information system should enable: solve a problem, make queries, reach answers, or examine possible solutions. Here, data are manipulated digitally, and not manually, because we manipulate the data on events and activities using digital cartographic objects. In other words, the points, lines and areas in this cartographic database are used for data management.

Therefore, the GIS is a general tool for problem solving. It is created for making a certain project. A successful GIS is built, not bought, and indented for analysts to draw out relevant data for forecasting and planning, as well as various pieces of information associated with a specific space, as well as the problem area which is a subject of analysis.

The role of GIS tools in waste management planning is dominant in landfill site selection process. In addition, GIS tools are also used for distribution and identification of locations for other elements of waste management system such as transfer station network, waste selection and processing centers, for defining transportation corridors, etc.

The method of multicriteria analyses and evaluation is used for identifying locations of elements of a waste management system in the GIS. This approach is inevitable in locating complex objects, such as, for example, regional municipal solid waste landfills. Its complexity is reflected both in the size and function of objects, as well as in relation to various possible spatial impacts, also in negative context.

The use of GIS in defining strategies, analyses and visualization of solutions and alternatives helps us consider and clearly represent various scenarios, as well as select the most suitable solutions through a prism of different relevant criteria (spatial, ecological, hydro-geological criteria, etc.) [7].

Therefore, in using the GIS in the selection of the most suitable landfill sites, two things of key importance are [8].:

1. Analysis of space, i.e. all of its physical-geographical and anthropogenetic characteristics. It is necessary to comprehensively consider the space on which the problem is to be solved or which can be useful for problem solving. In this process, because of social sensitivity associated with this issue, it is necessary to be impartial in considering a possible landfill site. This can only be achieved if the entire space is considered to the same level of detail and in the same manner;
2. Visualization of space and its characteristics and impacts. This is necessary so that all participants in the project could have equal chance to perceive and understand the subject problem area. This enables active participation in searching for solutions to an acceptable compromise [9]. All participants must consider the space, as well as its advantages and disadvantages for landfill site selection. This is precisely one of the most important advantages of using GIS tools in landfill site selection, as well as of choosing other elements of a waste management system.

Defining landfill site selection criteria is the main step in landfill site selection process. In the first phase, based on exclusiveness, the sites which do not satisfy these criteria are eliminated. Positive areas within which it is possible to search for the most suitable solutions are the result of this process. This phase represents an activity of microzoning. Using GIS tools, through overlapping cartographic presentations of a certain space carried out based on exclusion criteria, it is rather simple to eliminate unsuitable landfill sites.

After eliminating the unsuitable landfill sites, the attention is dedicated to the nomination of landfill sites within the remaining "conditionally suitable" zones. In this process, local governments and professional institutions can and must be of great importance, but soil investigations and collecting relevant data on physical-geographical and anthropogenetic characteristic of space are indeed of utmost importance.

Through nominating potential landfill sites, preconditions for the selection of the most suitable landfill site are created, which is followed by multicriteria analysis and evaluation of candidate sites. Site selection criteria are entered into tables and weighted for each candidate site based on the entered value scale. In this way, the evaluation process using GIS tools is carried out in an efficient manner and in a short period of time.

The role of GIS tools in the landfill site selection process is in that it enables faster singling out and clearer presentation of suitable and unsuitable sites based on previously given criteria.

In this context, it is evident that selection criteria and value scale for evaluation of candidate landfill sites are of key importance in this process, while GIS tools represent a powerful means which to a great extent facilitate and speed up the process. This refers not only to the landfill site selection process, but also to defining the spatial organization of the entire waste management system, as well as defining the transfer station network.

3.3. Landfill site selection criteria

3.3.1. Landfill site selection criteria

The most important step in this process is to define landfill site selection criteria.

There are two groups of criteria. The first group includes the so-called exclusion criteria that are used in the first phase of the landfill site selection process. Exclusion criteria are defined relative to the specific situation and they represent restriction criteria.

Some of exclusion criteria can be classified into a group of the following indicators:

- Distance from natural elements of space (watercourses, water sources, protected natural resources, etc.)
- Distance from anthropogenetic elements of space (infrastructure facilities, settlements, protected cultural structures, etc.)
- Terrain morphology
- Hydrological and geological characteristics of space
- Degradation of space
- Recommendations of local authorities in a form of intermunicipal corporation agreements, etc.

According to exclusion criteria, areas which should not be further analyzed are discarded, i.e. areas that will be analyzed and evaluated in consecutive phases singled out. In the elimination phase, a single-criterion method is mainly used.

After that, in cooperation with local institutions and experts, certain number of sites are nominated for which a multicriteria evaluation is carried out. In this context, criteria based on which each candidate site will be evaluated in the same way are defined. This is a second group of criteria.

Site evaluation criteria are mainly classified into several basic groups. Commonly, there are three basic groups of criteria whose definition varies from author to author:

- Ecological or environmental criteria,
- Socio-economic or social or spatial criteria,
- Technical and operational criteria (which usually also involve certain economic, spatial and ecological criteria).

Any variation of groups of basic indicators is possible. Regardless of the formulation of basic groups of criteria, they include approximately either the same or almost the same number of indicators and criteria that are analyzed and compared in the process of selection of the most suitable site for a landfill.

Number of landfill site selection criteria ranges from 20 to over 40. They are classified (or not classified) into groups of criteria to which they belong, which are also similar, but can be differently formulated.

A particularly sensitive and important step in landfill site selection that follows the choice of relevant criteria is to define value scales based on which each individual criteria is evaluated (valued, ranked). Each criteria is assigned its corresponding weight (value) which is determined based on expert's evaluation and evaluation of participants in the process of sanitary landfill site selection. Here, quantitative evaluation is commonly used (e.g. scores from 1 to 10, or from 1 to 5).

Qualitative/expert assessment can also be used, where criteria can be assessed as suitable, conditionally suitable or unsuitable. Qualitative assessment is today increasingly less used, because the use of new technologies enable more accurate and more sophisticated assessment under the principle of quantitative assessment. In this case, accurate and objective data are obtained that can be compared and used for making right decisions.

When a potential site is assessed according to all given criteria, it is possible to carry out the following two steps:

1. Adding up all obtained scores
2. Multiplying the obtained scores by importance values (weights).

The first step in evaluating candidate sites is the simplest one and will low requirements. The best score is obtained through adding up all obtained scores for each criterion. Evaluation of candidate site in this case does not have different scenarios that can be of great help to decision makers.

The second step is more complex as different scenarios can be used. For example, if criteria for locating candidate landfill site are classified into several basic groups, then the number of scenarios to be considered is consistent with the number of criteria groups. Criteria from one group are favored in the first scenario, the most important criteria in the second scenario are those from the second group, and so on. The final option is a situation when groups of criteria are multiplied by the same importance value, without favoring any of criteria group. By presenting the scenarios in synthesis Table, it is easy to identify which candidate sites are the most suitable in which scenarios. The PROMETHEE method [10] is an example of this approach.

The basic advantage of this procedure is in that decision makers have a clear idea of which is candidate site is the most suitable if criteria from a certain group of criteria (ecological or economic or spatial, etc.) are assessed as the most worthwhile criteria, and if basic criteria groups are dealt with equally. This greatly facilitates decision making. Regardless of which of the many methods for evaluation of potential landfill sites are used, the question of objectivity of the procedure arises taking into account that the selection of evaluation elements (criteria, weights), but also the very decision-making process, is a matter of objectivity of experts and decision makers. This can be considered as a common disadvantage of all methods for potential landfill site selection. Therefore, the subjectivity in this process must be minimized to the utmost limit, while objectivity must be maximized.

4. Case study: Regional landfill site selection in the Kolubara region in Serbia

4.1. Analysis of the present state in the Kolubara region

We have chosen the area of the Kolubara Region comprising 11 municipalities with 382,000 inhabitant as an example of theoretical knowledge presented in the first part of the present paper.

The Study on the Selection of Micro-location for the Regional Landfill with Recycling Centre and Regional Center for Municipal Solid Waste Management, Regional Plan for Solid Waste Management, as well as Strategic Environmental Assessment (SEA) for the same Plan have been elaborated for the Kolubara Region.

Municipal waste from the territory of 11 municipalities in the Region is disposed of to 10 unarranged sanitary city landfills and a certain number of illegal dumps. All existing landfills should be closed or remediated and recultivated in the shortest possible time. It is recommended to prolong the life-time of the existing landfills through the mentioned remediation projects. Recognizing the need for the final, contemporary waste disposal and management, 11 municipalities of the Region have united together in forming the regions for the development of a waste management. The initiatives that have been launched in this context have resulted in the elaboration of the "Study on the Selection of Micro-Location for the Regional Municipal Waste Landfill with Recycling Center for the Kolubara Region", based on which the location for regional landfill has been selected.

4.2. Locating the regional landfill site for Kolubara Region

The sanitary landfill construction implies carrying out activities in several phases, whereby it is necessary to observe a specific sequence. The process is mainly carried out in four phases as follows:

1. Identifying (selecting) the location
2. Determining the location (through the planning and design documentation) and creating conditions for bringing land to intended use
3. Elaborating the construction documents (technical documentation)
4. Landfill construction

The most sensitive and the most important step in making a concept of regional municipal solid waste management is a regional landfill site selection. Relative to the selected regional landfill site, other elements of a waste management system are also located and their spatial distribution carried out.

Once the selection of the most suitable landfill site is made, it is necessary to incorporate it in the planning solutions in order to create conditions for the elaboration of technical documentation, as well as for landfill construction.

4.3. Regional landfill site selection criteria

The elaboration of the Study on Landfill Site Selection represents the first step in making a concept of municipal solid waste management in the Kolubara Region.

The first step in landfill site selection is to define exclusion criteria.

Taking into account current legislation, Intermunicipal Agreement on Joint Waste Management, basic exclusion criteria used in practice, available data on the space, as well as relevant characteristics of a specific space, the following exclusion criteria have been defined, see [6]:

- Seismic activity over 9 MCS
- Distance of less than 500 meters from watercourses
- Distance of less than 500 meters, or 1.5 km from settlements, if not sheltered
- Distances of less than 500 meters from water supply sources
- Collision with the existing planning documents
- Distance of less than 500 meters from roads of the first category if the site is not sheltered
- Terrains with an inclination of over 30%
- Terrains of more than 300 meters above sea level
- Alluvial plains and karst terrains

Their corresponding areas have been identified using the GIS tools. Through overlapping the corresponding areas, the following is obtained:

1. Potentially suitable areas for landfill
2. Unsuitable areas within which it is not possible to locate the landfill.

Once the potentially suitable areas within which its is possible to search for the regional landfill site are singled out, the regional plans for waste management for the Kolubara District and Belgrade administrative area have been considered in which the area of the Kolubara lignite basin has been determined for a macro-location for the regional municipal waste landfill.

Besides, in the Intermunicipal Agreement on the Joint Waste Management it has been agreed that the landfill site will be located in the territory of the Ub municipality since Ub has agreed to accept the waste generated in the newly formed region for municipal solid waste management in its territory.

The location of Carić (within the territory of the Valjevo municipality) has been also considered taking into account that this location has been previously analyzed several times and assessed as a suitable site for a landfill.

In nominating the location, the following has been taken into account:

- Preliminary analyses of the entire area and possible central position of potential sites in the region
- Data collected during field visits
- Consultations and recommendations of relevant local institution and experts
- Guidelines set by the EU and the Waste Management Strategy of the Republic of Serbia
- Data and information from the existing planning documentation.

Based on the above mentioned, the following three potential landfill site locations have been proposed (Figure 1).

1. **Location KALENIĆ**, in the area of open pits in the Ub territory
2. **Location BOGDANOVICA ,** city landfill (dump) in Ub
3. **Location CARIĆ**, for which certain investigations have already been carried out which have indicated certain advantages for landfilling. The location is within the territory of Valjevo municipality.

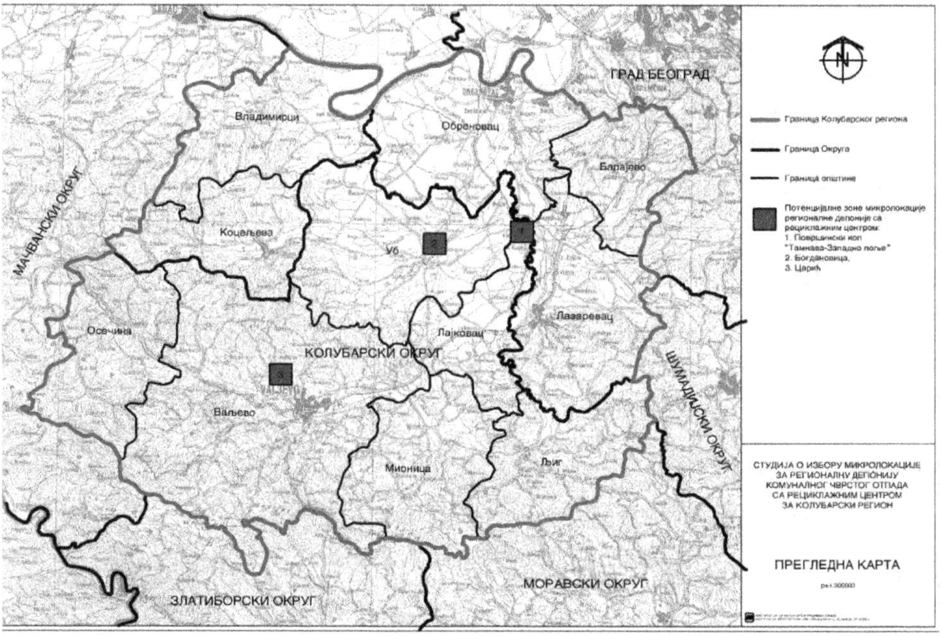

Figure 1. Position of potential landfill sites in the Kolubara Region

After the nomination of locations, the criteria for the evaluation and selection of the most suitable landfill site have been defined.

New criteria have been defined based on investigation and analysis of previous experiences of other countries and the EU guidelines, as well as on available relevant data for their evaluation. In this context, the following criteria for the selection of micro-location for regional landfill in the Kolubara Region for the municipal solid waste management have been defined:

1. Hydrogeological characteristics

a. Rock masses with fissure–cavernous porosity and a high water permeability (karstified rocks, limestones and dolostones)

b. Rocks with intergranular porosity, coarser grained rocks (coarse-grained gravel)

c. Rocks with low porosity (alluvial and glacial sediments)

d. Materials with a low water permeability, mainly impermeable complexes of 10-6≥ k ≥ 10-9m/s, or with a low water impermeability, but with small layer thickness of less than 1.0 m;

e. Water impermeable materials (clay, flysch) of k≤10-9m/s, the layer thickness ≥1.0 m.

2. Ground water

a. Aquifer is, over a brief period and at high water levels of greater frequency, above the bottom of the landfill in one part of its bottom area, while at other water levels, it is beneath the bottom; occasional flooding also occurs at the landfill site

b. Aquifer, at high water levels of small frequency, rarely rises above the landfill bottom; wetting of the landfill bottom is possible

c. Aquifer at water levels of 1 to 3 m is beneath the landfill bottom

d. Aquifer at high water levels > 3 m is beneath the landfill bottom

e. Aquifer does not exist

3. Distance from the boundaries of zones of sanitary protection of water supply sources

Distance from the boundary of

(a) narrower protection zone:(b) wider protection zone:

a. 0 to 0.2 kmbelt along the protection zone contours

b. 0.2 to 0.5 kmup to 0.5 km

c. 0.5 to 1.0 km0.5 to 1.0 km

d. 1.0 to 1.5 km1 to 2 km

e. more than 1.5 km> 2 km

4. Geological-tectonic characteristics

a. Pronounced fault zone

b. Fault carbonate rock masses with numerous surface and underground karst shapes or flat terrain

c. Flysch sediments, shales, marlsones, sandstones, etc.

d. Glacial sediments

e. Magmatic rocks

5. Distance from the closest settlements with concentrated development or residential zones of urban settlements

a. Distance 1.5 - 2 km, or 0.75 - 1 km if shelter

b. 2 -3 km, or 1 – 1.5 km with shelter

c. up to 4 km, or 1.5 - 2.0 km with shelter
d. up to 5 km, or 2.0 – 2.5 km with shelter
e. more than 5 km, or more than 2.5 km with shelter

6. Relief characteristics of the terrain

a. Broken relief, very uneven terrain, particularly pronounced in karst landscapes, incompact (scattered) spatial entity encompassing several valleys
b. Broken relief, uneven terrain, compact spatial entity
c. Incompact (scattered) spatial entity encompassing several valleys, naturally shaped terrain suitable for formation of valleys
d. compact waste entity, naturally shaped for locating a landfill site in a steep terrain or in natural depression
e. Mildly inclined or flat terrain, naturally shaped for locating a landfill site or possibly a landfill in excavated depressions or on earth fills

7. Available space for waste disposal and ancillary facilities

a. up to 5 yrs
b. up to 10 yrs
c. up to 15 yrs
d. up to 20 yrs
e. 20 yrs

8. Site acceptability

a. General landfill site disagreement
b. General agreement, but disagreement from local community
c. General agreement, but disagreement from certain individuals form local community
d. General agreement and somewhat moderate disagreement from local community
e. General acceptance of a landfill site

9. Engineering-geological characteristics

a. Incoherent rock masses, unstable slopes, slides and falls, active landslides
b. Complex of incoherent and semi—coherent rock masses (deluvial sediments), possible occurrence of landslides due to undercutting the foot of an existing slope
c. Semi-coherent rocks, possible occurrence of landslides due heavy falls
d. Coherent rocks, slightly stoned rock, stable slopes
e. Solid rocks, stable slopes even those of greater inclinations

10. Current land use

a. Cultivated agricultural land (ploughland, orchards), individual houses and other residential buildings within holdings, sportsgrounds, etc.
b. Quality tall forests;
c. Meadows
d. Pastures, shrub woods
e. Uncultivated land, thickets, barren land, excavations, quarries

11. Distance from individual water supply (wells)

a. 100 - 200 m, downstream of the landfill or approximately on landfill level
b. up to 500 m, downstream of the landfill or on the same level as the landfill
c. 500 to 1000 m, downstream or on the same level as the landfill
d. downstream of the landfill at the distance up to 200 m, downstream of the landfill at the distance of 1-1.5km;
e. downstream of the landfill at the distance of more than 200 m, downstream of the landfill at the distance of more than 1.5 km.

12. Landscape characteristics

a. Highly disturbed and completely changed natural ambience during landfill exploitation and after its closure
b. Highly disturbed natural ambience during landfill exploitation, and partly after the landfill closure
c. Natural ambience disturbed during landfill exploitation, and to a less extent after its closure
d. Natural ambience slightly disturbed during landfill exploitation, and undisturbed after its closure
e. Ambience not disturbed either during landfill exploitation or after its closure.

13. Linear distance from roads and railroads

more important roads | other roads

without shield | with shield | without shield | with shield

a. 500 m | 300 m | 300 m | 200 m
b. 600 m | 400 m | 400 m | 250 m
c. 800 m | 500 m | 500 m | 300 m
d. 1000 m | 600 m | 600 m | 400 m
e. >1000 m | >600 m | >600 m | >400 m

14. Distance to sacral structures, monuments of culture or protected natural resources

a. Distance 1.0 – 1.25 km, or 0.5 – 0.75 km where there is a shield
b. 1.25 -1.50 km, or 0.75 - 1,0 km with shield;
c. 1.5 – 2.0 km, or 1.0 - 1.25 km with shield;
d. 2 – 2.5 km, or 1.25 – 1.5 km with shield;
e. more than 2.5 km, or more than 1.5 km with shield

15. Seismic Activity

a. 9-8 MCS
b. 7 MCS
c. 6 MCS
d. 5 MCS
e. < 5 MCS

16. Existing site infrastructure

a. Absence of any infrastructure
b. Poor infrastructure
c. Only one infrastructure segment (access road, water supply line, electricity);
d. Several infrastructure segments
e. All or most of the infrastructure segments

17. Distance from surface watercourses

a. Permanent rivers or standing waters at the distance of 500 to 1000 m, there is a risk of flooding during high waters, defense measures against high waters required
b. Small watercourses, permanent or periodic ones (brooks, torrents), there is a flood risk, it is necessary to displace or channel these waters
c. Heavy inflow of rain waters from immediate catchments, defense against these waters requires more complex facilities; there is no flooding
d. Permanent watercourses at the distance greater than 1 km, no risk from flooding; defense standard solutions applicable
e. Great distance from watercourses, no risk from flooding, very low inflow of rain waters, simple protection against these waters possible

18. Terrain preparation

a. Very complex terrain leveling works, including intensive blasting on the greatest part of the site
b. Complex terrain leveling works, blasting required only in some parts of the landfill site
c. Terrain leveling works on the greatest part of the landfill site using machines
d. Terrain leveling on the smaller part of landfill site using machines
e. Simple terrain leveling works on the smaller part of the landfill site

19. Earth for covering the disposed waste – distance from the borrow site

a. greater than 5 km,
b. 2-5 km,
c. 1-2 km,
d. up to 1 km,
e. on site.

20. Position of the site in the Region

a. Completely dislocated relative to the central position in the Region; at the edge of the Region
b. Within the radius of 20 km relative to the central point in the Region,
c. Within the radius of 10 km relative to the central point in the Region,
d. Centrally positioned relative to the Region,
e. Within the radius of 10 km relative to the central point in the Region, but closer to the municipalities with the largest amounts of municipal solid waste.

21. Ownership of land

a. 100 % of land under private ownership, greater number of smaller plots
b. 100 % of land under private ownership, greater plots

c. About 75 % of land under private ownership, about 25 % of land under state ownership
d. About 50 % of land under private ownership and about the same amount of land under private ownership
e. 100 % of land under state ownership

22. Precipitations

a. 1500 mm
b. 1000 to 1500 mm
c. 600 to 1000 mm
d. 300 to 600 mm
e. < 300 mm

23. Air temperature

a. < 6° C
b. 6-9° C
c. 9-12° C
d. 2-15° C
e. > 15° C

24. Air flow

a. Very frequent high intensity winds, with prevailing wind direction towards settlements and other localities where people stay and work
b. Less frequent lower intensity winds with prevailing wind direction towards relevant facilities
c. Prevailing winds of changeable direction towards relevant facilities
d. Dominant winds blowing in the opposite direction, from settlements and other places where people stay and work, as well as low intensity winds blowing in direction towards the settlements
e. Most of the winds blowing in opposite direction, from settlements and other places where people stay and work

25. Distance to individual houses outside settlements

a. < 250 m
b. 500 m
c. 1000 m
d. 1500 m
e. 500 m

26. Site shelterness

a. Visible from all distances and all angles
b. Locality sheltered to a smaller extent
c. Locality sheltered to a greater extent
d. The glimpse of the locality can be caught in the great distance
e. Not at all visible, except when you come in the locality itself

27. Access road – reconstruction, or construction of a new road

New roadRoad reconstruction

a. > 1000 m,>1500 m
b. 500-1000 m800 – 1500 m
c. 200-500 m300 – 800 m
d. <200 m< 300 m
e. There is an access road of satisfactory characteristics

28. Providing electricity supply via the distribution network at the distance of:

a. > 2 km
b. 1 - 2 km
c. 0. 5 - 1 km
d. 300 - 500 m
e. < 300 m

29. Water supply in the locality

a. From the public water supply system via connection longer than 4 km, or from a local water supply via a connection longer than 3km
b. From the public water supply system via connection 2 to 4 km long, or from a local water supply via a connection up to 3 km long
c. From the public water supply system via connection from 1 to 2 km long, or from a local water supply via connection up to 1 km long
d. From the public water supply system via connection from 0.5 to 1 km long, or from a local water supply via connection up to 500 m long
e. From the public water supply system via connection up to 500 m long

30. Distance to agricultural land

a. < 100 m
b. 100 - 300 m
c. 300 - 500 m
d. 500 - 1000 m
e. > 1000 m

31. Distance from the main transmission line, gas pipeline, crude oil pipeline, drinking water pipeline

a. up to 100 m
b. 100 - 200 m
c. 200 - 300 m
d. 300 - 500 m
e. 500 m

32. Possibility of construction in phases and extension

a. No possibility of construction in phases or of extension
b. Limited possibility of construction in phases, but not of extension
c. Possibility of construction in phases, but not ofextension

d. Possibility of construction in phases and of limited extension
e. Possibility of construction in phases and of unlimited extension

Criteria are presented under the principle of exclusion criteria. More precisely, no detailed guidelines for evaluation have been given for criteria save for exclusion criteria which define requirements which a potential site MUST meet in locating the municipal solid waste landfill site.

4.4. Implementation of multicriteria evaluation method in landfill site selection

Potential micro-location for regional landfill in the Kolubara Region has been determined through multicriteria analysis and evaluation. Chosen criteria have been evaluated by assigning scores from 1 to 5 for each candidate site.

At the same time, depending on their importance in evaluating the locality quality, criteria have been classified into 3 pondering categories (PC). Each weight category has its specific value – weight, which is multiplied by the score of corresponding criteria. In this way, a final score is obtained for each criterion. Values by pondering categories are:

- PC1 = 1
- PC2 = 1.5
- PC3 = 3

The relation between pondering categories (PC) is: $K_{i+1} = K_i/1.5$

PC 3	PC 2	PC 1
	Landfill site selection criteria	
1 - 8	9 - 20	21 - 32

Table 1. Grouping the criteria by pondering categories (PC)

Table 2 indicates that after assigning a score to each criterion, the Kalenić location has been singled out as the most suitable one. The other two locations (Bogdanovica and Carić) have been assigned much poorer scores compared to the Kalenić location. However, in cases when the difference in ranks between candidate locations is extremely small at the end of evaluation process, it is difficult to make a final decision on which site is the most suitable. In this case, it necessary to carry out an additional evaluation which implies the evaluation of candidate sites by different scenarios. The chosen site selection criteria are then grouped into basic groups, while in the "additional" valuation process, the criteria from one of the basic groups are favored in each scenario, see [6].

There are so many scenarios as groups, plus one for the scenario according to which each basic criteria group is evaluated equally (for the last scenario, the data taken from basic evaluation or criteria are multiplied by weight value). In this way, decision makers are given opportunity to choose the option based on their policy, and thus select the most suitable site.

In regional landfill site selection for Kolubara Region, no "additional" evaluation has been required due to evident advantages of location Kalenić.

Criteria	PC	Kalenić	Bogdanovica	Carić
1. Hydrogeological characteristics	**PC 3**	12	9	9
2. Groundwater	**PC 3**	15	6	12
3. Distance from the boundaries of zones of sanitary protection of water supply sources	**PC 3**	12	15	15
4. Geological-tectonic characteristics	**PC 3**	12	9	9
5. Distance from the nearest settlements with concentrated development or residential zones of urban settlements	**PC 3**	12	3	12
6. Relief characteristics of the terrain	**PC 3**	15	15	12
7. Available space for waste disposal and ancillary facilities	**PC 3**	15	3	6
8. Site acceptability	**PC 3**	15	6	6
9. Engineering-geological characteristics	**PC 2**	3	3	3
10. Current land use	**PC 2**	7.5	7.5	1.5
11. Distance from individual water supply (wells)	**PC 2**	7.5	7.5	7.5
12. Landscape characteristics	**PC 2**	7.5	4.5	1.5
13. Linear distance from roads and railroads	**PC 2**	7.5	1.5	7.5
14. Distance to sacral structures, monuments of culture or protected natural resources	**PC 2**	7.5	7.5	7.5
15. Seismic Activity MCS	**PC 2**	3	3	3
16. Existing site infrastructure	**PC 2**	7.5	3	1.5
17. Distance from surface watercourses	**PC 2**	6	1.5	7.5
18. Terrain preparation	**PC 2**	4.5	4.5	4.5
19. Earth for covering the disposed waste – distance from the borrow site	**PC 2**	7.5	3	1.5
20. Position of location in the Region	**PC 2**	7.5	6	3
21. Ownership of land	**PC 1**	5	4	1
22. Precipitation	**PC 1**	3	3	3
23. Air temperature	**PC 1**	2	2	2
24. Air flow	**PC 1**	4	4	3
25. Distance to individual houses outside the settlement	**PC 1**	3	2	1
26. Location shelterness	**PC 1**	5	3	4
27. Access road - reconstruction or construction of new road	**PC 1**	3	3	1
28. Providing electricity supply via the distribution network at the distance of	**PC 1**	3	3	2
29. Water supply in the locality	**PC 1**	4	3	1
30. Distance to agricultural land	**PC 1**	5	4	1
31. Distance from main transmission line, gas pipeline, crude oil pipeline, drinking water pipeline	**PC 1**	5	2	5
32. Possibility of construction in phases and extension	**PC 1**	5	3	2
Total sum of criteria scores		**231.5**	**154.5**	**154**

Table 2. Evaluation of potential site by chosen criteria, see [2].

Had the final results of evaluation for all candidate locations been equal, the "additional" evaluation process would have been carried out in a manner as described in the text that follows. Namely, chosen criteria for additional evaluation would be classified into three basic groups (Table 3).

ECOLOGICAL	SPATIAL	SOCIO-ECONOMIC
Landfill site selection criteria		
1, 2, 3, 4, 6, 9, 10, 12, 15, 22, 23, 24, 26	5, 7, 11, 13, 14, 17, 19, 20, 25, 30, 31	8, 16, 18, 21, 27, 28, 29, 32

Table 3. Classification of chosen criteria into basic criteria groups

The scores of each criteria obtained in the basic evaluation process would then be multiplied by weight values for criteria groups according to different scenarios (Table 3). Weight values would actually be the percentage values whose sum is 100%.

Scores of each criteria from the basic evaluation process would be then multiplied by weight values for criteria groups according to different scenarios (Table 4).

Weight values would actually be the percentage values whose sum is 100%.

Scenario / Basic criteria group	SC 1	SC 2	SC 3	SC 4
EKOLOGICAL	**0.50**	0.25	0.25	**0.33**
SPATIAL	0.25	**0.50**	0.25	**0.33**
SOCIO-ECONOMIC	0.25	0.25	**0.50**	**0.33**

Table 4. Criteria weight values according to different scenarios

After multiplying the criteria values from basic evaluation by weight criteria according to different scenarios and their sum for each candidate site, the ranking of candidate sites according to different scenarios would be obtained (Table 5).

Scenario	SC 1	SC 2	SC 3	SC 4
	Site ranks			
Candidate site	Kalenić (83.12)	Kalenić (77.50)	Kalenić (73.32)	Kalenić (76.23)
	Bogdanovica (63.12)	Carić (53.25)	Bogdanovica (47.40)	Bogdanovica (50.98)
	Carić (58.62)	Bogdanovica (48.87)	Carić (45.07)	Carić (50.82)

Table 5. Ranking of candidate sites according to different scenarios

Through multicriteria evaluation according to different scenarios, several options and different arguments for the selection of the most suitable site are made available to decision makers. Implementation of different scenarios is based of the PROMETHEE method.

In this case, it has been shown that the location Kalenić has the best values in all four scenarios, while it is evident that the remaining two locations differ depending on scenario. The location Carić is better valued for the scenario 2, while location Bogdanovica is better valued in other three scenarios.

4.5. Multicriteria analysis and evaluation using GIS tools

In selecting the landfill site in the Kolubara Region for municipal solid waste management, the GIS tools have been implemented in singling out areas to be eliminated. The areas to be eliminated have been singled out based on defined exclusion criteria. Each of exclusion criteria has been presented graphically (cartographic presentation), and corresponding areas have been identified using GIS technology. Through overlapping maps of each exclusion criteria, negative areas have been singled out that should not be further analyzed in the landfill site selection process for the municipal solid waste management. Negative areas are shown in the synthesis map (Figure 2).

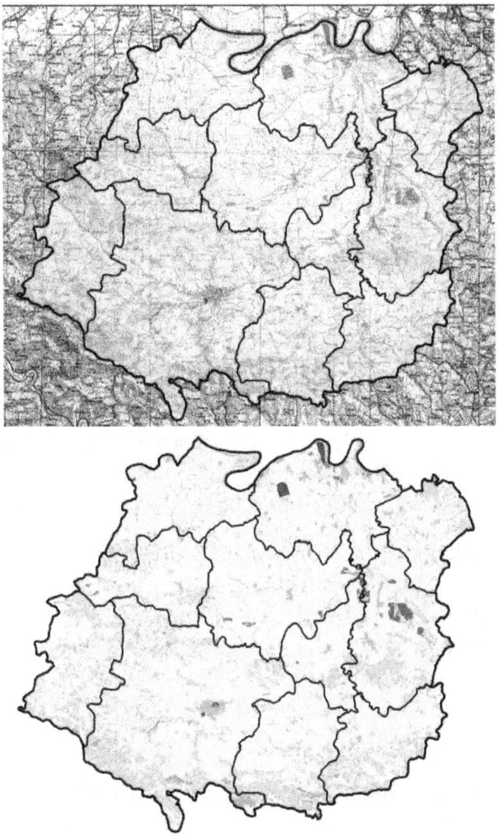

Figure 2. Analsys of land use structure based on CORINE information system [11].

Figure 2. depicts the structure of land use in Kolubara Region based on the CORINE (Coordination of Information on the Environment). The CORINE program is the European information base as support to the sustainable development policy of the European Union. The database contains data on: urban areas, crop yield, meadows, forests and natural vegetation, waters,. as well as other dynamic processes in the environment. All mentioned data are cartographically presented, which enables a more simple analysis of the subject area. The CORINE program was initiated in 1985. At the beginning, the program was developed and tested on 10 regions of the European Union by demonstrating the feasibility of the approach. Satellite photographs on which the CORINE database is based have been geometrically and radiometrically supplemented and with abundance of data which are in the CORINE Land Cover organized hierarchically in three levels classified in 44 classes (correspondingly presented spatial features and data). After showing positive results, in 1994 the European Environment Agency based in Copenhagen undertook the maintenance and use of the CORINE Land Cover database. Since then, the CORINE Land Cover (CLC) has been affirmed which is reflected in the fact that an increasing number of European countries are involved in the CLC project which has provided them with an opportunity to more efficiently pursue their environmental protection and sustainable development policies, as well as to carry out analyses for various needs and development strategies. Today, 64 European countries are involved in the CORINE Land Cover 2000 Project, with clearly defined and synchronized methodology for collection, processing, as well as presentation of data, in the function of the elaboration of environmental management plans [12].

In the elaboration of the "Study on the Selection of Micro-Location for the Regional Municipal Waste Landfill with Recycling Center for the Kolubara Region", the CORINE information base has not been available for Serbia, thereby for the Kolubara Region either. However, once the information on the environment from the CORINE program have become available for users in Serbia, all results from the elimination phase of landfill site selection contained in the Study have been checked and, what is even more important, confirmed. By using the CORINE program in accordance with the defined eliminating criteria for Kolubara Region, the selection of “negative” areas has been much easier and faster. The CORINE Information base to a great extent meets the needs of elimination phase in landfill site selection, thus this phase should be used as much as possible.

On the synthesis map (Figure 3), which is a final phase in the process of elimination of “negative” areas, the areas which do not satisfy basic conditions relative to the established exclusion criteria are denoted by red color. These are mainly corridors along watercourses, first category roads, distances to settlements, areas at over 300 meters above sea level, water supply sources, etc. Thus, it is the matter of exclusion criteria represented by minimum required distance of the future landfill site relative to them [13].

In the elimination phase, it is also possible to use some other criteria such as, for example, central position of a landfill relative to the Region. This means that, because of the cost-effectiveness of the waste management system, i.e. transportation costs, it is necessary to position a landfill within the radius of 20 or 30 km relative to the central point of the Region. However, in such case, a great number of areas that merit further analysis by their characteristics can be excluded, while the problem of central positioning of a landfill can be overcome through a good organization of transfer station network in the Region. In this context, it is important to emphasize that it is not necessary to introduce a great number of exclusion criteria, but to limit the choice of exclusion criteria to the most relevant ones, as shown on the example of the Kolubara Region.

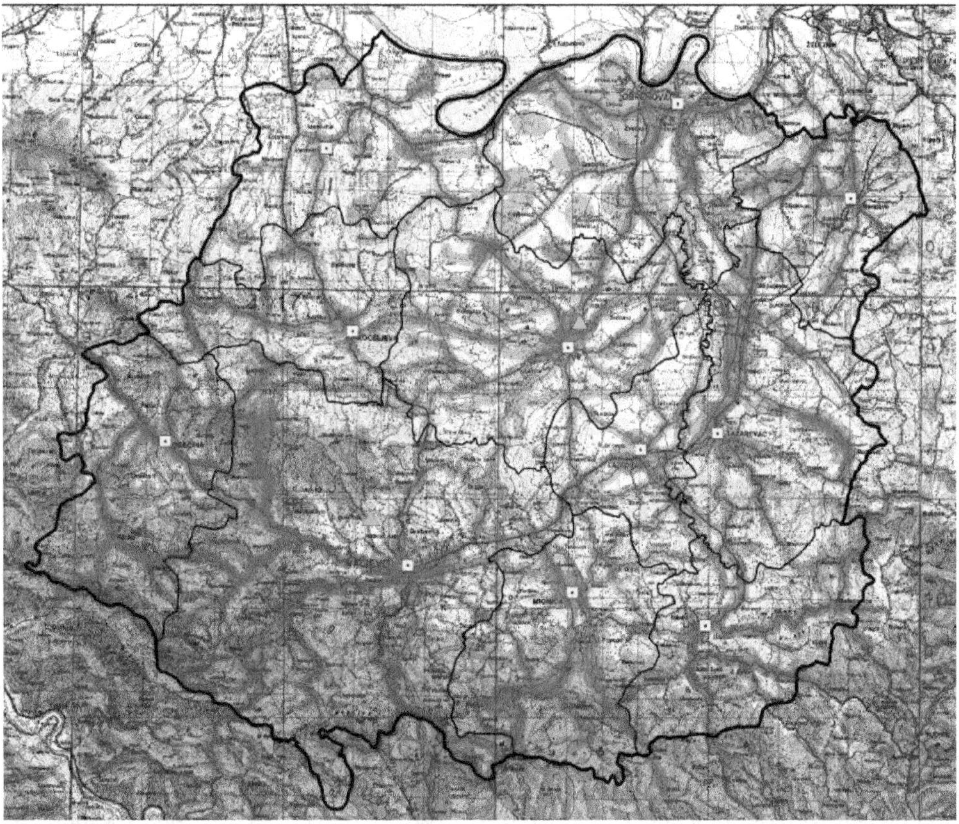

Figure 3. Suitability/elimination map

Once the consultations with relevant entities have been carried out, as well as preliminary analysis of areas that have not been eliminated in the first phase of the landfill site selection process, three sites have been singled which have been included in the process of detailed analysis and multicriteria evaluation.

The use of GIS information base in this phase of landfill site selection has considerably accelerated the process of evaluation according to 32 given criteria. Once the location Kalenić has been assessed as the most suitable one and singled out as the most acceptable one, the landfill site selection process has been completed. However, this is not where the use of GIS tools ends. Their role is also in implementation of a uniform information system for waste management which consolidates data on landfills, transfer stations, waste generation, waste flows, as well as other data important for an efficient waste management.

In this sense, GIS tools represent an information support in the functioning of the waste management system.

5. Conclusions

Landfill site selection is the most sensitive task placed before the participants in the process of planning spatial organization of a waste management system, particularly in countries in which there is insufficient awareness and lack of information in the population, and, consequently, there is a resistance to plans to locate a landfill in their area, known as NIMBY (not in my backyard) syndrome. Out of these reasons, this problem is overcome through defining the elimination and basic criteria for landfill site selection based on which an multicriteria evaluation is carried out, along with mandatory inclusion of all relevant stakeholders in the process of selecting the most suitable landfill site. In the present paper, such exclusion criteria have been chosen that are appropriate to the specific space which has been the subject of our investigation, as well as according to available spatial data. In this context, the paper emphases the fact that the choice of exclusion criteria is conditioned by a specific physical properties of space. After the phase of elimination of "negative" areas, a multicriteria analysis of sites that have been nominated based on a set of basic criteria has been carried out. Altogether 32 criteria have been defined that are based on efficient functioning of a landfill, as well as on efficient environmental protection at specific landfill site and its surroundings. A multicriteria evaluation model has been offered and value scale for evaluation of each criteria defined. The multicriteria evaluation model has been also used for different scenarios. In this context, basic criteria for landfill site selection have been grouped into several basic groups, while in the evaluation process for each criteria only one of the basic groups has been evaluated. Such approach enables decision makers to choose the most suitable option and to make best decision according to their policy.

Comprehensive consideration of the problem associated with landfill site selection for physical elements of waste management system implies the use of GIS tools, thus providing a more sophisticated process of spatial analysis and searching for better options, as well as accelerating and visually enriching the process. Advantage of using GIS tools is in that it enables faster singling out and clearer presentation of suitable and unsuitable landfill sites based of previously given criteria. The paper shows the example of advantages and disadvantages, as well as possibilities of implementing GIG in regional landfill site selection for municipal solid waste management in the Kolubara Region. The GIS applications are particularly suitable for elimination phase where, based on the given exclusion criteria and spatial data, the "negative" areas within which potential landfill sites are not be searched are very quickly and easily eliminated. The entire process is presented cartographically. The possibility of implementing the CORINE Program – a uniform European information base on the environment and space usage, which is particularly suitable for elimination phase in landfill site selection as it provides abundance of geospatial data, has been highlighted. Furthermore, the possibility of efficient waste management using database in the GIS is highlighted. The system supported by such data enables quality and fast waste management, monitoring, waste data updating, as well as the best basis for planning waste management strategy at regional level.

Author details

Boško Josimović and Igor Marić
Institute of Architecture and Urban & Spatial Planning of Serbia, Serbia

Acknowledgement

This work has resulted from research within the scientific project: "Sustainable spatial development of Danube area in Serbia " (TR 036036), which was financed within the program Technological development by the Ministry of Education and Science of the Republic of Rerbia from 2011 to 2014.

6. References

[1] Ilić M (2005) Waste Management in the Focus of Controversial Interests. In: Lechner P, editor. Waste Management. Vienna: Boku University. Pp. 95-96.

[2] Boško Josimović, Marina Ilić, Dejan Filipović (2009) Planning of Municipal Waste Management. Belgrade: Institute of Architecture and Urban & Spatial Planning of Serbia. 157 p.

[3] Waste Management Strategy of the Republic of Serbia 2010-2019.

[4] Boško Josimović (2003) Implementation of Environmental Management System in spatial planning – master's thesis. Belgrade: Faculty of Geography. 146 p.

[5] UK Department for the Environment (1995) Landfill Design, Construction and Operation Practice.

[6] Josimović B, Marić I, Manić B (2011) Methodological approach to the determination of landfill location for municipal solid waste - Case study: Regional landfill in Kolubara region. Architecture and Urbanism. 32: 55-64.

[7] Catalano A, Zhang M, Rice J (2006) The Use of GIS to Manage, Analyze, and Visualize Data Collected During an Investigation of a Proposed Landfill. Available: http://gis.esri.com/library/userconf/proc98/.

[8] Margeta J, Prskalo G (2007) Sanitary Landfill Site Selection, Građevinar. 58. No. 12.

[9] Higgs G (2006) Integrating multi-criteria techniques with geographical information systems in waste facility location to enhance public participation. Waste Management & Research: pp 105-117.

[10] Brans J.P, Vincke Ph (1985) Preference Ranking Organisation Method for Enrichment Evaluations – The Promitee Method for Multiple Criteria Decision Making. Management Science. 31: 647-656.

[11] Josimović B, Crnčević T (2010) Implementation of Strategic Environmental Assessment in Serbia with special reference to the Regional Plan of Waste Management. In: Santosh Kumar Sarkar, editor. Environmental Management. Rijeka: SCIYO. pp. 95-113.

[12] CORINE Land Cover 2000, http://terrestrial.eionet.europa.eu/CLC2000.

[13] Josimović B, Krunić N (2008) Implementation of GIS in the selection of locations for regional landfill in the Kolubara Region. SPATIUM. 17: 72-77.

Chapter 11

Developing a South-European Eco-Quarter Design and Assessment Tool Based on the Concept of Territorial Capital

Stella Kyvelou, Maria Sinou, Isabelle Baer and Toni Papadopoulos

Additional information is available at the end of the chapter

1. Introduction

Many studies have been undertaken with regard to eco-neighbourhoods in Europe. However, most of the projects that have been completed and are being analysed in relevant studies are located in Northern Europe, i.e. the BedZed in England, Hammarby (Stockholm) and BO01 (Malmö) in Sweden, Kronsberg (Hannover) in Germany, Vesterbro (Copenhagen) and Kolding in Denmark, Vauban in Freiburg and others. Findings from these projects permit nowadays to speak about a Northern European eco-neighborhood model (Souami, 2009). However, it would be interesting to investigate eco-neighborhood projects in Southern Europe that are either already realised or still in the design phase. The questions that rise are on one hand the sustainability approach that was followed for these projects and on the other hand the specific criteria involved in each case. To this end, an investigation is undertaken regarding the tools that are being used in terms of the environmental principles-criteria which are taken into account by each of them and how easy these tools are to use. Finally, a comparative analysis follows regarding the different Southern European projects and the environmental criteria involved in their implementation.

The present chapter can be summarised in three fundamental objectives: a) the investigation of contemporary tools and methods of planning and assessment of eco-neighborhoods aiming at identifying similarities and differences but also issues that can lead to an efficient Mediterranean methodology, b) the study of examples of Mediterranean eco-neighborhoods in order to create a good and bad practice guide and g) the proposal of a new assessment tool for the Mediterranean eco-neighborhood, based on the concept of territorial capital (OECD, 2001).

The methodology that was followed regarding the first objective focuses in the parametric analysis of basic criteria of existing tools seeking common ground and differences. As for the second objective the environmental criteria of examples under consideration are investigated with the use of the One Planet Living framework, while the new "SDMed eco-neighborhood tool" was based on:

1. the research and parametric analysis among the tools that concern the development of eco neighborhoods;
2. the SDMed building performance assessment tool (Sinou & Kyvelou, 2006);
3. the concept of territorial capital (OECD, 2001) and its expoitation at local level and the approach of territorial cohesion (both internal and external) that should govern an agglomeration or urban development, even at the scale of a neighborhood.

2. Definition of eco-neighbourhoods

Different approaches and perspectives can be identified regarding the definition of the term "neighborhood" and therefore "eco-neighborhood". One of the most common is to do with density and population. The link between the levels of density and land take in a typical neighborhood of 7500 people. The message is clear: the lower the density, the larger the amount of area that is occupied by buildings, roads and open space. Density per se is not an indicator of urban quality. An interesting definition of eco-neighborhood is given by Barton where he categorises according to different spatial scales. The smallest scale is the building scale; the next one is the home place scale, then the neighborhood scale, the small town scale and finally the city scale. The key sustainability and health issues identified by Barton are: plan for local facilities with attractive walking routes, local hubs to support healthy lifestyles and development of local food, waste, water and energy capture systems (Barton, 2010).

Neighborhood is defined as a residential or mixed use area around which people can conveniently walk. Its scale is geared to pedestrian access and it is essentially a spatial construct, a place. It may or not have clear edges. It is not necessarily centred on local facilities, but it does have an identity, which local people recognize and value (Barton, 2000). Moreover, it is interesting to note the three different facets of neighborhood that Barton distinguishes. Firstly, the neighborhood perceived as the base for home life, education, leisure and employment activities. Secondly, seen as a place, as an aesthetic experience and thirdly as the locus for community (Barton, 2000). All these three facets can be identified in the tools that will be presented in section three; however each tool seems to have a slightly different orientation according to whether it will be used from town planners, or urban designers, or even at an earlier stage from stakeholders during decision making. Moreover, it is significant to note at this point that the most coherent and complete approach is the one that takes into consideration in planning not only issues of energy, transport and resources but also social and economic parameters.

About the model of eco-neighborhood or eco-district, we should mention that a meaningful typology has been drafted by Souami (Souami,2009). He distinguishes three phases of eco neighborhoods' creation, in less than two decades. According to Souami, each phase is corresponding to one of the three different types of eco-neighborhood and it is the second type that permits more performances compared with the other two types, thus representing in a certain way 'the very model of the so-called 'eco-neighbourhood'.

2.1. The initial eco-neighbourhood type of the '80s

According to Souami, the initial eco-neighborhood type was most often a small pool of buildings located in the periphery of cities or in rural areas. The initiators of such projects were usually professionals and experts, politically active, enrolled in so-called alternative movements. Convinced about the importance of 'green' development and construction, the founders of eco-neighborhoods adapted both the idea and the process before choosing the site to accommodate and implement their ideas which was chosen later on. During the 1980s, we meet neighborhoods of this type in Austria, in the Netherlands and in Germany. They are, in fact, eco-villages transformed into neighborhoods and the organisation in a community or associative form of development is often used to group the inhabitants, in order to organise the public areas and implement the project.

2.2. The 'prototype' of eco-neighbourhood of the '90s

In this case, some communities have taken advantage of exceptional urban events to initiate sustainable districts on their territory: World's Fair in Hanover, B01 exhibition in Malmö, London Olympics, Olympic Games bid in Paris, Zaragoza's candidature for EXPO 2008 that initiated the 'Ecociudad Valdespartera' etc. These events have been all opportunities to initiate positive processes that go beyond conventional practices, showing ambitious environmental goals.

The projects are accompanied by an important work of communication, especially internationally. They are developed as exemplary neighborhoods, particularly successful demonstration projects addressing both to technicians and local politicians. Nevertheless, some of them constitute events by themselves. As far as governance and investment schemes are concerned, we should note that:

- partnerships involve communities, private and public developers, social housing agencies, several operators of urban services and facilities (energy, water, etc.),groups of experts and many contractors
- funds are cumulative and come from various sources: local (municipalities, public and private developers), national (sectoral programmes, exceptional ministerial grants or subsidies) and international (various European programmes).

Leaders of local projects show innovative and mainly broadly applied technical solutions (systematic recycling of rainwater, deployment of extended solar panels, photovoltaic panels, etc.) and technicians and policy-makers have the opportunity to test, validate and correct certain choices. These projects are also considered as places that promote a learning procedure for stakeholders and citizens (Kyvelou, 2010).

2.3. From the mid-90s a new type of eco-neighbourhood appears based on environmental quality criteria

In the third category, eco-neighborhood projects are initiated in a conventional manner since they mobilise ordinary tools of development and construction but they integrate environmental quality objectives. In other words, these districts adopt common and unexceptional production methods in order to integrate sustainable development perspectives. Some of them clearly refer to the achievements of the 'prototype' sustainable neighborhoods (the so-called European 'vedettes'). These projects, often modest in size, are being planned in a long-term period and they are sometimes considered as resulting from the dissemination procedure of the proto-neighborhoods and the 'prototype' ones. Nevertheless, Souami argues that, according to his investigations and research, there is no systematic genealogy and explicit influences between these neighborhoods (Souami, 2009).

Furthermore, history shows that eco-neighborhood projects concern issues arising from territorial, urban, social and economic aspirations that go beyond environmental considerations. These socioeconomic and urban contexts are different in each case: economic prosperity for some of them, acute socioeconomic crisis for others, reshaping of political and institutional context, poverty, etc. This diversity proves that there is no common profile to serve as a basis for the creation of an eco-neighborhood. In all cases, eco-neighborhoods are implemented in order to enhance image of the city and local identity. The objective is to protect the sites concerned and therefore the cities where they belong, from their prior image. This image is part of the elaboration of public policies across the whole city. The classification of a district based on its environmental performance depends on a long-term work on the construction of place identity and the deepening of the feeling of belonging to this neighborhood. The place would no longer be identified by its history, its people, its animation or attendance. It would initially live through the image of environmental performance that circulates outside. The existence of prior approaches and actions on environmental issues is not always an asset for the development of operational projects aimed at sustainable development. In some cases, certain environmental policies have applications in connection with sustainable urban projects. In other cases, affected communities have not been able to mobilise their achievements of past policies to initiate and carry out operational projects of environmental quality. Sometimes, sustainable neighbourhoods are not preceded by pre-existing environmental policies. Summing up, we should note that:

- there is not a prerequisite for environmental policy or a prerequisite for sustainable development to achieve eco-neighbourhoods;
- eco-neighborhoods are often conducted in parallel with the establishment of local policies for sustainable development and these projects may contribute to developing local sustainable development policies;
- eco-neighborhoods are not the operational implementation of existing policies for sustainable development which prove and demonstrate their effectiveness.

3. Eco-neighborhood tools

There are numerous tools developed so far, to assess performance of eco-neighborhoods and provide guidance for their planning and design. The following are the most popular ones.

3.1. To BioRegional one planet living framework

This framework consists of ten principles which should govern sustainable communities, namely, zero carbon, zero waste, sustainable transport, sustainable materials, local and sustainable food, sustainable water, land use and wildlife, culture and heritage, equity and local economy and lastly health and happiness. This approach is a very simple one and can be easily used to help individuals and local stakeholders to examine the sustainability challenges and develop appropriate solutions (Table 1).

Zero carbon	Making buildings more energy efficient and delivering all energy with renewable technologies.
Zero waste	Reducing waste, reusing where possible, and ultimately sending zero waste to landfill.
Sustainable transport	Encouraging low carbon modes of transport to reduce emissions, reducing the need to travel.
Sustainable materials	Using sustainable healthy products, with low embodied energy, sourced locally, made from renewable or waste resources.
Local and sustainable food	Choosing low impact, local, seasonal and organic diets and reducing food waste.
Sustainable water	Using water more efficiently in buildings and in the products we buy; tackling local flooding and water course pollution.
Land use and wildlife	Protecting and restoring biodiversity and natural habitats through appropriate land use and integration into the built environment.
Culture and heritage	Reviving local identity and wisdom; supporting and participating in the arts.
Equity and local economy	Creating bioregional economies that support fair employment, inclusive communities and international fair trade.
Health and happiness	Encouraging active, sociable, meaningful lives to promote good health and well being.

Table 1. Principles of BioRegional One Planet Living framework

The One Planet Communities programme uses a set of Common International Targets against each of the 10 One Planet principles to ensure that international partners' projects are guided towards a shared end-point by 2020 and to determine what level of performance is required for a development to be endorsed. Behind the One Planet initiative there are three overarching environmental drivers:

- sustainable ecological footprint;
- sustainable carbon footprint; and
- clean (non-polluting) activities.

3.2. The Eco Town framework by the Cambridge quality charter of growth

The Eco-Town framework focuses on state of the art green building, energy and transport technologies and materials to be used in an urban development context. The task is to ensure zero-carbon housing and that energy efficiencies are achieved through waste reduction, energy conservation technologies and use of more sustainable sources of energy.

The Eco Town approach refers to new settlements with a minimum of 5000 homes where the developments should reach zero carbon standards, should provide good range of facilities and affordable housing. The framework consists of four fields, the four Cs, namely, climate, connectivity, community and character. Each one of the four is subdivided in several criteria (Table 2).

UK Eco-Towns	
Climate	Energy
	Water
	Environment
	Planning for low carbon
	Low-environmental
Character	Place-making
	New design and High Design Standards
	Attractiveness and desirability
	Investment
	Locally-based facilities
Connectivity	Employment opportunities
	Transport
	Services
Community	Social mix
	Sustainable community principles
	Governance
	Delivery organisation

Table 2. Principles of Eco-Town framework

3.3. The DPL approach

DPL is an approach for sustainable urban planning that attempts to quantify and measure sustainability of urban areas (districts) based on 25 environmental, social and economic indicators (Planet, People, Profit) (Table 3). Planet indicators are subdivided into two categories, namely, stocks and local environment. People indicators are subdivided into four categories namely, safety, services, green space water and quality. Profit indicators are subdivided into three categories namely, economic vitality, sustainable businesses and capacity change. DPL was developed by IVAM in cooperation with TNO Environment and Geosciences and with financial support from the Dutch Ministry of Housing Spatial Planning and the Environment (VROM).

DPL – sustainability profile of a district		
PLANET	**PEOPLE**	**PROFIT**
stocks	safety	economic vitality
1. Material use	10. Social safety	20. Local employment
2. Energy use	11. Traffic safety	21. Local economic activity
3. Land use	12. External safety	sustainable businesses
local environment	services	22. sustainable businesses
4. Water management	13. Quality of services	capacity to change
5. Soil contamination	14. Access to services capacity to change	23. Flexibility
6. Waste management	green space and water	24. Mixed use
7. Air pollution green space and water	15. Local green space	25. ICT infrastructure
8. Noise	16. Local water	
9.Smells	quality	
	17. Quality of the district	
	18. Quality of the dwellings	
	19. Social cohesion	

Table 3. The indicators of sustainability assessment according to DPL

3.4. The Med Eco-Quartier approach

The objective of the Med Eco-Quartiers Project was to define precisely the criteria and tools for creating eco-neighborhoods in the Mediterranean region, by studying different cultures, procedural approaches and environments. As part of the project four working tools were developed. They range from the phase of project design to the final realisation phase. The four tools are *Med Eco-urbanisme, Med Eco-constructibilité, Med Eco-gouvernance* and *Med Eco-sensibilisation* (Tables 4 & 5)

Criteria of MED ECO-urbanism
1. Government and organisation
2. Economic and social growth
3. Natural and cultural/archaeological heritage
4. Restriction of urban sprawl
5. Local services
6. Mediterranean natural and climatic characteristics

Table 4. The six criteria for the development of Mediterranean eco-neighborhoods by the MED Eco-urbanism

Preoccupations for the Mediterranean eco-neighborhoods
Preserving resources
Reducing pollution
Reducing waste
Managing natural and technological risks
Improving comfort
Preserving health
Culture and heritage
Integration of the public into the project
Maintenance evolutions
mastering cost excess
Local networks
Low nuisances building sites

Table 5. Preoccupations for the Mediterranean eco-neighborhoods, source : "MED-Ecoquartiers"

The Med Eco-planning tool consists of a grid of decision support and audience piloting development project. The overall approach, crossing the various themes of sustainable urban design leads to the fundamental objectives of the project. The study of the Med Eco-planning tool recommendations allows the specification of the characteristics of the new district. Even if the Med Eco-planning tool provides a framework for reflection and action, it remains a deliverable that has not proved its replication value since it has not been linked with implementation of eco-neighborhoods in the Mediterranean (Kyvelou & Papadopoulos, 2011).

3.5. The LEED for Neighborhood development project scorecard

The tool is subdivided into five categories. Each of them has several analysis criteria, which are either prerequisite or are being given a score (Table 6). The main categories are Smart Location and Linkage, Neighbourhood pattern and Design, Green Infrastructure and

Buildings, Innovation and Design Process and Regional Priority Credit. The tool similarly to LEED for buildings calculates a certification estimate and gives five total scores, namely certified, silver, gold and platinum.

LEED for Neighborhood Development	
Smart Location and Linkage	27 points
Neighborhood Pattern and Design	44 points
Green Infrastructure and Buildings	29 points
Innovation and Design Process	6 points
Regional Priority Credit	4 points

Table 6. The basic categories of LEED for Neighborhood tool

3.6. The HQE aménagement

Formalised in March 2010, the "HQE aménagement" has been subject of a Guide issued under the auspices of the HQE Association. It is, primarily, a pragmatic and ambitious methodology, mostly based on the feedback from concrete operational projects of development. It is based on business and professional logic, which is the one of the developers. The "HQE aménagement" also aims at equipping every stakeholder involved in the development with a reference framework and a common vocabulary for conducting eco-districts and improving professional practices.

HQE Aménagement	
Objectives	**Themes**
Ensure integration and cohesion in relation to the urban tissue and other territorial scales	1. Wider territorial unity and local frame 2. Density 3. Mobility and accessibility 4. Cultural heritage, landscape and identity 5. Adaptability and evolutivity
Preserve natural resources and promote environmental and health quality	6. Water 7. Energy and Climate 8. Materials and equipment 9. Waste 10. Ecosystem and biodiversity 11. Natural and technological risks 12. Health
Promote social life and support dynamics of local economy	13. Economy of the project 14. Mixture of uses and land uses 15. Atmosphere and public spaces 16. Integration and training 17. Local economy dynamics

Table 7. The basic topics of HQE Aménagement

It is a thematic approach that describes the objectives that are sought within the sustainable operation of development. Organised in 17 themes, it allows the direct choice of sustainable actions for the implementation of the following characteristics and subjects of interest (Table 7). The 17 themes are divided into three major sets: territorial analysis - technical and environmental analysis - socio-economic analysis.

- Territorial Analysis: Ensure integration and consistency of the eco-district with the urban area and other territorial levels.
- Technical and Environmental Analysis: Preserve natural resources and promote environmental quality and health.
- Socio-Economic analysis: Promote social life and strengthen the dynamics of local economy.

4. Is there a South-European model of eco-neighbourhood?

The north-european eco-neighborhood model is mainly described by its technical and environmental performance in terms of energy, saving water or recycling materials. They seem to be the main mechanism to move from principles to the effective implementation of sustainable urban development. The urban planning and design and the implementation of networks are resulting from this approach. This model governed by the environmental approach and its performance is mainly used by communities as a powerful tool of communication, promoting the region and even as leverage to reverse social and economic depreciation. However, these social and economic aspects are not sufficiently highlighted and are not explicitly included within the agreed content of the model. On the contrary, Southern European countries and especially the Mediterranean seem to prefix social, economic and governance issues and less attention is paid to environmental performances at least from the point of view of their initial definition and specification.

4.1. The case of France: Relative delay, centrally directed movement, focus on societal issues

In France, despite a relative delay, many cities have been engaged recently in the process of sustainable neighborhoods. Most of the projects are being actually studied and implemented, so their status does not allow to fully evaluate the results and present a meaningful analysis. In addition, the famous "Grenelle de l' Environnement" has proposed to initiate a plan of voluntary eco-neighborhoods driven by local governments: at least one eco-neighborhood in every municipality that intends to realise programmes of housing development until 2012 (in continuity with the existing urban texture and integrated in the city master plans) as well as fifteen large-scale projects of energy, architectural and social innovation, while the release of growth is planned to occur by 2012 through the creation of ten "Ecopolises", that is cities of at least 50,000 inhabitants integrating environmental quality and new information and communication technologies ("Attali" commission).

What is important in France is the strong political will and the consequent centrally directed generalisation: the "Sustainable City Plan" (Plan Ville Durable), presented to the Council of

Ministers on 22 October 2008, aimed at fostering the emergence of a new way to design, build, develop and manage the city. As defined by the Ministry of Ecology, Sustainable Development, Transport and Housing (MEDDTL), the eco-neighborhood is a sustainable operation of high demonstrative and exemplary value. Considered as key measure of the Sustainable City Plan of MEDDTL, it contributes to improving quality of life, while adapting to tomorrow's challenges: "preserving resources and landscapes, while preparing the conditions for the creation of a suitable housing supply". A first call for projects for the competition entitled "EcoQuartier" was launched in October 2008 with local communities to bring together stakeholders of quality operations within an operational Club, enhance their actions and allow the dissemination of good practices. 160 projects have been submitted by communities, coming from all French regions. In 2009, the following cross-cutting issues have enabled to distinguish the winning communities through a ranking of 28 projects:

- Relevance of the urban project, governance, management and structuring of the project, mixt uses and environmental aspects of the development.
- Water, waste, biodiversity, mobility, energy efficiency and renewable energy, density and urban forms, sustainable construction.

The National Award was delivered to the ZAC of Bonne in Grenoble (38). The second call for eco-neighborhood projects was launched in January 2011. A new eco-neighborhood grid was set up in order to serve the project analysis by experts but also to provide a "framework for thought" by any community seeking to implement an eco-neighborhood. In 2011, 393 applications were submitted. A double Grand National Award was delivered to:

- Nancy, and Laxou Maxéville (54 - Urban Community of Greater Nancy) - The Hague Plateau
- Roubaix, Tourcoing and Wattrelos (59 - Urban Community of Lille) - The Union.

The success of the two calls for proposals (2009 and 2011) demonstrated the enthusiasm of local communities to develop sustainable operations, whether in cities, towns or rural communities. A committee charged to form a proposal for an "écoquartier label" was formed. This committee issued, late 2011, recommendations addressing to the Minister of Housing and Urban Development for the establishment of a national eco-district label.

These centrally directed processes in cooperation with local authorities highlight already major projects and underline the fact that the phenomenon of eco-neighborhoods is widespread in France and building resilience is ongoing for these innovative projects of urban development.

Moreover, the HQE Association, a public character platform of stakeholders dealing with sustainable building and sustainable urban development, in collaboration with SNAL (Syndicat National des Aménageurs Lotisseurs) which is a federation of over 300 private developers who contribute to the production of more than 25,000 houses annually, have produced a guide on how to integrate the HQE process for buildings in urban projects. This was the concept that led to the creation of a new HQE process entitled "HQE Aménagement" which has been subject of experimentation since 2006 in 10 pilot community projects. This approach aims at ensuring that all environmental, social and economic concerns are handled by developers and should allow private or public developers to monitor the project so that it incorporates a variety of concerns and all stakeholders likely to be involved: professionals, residents, technical services of the community. The "HQE Aménagement" is fully compatible with the future label "écoquartier" planned by the Ministry of Ecology, Sustainable Development, Transport and Housing (MEDDTL) as far as both piloting, management system and process of the development project are concerned.

To note that the design of eco-neighborhoods in France gives emphasis on the social dimension of the outputs (Lefèvre & Sabard, 2009) even by means of the rebirth of the cooperative movement, which is evident since the beginning of 2000 in France.

4.2. A mediterranean eco-neighborhood model elaborated in the framework of an interregional cooperation programme

The overall objective of the MED-Ecoquartiers project, carried out in the frame of the Medocc-INTERREG IIB and based on a European regional partnership, was to produce a common methodology in order to serve as a set of principles for the creation of new neighborhoods that are consistent with the principles of sustainability (planning, building, energy, mobility, quality of life, activities, natural resources, historic buildings and landscape) in the countries of the Western Mediterranean. It grouped the cities of Pezenas (eco-neighborhood of Saint-Christol, 29 ha for 1700 inhabitants), Dos Hermanas (Spain, eco-neighborhood Montequinto for 12,000 inhabitants), Faenza (Italy, eco-district of San Rocco, 350 dwellings for 1,000 inhabitants), Elefsis (Greece, eco-neighborhood of 88 apartments on 3 hectares).

4.3. Strategic spatial planning prevailing in an Italian eco-district project

Under the hypothesis that "there is no a unique Mediterranean city, but only many different Mediterranean cities" and looking for describing as much as possible the Mediterranean diversity, the Italian partners of the Med-Ecoquartiers Project have elaborated one of the most significant examples in St Rocco neighborhood in the City of Faenza. The San Rocco neighborhood project in Faenza addressed the two fundamental issues for the construction of new neighborhoods; the first one related to land use and consequently town planning, while the second one related to experimental aspects, implementation techniques, the use of materials and innovative technology that can improve the overall quality of the ecosystem. The San Rocco neighborhood seems to be totally oriented towards the "relationship style of planning", even though a great deal of attention has been paid to the physical product, if only to ensure environmental sustainability. The experimental planning and building of the San Rocco neighborhood bridges the gap between product focused and relationship focused planning and is strongly Mediterranean in culture. Residents' involvement was also enhanced. Another useful conclusion deriving from the Italian Project is that a Mediterranean eco-neighborhood results from strategic planning (Nonni, Laghi, 2008). The project is a strategic planning one with people and their relationships at the centre.

4.4. Greek eco-neighborhood projects linked with workers' housing projects

Eco district ideas are still in their infancy in Greece, despite a boost in green development initiated recently by the government. There is no real "eco-neighborhood" implemented in Greece and many questions need to be answered on the way to implement eco-neighborhood projects in the country. Nevertheless, Greece has participated in the Med-Ecoquartiers through a project to construct 88 housing units in the city of Elefsis. The project was carried out by the Workers' Housing Organization, a public institution that since its inception in 1954, is responsible for the construction of social or workers housing dwellings, but the Elefsis project was already in construction phase which is a fact that has hampered the implementation of the Med-Ecoquartiers tools. They only contributed to a partial modification of the original design and improved the environmental performance of the project in some areas. However, the involvement of the Organization in the Project has contributed to a broader learning process to the extent that achieved awareness and knowledge around the question of eco-neighborhoods has had a replication effect through the adoption of the criteria introduced by the programme to other projects managed by the same organisation : a new settlement in Iasmos (Rodopi) has been designed as a pilot village for implementing as much as possible the methodological tools produced by Med-Ecoquartiers and major part of the sustainable planning was the consultation between the various stakeholders, the local government and the residents in order to build the necessary resilience (Kyvelou & Papadopoulos,2011). Moreover, the Organisation has announced a European architectural competition in collaboration to the Greek Institute of Architecture (EIA) regarding the environmental design of a new social housing settlement and is currently expanding its ecological action, by participating in the ELIH-MED Project dealing with energy refurbishment of low-income housing in the Mediterranean.

4.4.1. "Green neighborhood" projects in depreciated areas of western Athens

The most recent attempts to develop eco-neighborhood projects in Athens is the one led by the Ministry of Environment, Energy and Climate change and the Centre for Renewable Energy Sources (CRES) in the depreciated and low-income area of western Athens, namely the municipalities of Aigaleon and Aghia Varvara. Nevertheless the scale of the projects is too small and no real sustainable approach can be implemented. In fact, a block of 4 social housing buildings has been chosen to serve for the pilot implementation of both zero energy buildings and an interior urban oasis as well, to improve microclimate conditions. The ministry is also attempting to implement a particular public-private partnership scheme, through voluntary agreements with small construction enterprises which can provide construction materials and building products to affordable prices. The success of this scheme is crucial for accomplishing a major objective of the project, that is to stimulate local economy dynamics. Another remark that has to be made concerns the failure of a firstly launched project in Aigaleo due to the non capacity of the local authority to fully understand the project and find the necessary institutional tools to cope with the land and buildings ownership related constraints.

As fas as the Aghia Varvara project is concerned, both on-site visits, use of interactive questionnaires and data collection from the Public Power Corporation confirmed the need for interventions in buildings in three key areas: The first one concerns exterior insulation of the building shell, replacement of old simple glazed window frames with double glazing, replacement of old blinds and use of cold paintings. The second axis is to replace the various heating and cooling systems with an energy efficient central heating and cooling system. Finally, ensuring hot water through a central solar system is the third axis of the operation.

The programme is aiming at maximising energy efficiency of the neighborhood, achieving thermal comfort for residents and improving significantly their quality of life and at the same time minimise environmental impacts.

Figure 1. Social housing apartment buildings and the urban tissue of the Aghia Varvara (Western Athens) green neighborhood project.

4.4.2. Private eco-developments in high-income suburbs, focusing on marketing

Private projects can be often met in the Mediterranean area as they concern the creation of eco-villages either of secondary residences and touristic complexes or high income level housing in prestigious suburbs. In these projects, innovation and marketing for commercial reasons are usually prevailing. An example of such a development is the "Designer Village" developed by a private construction company in Dionysos, on the foothill of the Pendeli mountain. The project concerns the development of 85 plots in which 240 dwellings are being erected. Each plot has approx. an area of 1500m2. In an attempt to use green marketing tools, nine Greek architectural cabinets of different architectural perceptions and tendencies have been invited by the construction company, to put down their inspirations. "Designer Village" is already referred to as a kind of park of exemplary Mediterranean architecture, characterized as an "excellent project" by an EC programme competition. It focuses on energy efficient techniques and improvement of thermal and visual comfort and indoor air quality. Although the environmental objectives and targets associated with the rational use of resources (energy, water etc) are ambitious, the complex is far from being characterised as eco-neighborhood since it doesn't promote neither social mixing, nor economic efficiency, accessibility and affordability or cooperation among inhabitants. In conclusion we would say that an eco-neighborhood may be regarded as such only if it is the result of social dynamics and not a simple consumer product (Kyvelou & Papadopoulos, 2011).

5. Findings of the comparative study of the existing assessment tools

The investigation with regard to the existing tools showed that they address different parameters and have different focus. It is not easy to select the most appropriate tool, since there is none of them addressing all the issues and remaining, at the same time, easy to use. It was shown that most of the environmental, social and economic parameters are being implemented in different ways for each project.

However, in many of the cases, there are parameters that were neglected. Two of them is good design and place-making. It is strongly believed by the authors that the latter should be included to an eco-neighborhood approach and addressed at an earlier stage. The aspect of good design is not only related to environmental design and its many criteria but also to issues such as attractiveness and high design standards which are addressed by the Cambridge Quality Charter of Growth. Place-making, that is making of a real coherent neighborhood with social and territorial cohesion is also often neglected.

On the other hand, since current economic and financial crisis lead to fragmentation of society, special attention has to be paid, in these projects, on social equity issues as one of the aims of sustainable territorial development. The principle of social equity focuses on the right to work and housing but also on the access to services and public goods and on the fight against poverty and social exclusion. Each citizen must have access to a job and decent housing, but also to the essential public goods. The access to housing, medical care, education and information should lead to a more stable society capable of solidarity, tolerance and generating participation. It can develop its traditions while promoting a sustainable lifestyle. Finally, 'social equity' can be defined as the addition of the merit principle of equality. This aim can be divided into four targets: ensuring everyone an adequate housing, ensuring access to efficient public services, promoting access to employment, fighting against occupational and social exclusion. To achieve the first target, the issues in a territorial project are:

- promote the social mix
- promote diversity of housing supply
- promote affordable housing policy
- integrate urban projects in a social housing policy consistent with the entire city.

Apart from social equity principles, an eco-neighbourhood initiative should be based on adaptive and flexible governance schemes (Chouvet, 2007), that is bring together community stakeholders, property developers, utilities, and the city to solidify a shared sense of purpose and partnership through the following actions:

- establish municipal policy and organisational structures to support the eco-district development;
- create an engagement and governance strategy to build community support, set priorities and actions;
- develop an assessment and management to guide project development and track ongoing performance;
- identify commercialisation opportunities for the private sector to test promising products and practices;
- implement sustainability projects through technical and economic feasibility analysis, assembly of project financing, and establishment of public-private partnerships (Kyvelou & Karaiskou, 2007).

6. Comparing Mediterranean eco-neighbourhoods

This part of the research has identified examples of eco-neighborhoods in Southern Europe, either completed or in the design phase. These neighborhoods have various scales and cover regions approximately from 10 to 250 hectares (Table 8). This is an ongoing investigation aiming to identify and classify examples, which can be used as models in future eco-neighborhood design.

Amongst the 16 presented cases, one of them is located in Portugal, three in Spain, three in France and nine in Italy. In Table 8 some general characteristics are presented for each example while in Table 9 an analysis of the different environmental, economic and social parameters that were implemented in each project is undertaken. For methodological reasons, a single tool was selected for the purposes of this classification, the "One Planet Living framework" initiated by Bioregional. The 10 principles of the tool form the corpus of the analysis which will follow.

We shoud note here that the North-European model is mainly described by its technical and environmental components. The performance in terms of energy, saving water or recycling materials are particularly highlighted. They seem to be the main mechanism to move from principles to the effective implementation of sustainable urban development. The urban planning and design and the implementation of networks are resulting from this approach. This 'model governed by the environmental approach and its performance' is mainly used by communities as a powerful tool of communication, promoting the region and even as leverage to reverse social and economic depreciation. However, these social and economic aspects are not sufficiently highlighted and are not explicitly included within the agreed content of the model of sustainable urban development.

On the contrary, Southern European countries and especially as far as the examples examined in the Mediterranean, are concerned, seem to prefix social, economic and governance issues and less attention is paid to environmental performances at least from the point of view of their initial definition and specification. In Table 8, it is also clear that sustainable transport is also a field of Mediterranean interest together with health and hapiness issues.

6.1. What would be the model of a Southern European eco-neighborhood?

The study of a series of eco-neighborhoods in Southern European countries led us to the conclusion that a Southern European eco-neighborhood model is certainly emerging. Specifically, the experience of Southern Europe shows that eco-neighborhoods are nether merely expressions of integration of sustainable development in city planning nor only products integrating new technologies and alternative energies.

The eco-neighborhoods, as developed in Europe are, in our view, important local aspects of strategic spatial planning, as this is reborn and reshaped today in Europe, thus being products of a dynamic political and social process. Eco-neighborhoods are manifestations of the change of regulatory planning (based on physical planning) to a territorial management process where "territorial marketing" has a predominant role (Kyvelou, 2010).They reaffirm, moreover, that sustainable development is an exceptional unifying element and point of recasting spatial policies. Sustainable neighborhoods are also emphasizing the issue of scale which is nowadays one of the most important theoretical discussions in spatial and urban planning. Another important conclusion is that an eco-neighborhood mainly depends on the extent to which there is a tradition of strategic spatial planning, on the cultural tradition and level of collective and community involvement (Chiotinis, 2006) and on the presence of catalytic investments and the possibility of implementing public - private partnerships towards a process of stimulating local economy dynamics (Kyvelou & Karaiskou, 2006).

The Mediterranean countries usually lagging behind as far as the above conditions are concerned, present a spontaneity and a delay of implementation. Constraints of implementation show that an eco-neighborhood initiative should be based on adaptive and flexible governance schemes, and on engagement strategies that would bring together community stakeholders, property developers, utilities, and the city to solidify a shared sense of purpose and partnership through the following actions :

a. Establish municipal policy and organisational structures to support the eco-neighborhood development;
b. Create an engagement and governance strategy to build community support, set priorities and actions;
c. Develop an assessment and management toolkit to guide project development and track ongoing performance;
d. Identify commercialization opportunities for the private sector to test promising products and practices; and
e. Implement sustainability projects through technical and economic feasibility analysis, assembly of project financing, and establishment of public-private partnerships (Kyvelou, Marava & Kokkoni, 2011).

Both lessons from the North-European examples of eco-neighborhoods and observation from different types of eco-districts in Southern Europe, either public or private projects, can provide helpful criticism and a good framework to discuss issues of efficient implementation in the near future. This framework could help to modeling work that is both to the construction of a model that can predict future trends and to the affirmation of the model, meaning that it can be used in the reproduction and replication of certain practices and related projects.

Project name	plan	area	dwellings	inhabitants	total cost
Portugal					
Mata de Sesimbra, Lisbon	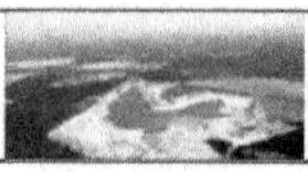	5300 he+4800he natural	8000 units tourist	25.000 inhabitants	800 mil Euros
Spain					
Entrenucleus, Dos Hermanas, Seville		39,5 hectares, 7, 2 hectares settlement 160,50 km2	13500 units homes and offices	126.000 inhabitants	244 mil Euros
Ecociudad Valdespartera, Zaragoza		243,3 hectares	9.678 social housing units	31.000 inhabitants	N/A
Logroño Montecorvo, Rioja Province		56 hectares	3.000 carbon neutral housing units	130.000 inhabitants	388 mil Euros
France					
Andromède in Blagnac		210 hectares	3.700 housing	1 million	N/A
Ville de Pézenas, Quartier Saint Christol		29 hectares	90.000 sq met.	250 famillies	31 mil Euros
Saint Lys-Haute Garonne		7 hectares	N/A	7500	N/A
Italy					
St.Rocco, Faenza	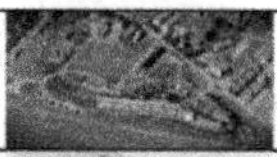	8,3 hectares	390 housing units	950	9,4 mil Euros
Malizia Ecologic quarter, Siena		5 hectares	150 housing units +commercial etc	N/A	N/A
Cognento project, Modena		11,7 hectares	220 housing units	770	5,5 mil Euros
BIOPEP		14, 7 hectares	340 housing units	N/A	N/A
Sanpolino quarter, Brescia		N/A	2.000 apartments	N/A	N/A
Parco Ottavi project, Reggio Emilia		53 hectares	131.000 sq m surface,1.500 housing units	4.500	N/A
Pietrasana, Vigevano, - Pavia		N/A	220 lodgings	N/A	5,7 mil Euros
Villa Fastiggi, Pesaro		15 hectares	333 dwellings	N/A	N/A
Bologne, quartier de la Bolognina ouest		30 hectares	N/A	64.000	N/A

Table 8. Examples of Mediterranean eco-neighbourhoods

Aghia Varvara	Iasmos	Elefsis	Greece	Quartier de la Bolognina	Villa Fastiggi, Pesaro	Public Buildingsr, Pietrasana	Parco Ottavi , Reggio Emilia	Sanpolino quarter, Brescia	BIOPEP, Modena	Malizia ,SienaCognento	St.Rocco, Faenza	Italy	Ville de Pézenas,Saint Christol	Andromède quarter, Blagnac	France	Logroño Montecorvo, Rioja	Valdespartera, Zaragoza	Entrenucleus, Seville	Spain	Mata de Sesimbra, Lisbon	Portugal	CATEGORIES IDENTIFIED FROM THE PROJECTS	ONE PLANET LIVING PRINCIPLES
•	•															•	•					Solar - panels, wind control	Zero Carbon
•	•				•			•									•					orientation	
•	•												•	•		•				•		Renewable energy	
						•				•			•									biomass	
	•						•				•							•		•		Photovoltaic systems	
•							•	•	•		•											Thermal stations	
	•	•					•		•				•	•						•		Waste management	Zero Waste
	•			•	•		•	•	•	•	•		•	•						•		Pedestrian routes	Sustainable Transport
		•		•	•		•	•	•		•		•	•						•		Bicycle network	
				•	•	•	•	•			•		•				•	•		•		Green belts	
								•						•				•		•		Urban bus lines	
						•	•															Covered parking	
•	•	•			•				•	•							•			•		Sustainable materials	Sustainable Materials
																	•			•		Local and sustainable foods	Local-Sust. Food
							•	•		•	•		•	•				•		•		Rainwater storing	Sustainable Water
							•			•	•		•	•								Water reutilisation	
							•	•	•							•						Water nets	
									•	•							•					Water devices	
	•	•			•			•									•			•		Land and wildlife	Land and Wildlife
				•							•		•							•		Connection with historical centres	Culture and Heritage
		•		•													•					New cultural centres	
					•		•			•	•		•	•						•		Social services	Equity and Local Economy
													•	•								Generation mix	
•																	•					Integrating urban environment	
						•	•	•									•					Sports facilities	Health and Happiness
					•	•	•	•	•	•				•						•		Public buildings	
							•	•														Kindergartens	
•					•		•	•	•				•	•				•		•		Urban parks	

Table 9. Parametric analysis of the sustainability criteria according to the One Planet Living framework.

6.2. What would be an efficient sustainability assessment tool for Southern European eco-neighborhoods?

The short analysis of the eco-neighbourhood design tools has shown that they have different orientations in terms of their focus on environmental, social or economic strategy. Some of the tools seem to be more design oriented while others have a social resilience focus. The Eco-Neighborhood is more than a mere buzzword or local marketing tagline. The neighborhoods have ambitious targets that go well beyond load reduction. They draw upon new and often complex practices, from urban project design to construction, use and assessment and it is often complicated for the local authorities to implement these practices as new methodologies need to be addressed and many of them recoil at doing so, from the very start of the project.

The above described experiences in Southern Europe show that an assessment tool for eco-neighborhoods cannot be efficient if nor directly linked with the valorisation of the territorial potential and the territorial capital (OECD, 2001), at local level. We have therefore proceeded to the formation of a new tool, largely inspired by the territorial capital and the territorial cohesion concepts, which gives emphasis to the holistic approach of the city, its neighborhoods and its relationship and interdependence with its region. The so-called "SDMed eco-neighborhood" tool is structured around the main determinants of the territorial capital (geographical, cultural, political, material, social and intellectual capitals) and is analysed in a series of criteria capable to manage the complexity and diversity of the Mediterranean urban phenomenon (Sinou &Kyvelou, 2006)

The concept of territorial capital was first proposed in a regional policy context by the OECD in its Territorial Outlook (OECD, 2001), and it has been later on reiterated by DG Regio of the European Commission: "Each Region has a specific 'territorial capital' that is distinct from that of other areas and generates a higher return for specific kinds of investments than for others, since these are better suited to the area and use its assets and potential more effectively. Territorial development policies (policies with a territorial approach to development) should first and foremost help areas to develop their territorial capital" (European Commission, 2005).Territorial capital is referring to the following elements: a/ a system of localised externalities, both pecuniary and technological; b/ a system of localised production activities, traditions, skills and know-how; c/ a system of localised proximity relationships which constitute a 'capital' – of a social psychological and political nature – in that they enhance the static and dynamic productivity of local factors, d/ a system of cultural elements and values which attribute sense and meaning to local practices and structures and define local identities; they acquire an economic value whenever they can be transformed into marketable products – goods, services and assets – or they boost the internal capacity to exploit local potentials; e/ a system of rules and practices defining a local governance model. Accordingly, the OECD has rightly drawn up a long list of factors acting as the determinants of territorial capital, and which range from traditional material assets to more recent immaterial ones. All the above have been used to draft the SDMed Eco-neighborhood tool.

Territorial capital	Territorial cohesion components
Geographic	Emmissions
	Landscape resources
Cultural	Cultural heritage
Political	Governance
Material	Economic growth-wealth
	Resources
	Sustainable transports-mobility
Social	Health and safety
	External accessibility
	Internal connectivity
Intellectual	Creativity

Table 10. The conceptual elements behind the SDMed Eco-neighborhood tool.

7. The SDMed Eco-Neighborhood tool

As it was mentioned in the introduction the SDMed Eco-Neighbourhood tool was based on:

1. the research and parametric analysis between the existing eco neighborhood tools;
2. the SDMed building performance assessment tool (Sinou & Kyvelou, 2006);
3. the concept of territorial capital (OECD, 2001);
4. The current economic and financial crisis and the consequent limitation of public funds.

By its definition, the concept of territorial capital can be divided in geographic, cultural, political, material, social and intellectual capital, while approaching territorial cohesion we can categorise actions in emissions, local resources, cultural heritage, governance economic growth, resources, sustainable transport-mobility, health and safety, external accessibility, internal connectivity and creativity (Kyvelou, 2010).

Thus, the structure of the new tool is constituted by the main subdivisions of territorial capital, eleven objectives-targets linked to territorial cohesion and in 39 accordingly sub-targets (Table 11). These sub-targets are further divided in more criteria in order to include all the parameters that can influence sustainable neighborhood. The final depiction of the tool is under study; however a first attempt is presented in the diagram of Table 10.

SDMed ECO NEIGHBORHOOD tool				
Territorial capital	11 Targets territorial cohesion	39 Sub-Targets		
Geographic	Emissions	1.1	Emissions (CO2)	Reduction of CO2 emissions
		1.2	Water waste	Management of water waste
		1.3	Production of solid waste	Management of domestic waste Management of construction waste Management of solid waste infrastructure
		1.4	Management of litter and waste	Network of sewage
		1.5	Urban heat island effect	Measures to reduce heat island effect
		1.6	Night-time light pollution	Reduction of light pollution
	Landscape resources	2.1	Natural and technological risks	Local management of natural risks (Earthquake, heat wave, tsunami) Local management of technological risks
		2.2	Influence to the urban form	Optimisation of land use Compact growth – Density Intelligent planning Integration of environmental issues in urban planning
		2.3	Adaptive opportunity	
		2.4	Quality of public spaces	Tree-lined and shaded streets Cohesion and linkage among urban spaces Design with bioclimatic criteria
		2.5	Comfort (thermal, visual, acoustic)	Nuisance linked to the neighborhood Noise pollution in the neighborhood from vehicles or activities Minimisation of construction nuisance

SDMed ECO NEIGHBORHOOD tool				
Territorial capital	11 Targets territorial cohesion	39 Sub-Targets		
				Visual quality of natural environment - view Visual quality of built environment Preferred locations Water efficient landscape Steep slope protection
		2.6	Air quality	Interior air quality Exterior air quality
		2.7	Spatial comfort and comfort of activity	Quality of building Quality of housing Variety in housing Satisfaction of users and residents Support of architectural quality Innovation and exemplary performance
Cultural	Cultural heritage	3.1	Maintenance of natural heritage - biodiversity	Maintenance of wetlands and natural beauty landscapes Protection of rural land Extinction species and ecological communities
		3.2	Maintenance of cultural heritage	Conservation and re-use of cultural heritage
		3.3	Maintenance of built environment	Use of existing buildings Preservation of historical resources and adaptive reuse Brownfield redevelopment
Political	Governance	4.1	Functionality and possibility of services control	
		4.2	Adaptability and flexibility of services	
		4.3	Robustness and maintenance of	

SDMed ECO NEIGHBORHOOD tool				
Territorial capital	11 Targets territorial cohesion	39 Sub-Targets		
			services	
		4.4	Community involvement	
		4.5	Ownership of land and buildings	
		4.6	Public Private Partnerships	
Material	Economic growth-wealth	5.1	Cost of land and construction cost	Local economic dynamic Creation of social economy
		5.2	Cost of of life cycle (€/year) (maintenance, exploitation et deconstruction)	
		5.3	Cost of waste management and cost of emissions (€/year)	
		5.4	Support of local economy	Presence economic activities Presence of retail trade Local food production Mixed use neighborhood centers
	Resources	6.1	Effect in the energy resources	Improvement of energy efficiency for heating, cooling and electricity (buildings and infrastructures) Use of renewable energy (locally) Orientation Heating and cooling of district Certified green buildings
		6.2	Exhaustion of raw material	Integration of recycled and reused materials, constructions and process of demolition in buildings and the public spaces

SDMed ECO NEIGHBORHOOD tool				
Territorial capital	11 Targets territorial cohesion	39 Sub-Targets		
		6.3	Use and management of water	Consumption of potable water Use and management of rain water of Stormwater management
	Sustainable transport-mobility	7.1	Transportation cost	Development ICT Proximity of housing and job
		7.2	Improvement of transportation and mobility	Improvement of transport Safe and convenient paths for pedestrians and bicycles Areas with decreased dependence automotive Transit centre Network bicycle path and storage Reduction of parking footprint Street network
Social	Health and safety	8.1	Health and productivity	Improvement of cleanliness in the neighborhood and communal spaces Right and acess to care and health
		8.2	Safety of users	Improvement of safety of people and goods Improvement of road safety
	External accessibility	9.1	Accessibility for people with special needs	Accessibility
		9.2	Accessibility in open spaces	Access in public spaces Access in recreation facilities
	Internal connectivity	10.1	Involvement of users	Involvement of residents and users in the process of sustainable urban development Participation of residents in decision making and projects related to the community Strengthening of community

SDMed ECO NEIGHBORHOOD tool				
Territorial capital	11 Targets territorial cohesion	39 Sub-Targets		
				Promotion and participation of community
		10.2	Creation of work places	
		10.3	Social diversity	
Intellectual	Creativity	11.1	Support of education - Levels of education and professional skills	Diversity in age distribution Mixed income communities Connected and open community Collaborations Promotion of academic success Reinforcement of the role of school in the community International cultural contacts - connectivity

Table 11. The new SDMed ECO-NEIGHBOURHOOD tool

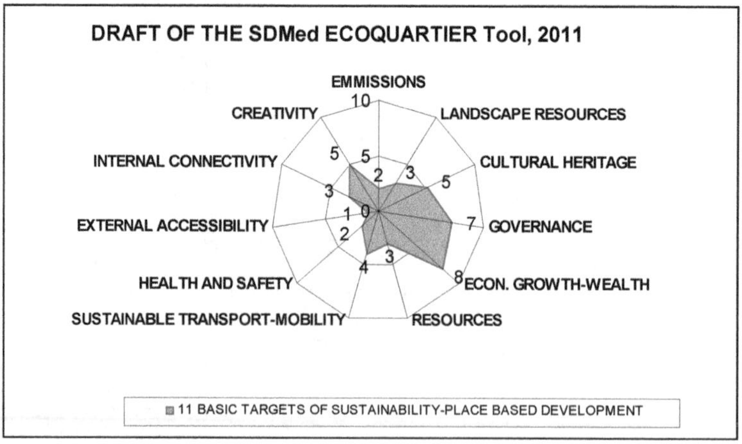

Figure 2. Draft illustration of the SDMed Eco-neighbourhood tool.

8. Conclusions

The climatic, social, cultural, governance and spatial planning related specificities of the Mediterranean region and the delay of eco-neighborhood development in the Mediterranean countries combined with current need of economic and social regeneration lead to the necessity to develop a tool adapted in these specific needs and particularities. The comparative research showed that the assessment tools for eco-neighborhoods have been structured around different principles and ideas. There is an abundance of tools and their evaluation constituted an aid for the creation of a new proposal. The parametric analysis of tools that were selected created a base for the creation of the new tool that has to be enriched in order to face the particular needs of the Mediterranean region, with sensitivity and taking into account current ecomonic and financial crisis.

Furthermore, the investigation of existing Mediterranean examples gives important information with regard to the sustainability criteria that are used in most of the cases.

Finally, the concept of place-based development and contemporary planning criteria like the one of territorial capital have been explored in order to form a proposal for a new methodological tool based on the need of place-making and urban regeneration under economic crisis and lack of public funds. Sustainability in the scale of a neighborhood cannot be a static process, has to do with innovation and has to ensure the creation of wealth at local level and the stimulation of local economy dynamics. Planning a contemporary eco-neighborhood is a complex procedure and addresses both economic, social, environmental and governance related challenges. Achieving an eco-neighborhood assumes the development and implementation of a process of project management and an action plan involving local actors at the different phases of a project. Furthermore, it involves setting objectives and implementing actions to achieve environmental, economic and social performances.

Author details

Stella Kyvelou
Panteion University of Social ad Political Sciences of Athens , Greece

Maria Sinou
Technological Educational Institution, Greece

Isabelle Baer
Syndicat National des professionels de l'aménagement et de lôtissement, France

Toni Papadopoulos
Workers' Housing Organisation, Greece

9. References

Barton H., (Ed.). (2000). *The Potential for Eco Neighbourhoods*, Earthscan Publications Ltd, ISBN 1 85383513 7, London.

Barton H.; Grant M. & Guise R. (2010). *Shaping Neighbourhoods, for local health and global sustainability*, Routledge, ISBN10: 0-415-49549-0; ISBN13: 978-0-415-49549-13, London and New York.

Chiotinis, N. (2006). The request of sustainability and architecture as cultural paradigm, *Management of Environmental Quality : An International Journal*, Vol. 17, No. 5, Emerald Group publishing Limited, pp.593-598, ISSN 1477-7835.

Chouvet, C. (2007). *Les quartiers durables: un exemple de démarche intégrée et participative*, Comité 21/Angenius, pp.4–20, http://www.comite21.org/docs/territoires-durables/ville-durable/les-quartiers-durables.pdf

Kyvelou, S. & Karaiskou, E. (2006). Urban development through PPPs in the Euro-Mediterranean region, *Management of Environmental Quality: An International Journal*, Vol. 17, No. 5, Emerald Group Publishing Limited, pp.599–610, ISSN 1477-7835.

Kyvelou, S., Hetzel, J., Sinou, M. & Iwamura, K. (2007). *L'application du développement durable au cadre bâti dans l'espace Méditerranéen: La démarche SD-MED*, Pulim (Ed.), pp.121, ISBN : 978-2-84287-425-4, France.

Kyvelou, S. (2010). *From spatial planning to spatial management: the concepts of strategic spatial planning and territorial cohesion in Europe*, KRITIKI (Eds.), ISBN 978-960-218-671-8, Athens.

Kyvelou, S. & Papadopoulos, A. (2010). *'Sustainable neighborhoods: lessons from Northern Europe – issues arising from a Mediterranean paradigm'*, in: I. Beriatos and M. Papageorgiou (Eds.): Spatial Planning-Urban Planning-Environment in the 21st Century, pp.315–326, University Editions of Thessaly, Volos, Greece, Mediterranean.

Kyvelou, S. & Papadopoulos T. (2010). Exploring a South-European eco-neighbourhood model: Planning forms, constraints of implementation and emerging resilience practices, *International Journal of Environment and Sustainable Development*, Vol. 14, Nos. 1/2, 2011, pp. 77-94, ISSN(print) 0960-1406.

Kyvelou, S., Sapounaki-Dracaki, L. & Papadopoulos, A. (2010). *«Eco-quartiers en Europe: leçons obtenues des pays du Nord, perspectives et politiques dans les pays de l'Europe du Sud»*, oral presentation at the 10th International EAUH Conference, Gand, August-September 2010.

Kyvelou, S., Marava, N. & Kokkoni, G. (2011). Perspectives of local public-private partnerships towards urban sustainability in Greece, *International Journal of Environment and Sustainable Development*, Vol. 14, Nos. 1/2, 2011, pp.95-111, ISSN(print)0960-1406.

Kyvelou, S., Dracaki, L., Sinou, M., & Papadopoulos A. (2011). Planning and building a South-European eco-neighborhood: from concepts and strategies to practice and assessment tools, *Review of decentralization, local government and regional development*, Vol.66, Athens, 2011, ISSN 1106-91-71.

Lefèvre, P. & Sabard, M. (2009). *Les écoquartiers*, Editions Apogée, p.261, ISBN: 978-2-84398-325-2, Paris.

Nonni, E. & Laghi, S. (2008). *Un eco quartiere mediterraneo: Il Quartiere Residenziale S. Rocco a Faenza'*, Comune di Faenza. http://www.avecnet.com/publications/files/page51_15.pdf

OECD Terrirorial Outlook (2001), ISBN: 9789264189911, Paris, p.300

Sinou, M. & Kyvelou, S. (2006). Present and future of building performance assessment tools, *Management of Environmental Quality: An International Journal*, Vol. 17 Iss: 5, pp.570 – 586, ISSN 1477-7835.

SNAL (2011), *"Ecoquartiers"*, ISBN 978-2-912683-84-7, Editions PC.

Souami, T. (2009). *Éco-quartiers, secrets de fabrication: analyse critique d'exemples européens*, Les Carnets de l'info (Eds.), ISBN 978-2-9166-2844-8, Paris.

Permissions

The contributors of this book come from diverse backgrounds, making this book a truly international effort. This book will bring forth new frontiers with its revolutionizing research information and detailed analysis of the nascent developments around the world.

We would like to thank Sime Curkovic, for lending his expertise to make the book truly unique. He has played a crucial role in the development of this book. Without his invaluable contribution this book wouldn't have been possible. He has made vital efforts to compile up to date information on the varied aspects of this subject to make this book a valuable addition to the collection of many professionals and students.

This book was conceptualized with the vision of imparting up-to-date information and advanced data in this field. To ensure the same, a matchless editorial board was set up. Every individual on the board went through rigorous rounds of assessment to prove their worth. After which they invested a large part of their time researching and compiling the most relevant data for our readers. Conferences and sessions were held from time to time between the editorial board and the contributing authors to present the data in the most comprehensible form. The editorial team has worked tirelessly to provide valuable and valid information to help people across the globe.

Every chapter published in this book has been scrutinized by our experts. Their significance has been extensively debated. The topics covered herein carry significant findings which will fuel the growth of the discipline. They may even be implemented as practical applications or may be referred to as a beginning point for another development. Chapters in this book were first published by InTech; hereby published with permission under the Creative Commons Attribution License or equivalent.

The editorial board has been involved in producing this book since its inception. They have spent rigorous hours researching and exploring the diverse topics which have resulted in the successful publishing of this book. They have passed on their knowledge of decades through this book. To expedite this challenging task, the publisher supported the team at every step. A small team of assistant editors was also appointed to further simplify the editing procedure and attain best results for the readers.

Our editorial team has been hand-picked from every corner of the world. Their multi-ethnicity adds dynamic inputs to the discussions which result in innovative outcomes. These outcomes are then further discussed with the researchers and contributors who give their valuable feedback and opinion regarding the same. The feedback is then collaborated with the researches and they are edited in a comprehensive manner to aid the understanding of the subject.

Apart from the editorial board, the designing team has also invested a significant amount of their time in understanding the subject and creating the most relevant covers. They scrutinized every image to scout for the most suitable representation of the subject and create an appropriate cover for the book.

The publishing team has been involved in this book since its early stages. They were actively engaged in every process, be it collecting the data, connecting with the contributors or procuring relevant information. The team has been an ardent support to the editorial, designing and production team. Their endless efforts to recruit the best for this project, has resulted in the accomplishment of this book. They are a veteran in the field of academics and their pool of knowledge is as vast as their experience in printing. Their expertise and guidance has proved useful at every step. Their uncompromising quality standards have made this book an exceptional effort. Their encouragement from time to time has been an inspiration for everyone.

The publisher and the editorial board hope that this book will prove to be a valuable piece of knowledge for researchers, students, practitioners and scholars across the globe.

List of Contributors

Fabio Iraldo and Marco Frey
Sant'Anna School of Advanced Studies, Pisa, Italy
IEFE - Institute for Environmental and Energy Policy and Economics, Milano

Francesco Testa and Tiberio Daddi
Sant'Anna School of Advanced Studies, Pisa, Italy

Antonio Zanin and Fabiano Marcos Bagatini
Universidade Comunitária da Região de Chapecó - UNOCHAPECÒ, Brazil

Ray-Yeng Yang and Ying-Chih Wu
Tainan Hydraulics Laboratory, National Cheng Kung University, Taiwan
Department of Hydraulic and Ocean Engineering, National Cheng Kung University, Taiwan

Hwung-Hweng Hwung
Department of Hydraulic and Ocean Engineering, National Cheng Kung University, Taiwan

Mansoor Zoveidavianpoor and Ariffin Samsuri
Universiti Teknologi, Faculty of Petroleum & Renewable Energy Engineering, Malaysia

Seyed Reza Shadizadeh
Petroleum University of Technology, Abadan Faculty of Petroleum Engineering, Iran

Danica Fazekašová
University of Prešov, Slovakia

Felipe de Jesús Escalona-Alcázar, Bianney Escobedo-Arellano, Brenda Castillo-Félix, Perla García-Sandoval, Luz Leticia Gurrola-Menchaca, Carlos Carrillo-Castillo, Ernesto-Patricio Núñez-Peña, Jorge Bluhm-Gutiérrez and Alicia Esparza-Martínez
Unidad Académica de Ciencias de la Tierra, Universidad Autónoma de Zacatecas Calzada de la Universidad, Fracc. Progreso, Zacatecas, México

Beatriz Gómez-Muñoz
Ecology Section, University of Jaén, Campus Las Lagunillas s/n, Jaén, Spain

David J. Hatch
Sustainable Soils and Grassland Systems Department, Rothamsted Research, North Wyke, Okehampton, Devon, UK

Roland Bol
Institute of Bio-and Geosciences, Agrosphere (IBG-3), Forschungszentrum Jülich GmbH, Jülich, Germany

Roberto García-Ruiz
Ecology Section, University of Jaén, Campus Las Lagunillas s/n, Jaén, Spain

Nils McCune
Ghent University, Belgium
Red de Estudios para el Desarrollo Rural AC, Mexico

Francisco Guevara-Hernandez
Facultad de Ciencias Agronomicas, Universidad Autonoma de Chiapas, Mexico

Jose Nahed-Toral
El Colegio de la Frontera Sur, San Cristobal de las Casas, Mexico

Paula Mendoza-Nazar
Facultad de Medicina, Veterinaria y Zootecnia, Universidad Autonoma de Chiapas, Mexico

Jesus Ovando-Cruz
Red de Estudios para el Desarrollo Rural A.C., Mexico

Benigno Ruiz-Sesma
Facultad de Medicina, Veterinaria y Zootecnia, Universidad Autonoma de Chiapas, Mexico

Leopoldo Medina-Sanson
Idem

Richard Y.M. Kangalawe
Institute of Resource Assessment, University of Dar es Salaam, Dar es Salaam, Tanzania

Boško Josimović and Igor Marić
Institute of Architecture and Urban & Spatial Planning of Serbia, Serbia

Stella Kyvelou
Panteion University of Social and Political Sciences of Athens, Greece

Maria Sinou
Technological Educational Institution, Greece

Isabelle Baer
Syndicat National des professionels de l'aménagement et de lôtissement, France

Toni Papadopoulos
Workers' Housing Organisation, Greece

Printed in the USA
CPSIA information can be obtained
at www.ICGtesting.com
JSHW011457221024
72173JS00005B/1105

9 781632 395788